Recent Advances in Weed Science

Bhagirath Singh Chauhan • Gulshan Mahajan
Editors

Recent Advances in Weed Science

Springer

Editors
Bhagirath Singh Chauhan
Queensland Alliance for Agriculture and Food Innovation (QAAFI)
The University of Queensland
Gatton, QLD, Australia

Gulshan Mahajan
Queensland Alliance for Agriculture and Food Innovation (QAAFI)
The University of Queensland
Gatton, QLD, Australia

ISBN 978-3-031-91013-5 ISBN 978-3-031-91014-2 (eBook)
https://doi.org/10.1007/978-3-031-91014-2

This Springer imprint is published by the registered company Springer Nature Switzerland AG
The registered company address is: Gewerbestrasse 11, 6330 Cham, Switzerland

Preface

Feeding the rapidly growing population will depend on agriculture remaining the fundamental provider, as this represents a substantial and challenging responsibility. Research scientists have worked relentlessly to change agricultural production from basic sustenance into commercial farming through advanced technologies that help farmers build profitable operations. Efficient weed management approaches are essential for maintaining and boosting agrarian profitability.

Farmers struggle with weed problems because they must increase crop yields to satisfy human population growth demands. Weed control methods have evolved from essential hand weeding to advanced, innovative techniques. Herbicide application helps solve labour shortages yet creates problems through weed resistance and environmental damage alongside herbicide residue. Scientists work to reduce agricultural challenges by building more potent molecules that work at reduced doses and researching innovative solutions such as nano-herbicides.

Aside from chemical methods, harvest weed seed control provides a non-chemical late-season approach to diminish weed seedbanks. Using unmanned aerial vehicles (UAVs) and sensor technologies, precision agriculture benefits from enhanced weed management through site-specific targeted actions. Current regenerative agriculture systems now include adaptive weed management practices, creating new opportunities for sustainable farming.

Scientists gain valuable insights into weed biology and ecology through molecular biology and omics approaches, enabling them to develop innovative management strategies. Integrated weed management frameworks utilize cultural practices and biological and technological methods to reduce the environmental impact of herbicides while increasing crop yields. Aquatic plant management commands attention because of its distinct challenges, which necessitate customized control methods.

This book combines recent advancements to tackle urgent problems, including herbicide resistance and ecological weed management, while promoting sustainable agricultural practices. It provides extensive insights that enable both growers and researchers to access collective expertise instead of depending on trial-and-error methods. By offering comprehensive insights, the new edition of this book is a strategic guide for students, weed scientists, and policymakers in creating research priorities and executing sustainable farming strategies.

Gatton, QLD, Australia

Bhagirath Singh Chauhan
Gulshan Mahajan

Contents

About the Editors

Bhagirath Singh Chauhan The University of Queensland, Queensland Alliance for Agriculture and Food Innovation, Gatton, Queensland, Australia.

Professor Chauhan received his PhD in Weed Science from the University of Adelaide, South Australia, in 2007. He leads weed research in the northern grain region of Australia. He has worked in several countries, including Australia, Bangladesh, India, the Philippines, Sri Lanka, and Vietnam. He is on the editorial boards of several highly regarded journals, including *Weed Science*. He has published over 500 articles and edited several books, including *Rice Production Worldwide*, *Crop Protection Under Changing Climate*, and *Non-chemical Weed Management*.

Gulshan Mahajan The University of Queensland, Queensland Alliance for Agriculture and Food Innovation, Gatton, Queensland, Australia.

Gulshan has vast experience working on the agronomy of rice in India. He is currently a Research Fellow at the University of Queensland. Gulshan has published over 150 articles and edited several books, including *Rice Production Worldwide*.

Chapter 1
Future Challenges and Opportunities in Weed Science

Robert L. Zimdahl

Abstract Agricultural scientists, farmers, ranchers, and the agriculture industry remain confident of their basic faith in the possibility of continued increasing production through the intelligent use of ever more efficient agricultural technology and research. Increasing production has been and remains the accepted way to achieve the moral obligation of feeding a growing population. This brief essay questions if agriculture's moral justification will hold as widespread, rational scientific, and moral arguments about human and environmental harm, public fear of technology, and concern about food quality dominate.

Keywords Herbicide dependency · Climate change · Weed challenges · Food security · Remote sensing · Sustainable agriculture

The Danish physicist Niels Bohr was the first to remind us—"It is difficult to predict, especially the future." I agreed to write this chapter with some trepidation and a modicum of fear about expressing eccentric, wild, and potentially wrong opinions. However, when the past is known and the present is understood, it is tempting to try to glimpse the future.

I assert that weed science, although young among the agricultural sciences, has an enviable, rich, productive history and will continue to contribute to agriculture, other disciplines, and food production. Weed control was recognized as a necessity by farmers who had been controlling weeds long before herbicides were invented. Herbicides changed the way control was done, but not its fundamental purpose—to improve the yield of desirable species. The chemical energy of herbicides replaced human, animal, and mechanical energy. No other method of weed control was as efficient at reducing the need for labor or as selective. People with hoes could distinguish weeds from crops and weed selectively. Mechanical and cultural methods, while effective, were not selective enough. Herbicides enabled prevention, reduced

R. L. Zimdahl (✉)
Agricultural Biology, Colorado State University, Fort Collins, CO, USA
e-mail: r.zimdahl@colostate.edu

B. S. Chauhan, G. Mahajan (eds.), *Recent Advances in Weed Science*,
https://doi.org/10.1007/978-3-031-91014-2_1

weed populations, and selectively removed weeds from crops. Weed control in the world's developed countries now depends on herbicides. This situation will prevail well into the twenty-first century.

1 Problems

There are seven important problems that have and may continue to hinder progress as follows:

1. The assumption that anyone can control weeds is made by those who do not understand either the complexity of agriculture or weed management. Marshall (2010) reinforced this assumption with the misconception that weed science is easy and, more importantly, has all the answers to weed problems, which it does not. Environmental and production demands will require significant adjustments in weed management and agricultural practice.
2. Although weeds have been and will continue to be components of agriculture and the environment, they lack the attention, appeal, and urgency of sudden infestations of other pests.
3. Weed science lacks foundational hypotheses "linked to established bodies of ecological and evolutionary theory to provide deeper theoretical justification, a broader vision, and increased collaboration across diverse disciplines" (Ward et al., 2014).
4. There is a lack of people and research funds (Davis et al., 2009). Research on weed biology, ecology, seed dormancy, and other problems leading to basic understanding rather than immediate control is done by too few scientists. Publicly funded interdisciplinary agricultural research has lacked adequate funding and, it seems, will remain so (Davis et al., 2009).
5. Underlying all agricultural issues there is always an unexamined ethical position (Zimdahl, 2022). Thompson (1995) pointed out there is only one imperative to produce as much as possible, regardless of the environmental/ecological costs and perhaps even if it is not profitable. Agricultural people cannot escape responsibility for societal views of its effect on the environment, other species, and themselves. Agriculture's views on ethical issues have not been and should be examined.
6. All in agriculture know farming is crucial to all economies (Economist, 2022) and important to the welfare of all. The public in most societies is certain that food is important but is abysmally unaware of the complex processes and the people who provide their food.
7. Climate change and lack of appropriate weed control practices will affect farmer's ability to produce. Modern agricultural technology that developed country farmers rely on is beyond the reach of poor farmers in the developing world. More than 90% of farmland in Africa has no irrigation, one-third of the world's people, and 60% of Africans do not receive warnings of impending natural disas-

ters or routine weather forecasts. Agriculture's admirable goal of feeding an expanding world population in a warmer, dryer climate would benefit from expanding its horizons beyond developed country farmers.

A few conflicting claims illustrate future challenges, which are as follows:

- Moss (2008) charged that the overall direction of weed research was wrong. There was too much emphasis on scientific effect at the expense of practical application. Moss argued that weed science was weed technology. He suggested his colleagues lacked an awareness of the complexities and resources needed to translate research results into actions for farmers.
- Ward et al. (2014) claimed two broad aims have been driving weed science research: improved weed management and improved understanding of weed biology and ecology. Research has developed a very high level of repetitiveness, a preponderance of purely descriptive studies, and has failed to clearly articulate novel hypotheses linked to established bodies of ecological and evolutionary theory. Although Ward et al. (2014) noted that studies of weed management remain important, they urged weed scientists to recognize the benefits of deeper theoretical justification, a broader vision, and increased collaboration across diverse disciplines (especially ecology). One might conclude that weed science research has not been as good (weeds, like the poor, are still with us) as many colleagues think it has been.
- Swanton (2022) accused weed science of being primarily reactive. Scientists responded to current needs and worked to solve on-farm problems. He recommended the discipline make long-term thinking automatic and common instead of rare. Long-term thinking is required because weed science, a sub-discipline of agriculture, must begin to answer complex questions regarding cropping systems and environmental challenges.
- A former Editor-in-Chief of Weed Research (Marshall, 2019) introduced "the post-herbicide era of weed science." He argued this was increasingly prescient as herbicides continue to face ever-increasing legislative restrictions and the challenge of evolved resistance. They are key influences on the practice of intensive agriculture, whose success is intimately linked to the heart of the planetary crises: climate change, global warming, loss of biodiversity, environmental harm, etc.
- Buhler (2017) argued that weed scientists must develop integrated cropping systems and weed control strategies in a comprehensive, environmentally, and economically viable system. This approach would "help reduce economic effects and improve weed control practices." Herbicides will continue to be an essential part of integrated cropping systems.
- Westwood et al. (2018) claimed that weed science was at a "critical juncture" because decades of chemical control have dramatically increased herbicide-resistant weed populations. The problems were critical because there were few new herbicides, new modes of action, and no economically acceptable alternative to herbicides in large acreage crops. They suggested new modes of action could be discovered using genetic engineering, computing power, automation,

employment of artificial intelligence, and machine vision to improve weed management.
- Gould (2002) portrayed the situation well by contrasting "immediate and practical" with "distant and deep" issues. Immediate and practical issues are about potent and unanticipated effects (e.g., herbicide resistance). Distant and deep issues include legislative, ethical, aesthetic, and practical consequences of altering agriculture's fundamental geometry and permitting scientists in the developed world to change the way agriculture is and ought to be practiced. The author advocated proper development and use while giving adequate, proper consideration to human and environmental health, agricultural progress, and sustainability.

This chapter deals with thoughts about future weed science research, but not in terms of what will be accomplished. It is conjecture, not prophecy. It might be best conceived as a proposal of what ought to be done. It may not be what will be done because research does not always follow a straight path, and other developments may change what is desirable and possible. For example, environmental legislation mandating reduced herbicide use could rapidly change the way agriculture is practiced. A description of research needs is a safer prophetic stance. It describes what could be done rather than describing what the situation will be like in several years. This approach, of course, reduces the possibility that the prophet may be wrong.

2 Research Needs

Dependence on herbicides for weed control is equivalent to treating the symptoms of a disease without actually curing the disease. Agriculture would be far better served if weed scientists learned how to control weed seed dormancy and seed germination so weeds could be prevented rather than controlled after they appear. No one knows enough about weed seed dormancy, and much research remains to be done to reach the prevention goal. Here and throughout this chapter, the emphasis will be on the two major goals put forth by Ward et al. (2014).

1. Discussion and debate of appropriate goals and the pathways necessary to achieve the goals.
2. Rediscovery of the ability to pose critical research questions rooted in and designed to advance the theoretical underpinnings of weed science.

Weed science began when herbicides (e.g., 2,4-D) made control possible without studying the weeds. Those who controlled had to know what weeds were to be controlled and where they were growing. That is, control was not blind. There are objects to be controlled, and they are known. But, with herbicides, it has not been necessary to know much more.

In general, herbicide development has neither exploited weak points in a plant's life cycle nor used specific physiological knowledge for control purposes. The safest approach has been to aim for complete control of weeds in a crop. As knowledge grows, scientists find that some plants may be beneficial and should not be controlled (Chandrasena, 2023). A series of projects could be developed to study the regulation of seed and bud dormancy of perennial weeds and the development and life of reproductive propagules (Wyse, 1992). Population genetics and modeling of crop-weed systems will contribute to improved weed management.

2.1 A. Weed Ecology

2.1.1 Future Role

Important insights on the future role of weed ecology are found in two papers—Neve et al. (2018) (35 authors) and MacLaren et al. (2020) (6 authors). Both papers support the increasingly dominant claim—the present weed management system is unsustainable because of its negative effects and dependence on chemical, capital, and petroleum energy. Both papers strongly advocate combining multiple known weed management techniques in a new integrated weed management (IWM) system. The creation of an integrated system based on agro-ecological approaches will require multi-disciplinary—trans-disciplinary—participation (Jordan et al., 2016). MacLaren et al. (2020) argue that "new herbicides, gene editing, and seed destructors do not address needed systemic challenges and are unlikely to provide sustainable solutions." Neve et al. (2018) advocate a better understanding of weed evolution, climate change, weed invasiveness" and, perhaps the greatest challenge, "disciplinary challenges for weed science." Neve et al. (2018) advocate "integration of agro-ecological weed management with socio-economic and technological approaches."

The scientific system that helped create these problems accepts credit but resists accepting blame for negative effects, which is part of the tragedy. It is an example of the agricultural mindset and justifies Mayer and Mayer's (1974) conclusion that the system is unsustainable. Their second claim is that the integration and isolation of the system have led to what they call—The Island Empire. Agriculture is a vast, wealthy, powerful intellectual, and institutional island. The Land Grant system created colleges of agriculture and allowed agriculture to be isolated from mainstream American life within the university. Mayer and Mayer accuse agricultural colleges of being separated from the university, mainstream of scientific thought, and national discussions about social policy. Agriculture does not ask for and only reluctantly receives outside criticism. Those who practice agriculture must move off their island.

2.1.2 Other Thoughts

Much of the basic information required to develop computer-based models of weed-crop systems and available control techniques has come from and will continue to be derived from weed biology and ecology research. What plants compete for and when competition is most severe between crops and weeds is known in sufficient detail to be useful in the development of weed management systems. The still-used (Dawson, 1965) period threshold concept of weed competition affirms that weed competition is nearly always time-dependent. Seedling weeds at crop emergence are less detrimental than those emerging later. This principle led to the timely use of herbicides and other techniques for weed management. Some crop cultivars are more competitive, and this needs to be considered when developing weed management systems. It is a basis for cooperative work with plant breeders.

Weed populations change with time and reasons are beginning to be understood. A major challenge presently dominating weed research is the appearance of herbicide resistance, often after only a few years of use in one field. Active research is coupled with the development of techniques to combat it. When resistance occurs, it does not lead to totally unmanageable weed populations because other weed-control techniques (e.g., cultivation and crop rotation) and other herbicides are available. Understanding why populations change and the management of population shifts is important to the development of successful, sustainable weed management. However, as Harker et al. (2012) note, the best way to reduce selection pressure for herbicide resistance is to reduce herbicide use, but dominant weed management programs advocate herbicide use.

Even casual observers of the world of weeds will recognize that weed problems have changed (see Van Wychen, 2016, 2020, 2022). Some of the most difficult weeds in most crops today were not important 10 or 20 years ago. This is evidence that weed scientists have developed successful solutions to some weed problems. It is also true that some common weeds (e.g., pigweeds, lambsquarters, velvetleaf, Canada thistle, and cheatgrass) have been targets of control programs for years. Thus, we have simultaneous evidence of success and continuing problems. It is also evidence that nature abhors empty niches. When successful control efforts have reduced the population of a species, they inevitably leave space unoccupied and resources unused. Other species move into empty niches created by successful weed control. Solutions to this dilemma take two forms. The first solution is to reduce the effectiveness of the niche. There are examples from other agricultural practice 1. Farmers typically over provide for crops. Fertilizer placement and precise rate recommendations have reduced surplus nutrients, but nitrogen runoff due to excessive application is a significant problem with notable externalities. 2. Whole fields are irrigated and light cannot be controlled. If water could be placed (e.g., drip irrigation) as precisely as fertilizer and only as much was provided, the attractiveness of the niche and the success of potential invaders could be reduced (i.e., preventive weed management).

The second approach has an element of prevention. Some of the important problem weeds of the next decade are already in fields or lurking on the edges. If they were identified and their weedy potential determined, weed scientists, cooperating with ecologists (see MacLaren et al., 2020), could try to predict those most likely to be successful invaders. They could be controlled or managed before their invasion. Invasive plant management is now a major area of weed science research, as indicated by the 2008 launch of the journal *Invasive Plant Science and Management*.

Basic biological–ecological knowledge is essential to either approach. Without it, weed scientists may be doomed to endure the Red Queen effect (a character in Lewis Carroll's classic book *Through the Looking-Glass*, 1871). The Red Queen tells Alice, "In this place, it takes all the running you can do to keep in the same place." Weeds and their control, especially with herbicides, seem to be evolving at about the same rate. In trying, and often succeeding, to eliminate weeds from fields, weed scientists have created, in a sense, better, more ecologically successful weeds while accepting herbicide's negative environmental effects.

A difficult and central issue for weed science is to understand the nature of weeds: What makes a weed a weed? How can weeds consistently come out ahead when matched up against the finest commercial varieties plant breeders have developed? Weeds persist, they spread, and they out-compete crop plants, reducing yields when left uncontrolled. Weeds are not conscious, but they seem to be clever. The nature of the competitive ability weeds possess seems an interesting target for research and an appropriate target for analysis through the generation of mutants.

Goethe's "The Sorcerer's Apprentice" and Mary Shelley's "Frankenstein," and, more recently, Michael Crichton's "Jurassic Park" reinforce the often inchoate fear of intelligent, rational concern about a powerful form of life manufactured with good intentions but excessive hubris, which might one day slip out of control (Specter, 2016). The 1950s gave us catchy phrases that still resonate—Better Living Through Chemistry and Atoms for Peace. We do not hear similar things now. Chernobyl/Fukushima nuclear reactors, agent orange, space shuttle crashes, thalidomide, ozone destruction, pesticides in food, and climate change dominate the public's thoughts. Scientists clearly solve problems, yet, in the public's view, untoward problems occur. These well-known problems, combined with human drug disasters, have made people suspicious of the efficacy and trustworthiness of science and scientists (Lemonick, 2006). It is in this context that public doubts about genetic modification of anything are raised and must be addressed. Weed scientists and others involved with genetically modified (GM) technology often think they could educate/tell people about what they do (William et al., 2001). Education is important, but careful listening followed by a conversation among equals may be better, especially in a time when science has made mistakes and is regarded with well-founded suspicion. Weed scientists should not regard themselves as the only acceptable arbiters of how developments in their science should be created and used. Because of public perceptions of greed, a bit of arrogance on the part of the developers, and a misunderstanding of science, and view genetic modification as a hazard, not salvation, and reject it (Specter, 2016).

2.1.3 Education

A review of some published articles on the future of weed science reveals few comments on the role of education. Research and appeals for more funding (Davis et al., 2009) dominate. There is at least one undergraduate weed science class at all US Land Grant universities, and several others are required of undergraduate and graduate students. The absence of discussion of what students ought to know among those who teach is disturbing. Surely, the education of the next generation of weed scientists with "innovative and diverse teaching practices" advocated by Chauhan et al. (2017) is as important to the collective future of weed science as biotechnology, invasive species, and new herbicides. If it is, why is education not closer to the top of the future agenda? We must integrate weed management and education.

2.2 Biological Weed Control

What if the agrochemical industry had never developed? Just suppose someone had discovered an organism in 1944 that selectively controlled some annual broadleaf weeds in small grains. What would weed science be like? Would we recommend all the herbicides we now do? Would our environmental problem be what to do with mutating organisms rather than polluting chemicals? Hypothetical questions lead to interesting discussions, but they do not always solve problems.

Biological control is still in its infancy compared to other control methods. Its theoretical potential, unrealized in all but a few cases, has had little effect on IWM systems for agronomic or horticultural crops. Future research may discover specific biological control organisms and combinations of organisms that are effective, safe, and can be integrated with other methods of weed management. There are only a few now. Research continues to discover and exploit natural mechanisms of predation and parasitism for weed management (Julien et al., 2012).

Those who understand the techniques of biological weed management readily acknowledge that it is not devoid of environmental concerns. However, the "massive accumulation of environmental problems—air pollution, acid rain, the greenhouse effect, ozone depletion, climate change—suggests few citizens and virtually no politicians have seriously considered the very meaning of progress if the future must be … different" (Bellah et al., 1991). What the public may regard as the thoughtless exploitation of the earth at the expense of future generations will not be tolerated, and weed managers may have to rethink what proper weed control is.

2.3 Weed Management (See Hall et al., 2000; Hamill et al., 2004)

Rotation of crops in a field is an effective way to reduce weed competition and may be a more important part of future weed science research. Future recommendations may include the use of more smothering and green manure crops to keep land covered and protected from wind and water erosion. Research may demonstrate that cover/protective crops may have economic value in addition to their environmental value. It is well known that crop rotations reduce soil erosion and manage weeds, but research has not been done, so one cannot say what weeds will be controlled effectively and how rotations can be used to the greatest advantage. Rotational research will include research on the relative competitiveness of crop cultivars and weeds.

A few herbicides control some perennial broadleaf or grass species. *Bromus tectorum*, *Cyperus esculentus*, *Sorghum halepense*, and *Convolvulus arvensis* are not generally controlled selectively by available herbicides. Few herbicides sufficiently active on perennials are not selective enough to be used in most crops, and several persist in soil. It is relatively easy to control emerged shoots of perennials but far more difficult to assure translocation throughout the extensive root, root runner, rhizome, or stolon systems. A continuing problem for turf managers is selective control of coarse-leaved perennial grasses in fine-leaved turf grass.

2.4 Soil Erosion

Cultivation, from plowing to interrow tillage, has been part of growing crops and weed management for years. Its benefits are well known. Evidence of negative environmental effects from tillage is growing (Logan, 1995; Montgomery, 2007). Cultivation will continue to be part of crop and weed management systems, but its potential to encourage soil erosion must be recognized. Worldwide, 2 billion hectares of soil (greater than the area of the United States and Canada combined) have been degraded (Cummings, 2006). Sir Albert Howard's (1940) book *An Agricultural Testament* criticized the rise of what he called "scientific agriculture." Following the work of Leibig in the mid-nineteenth century, scientists thought all plants needed from soil were the correct quantities of nutrients. Without Liebig's work and the development of nitrogen fertilizer, it is highly probable that 40% of humanity would not be here.

Howard was concerned with farmer's success and feeding people. He was confident both were dependent on soil health, which, in his view, the nation's health depended on. Haney (cited in Ohlson (2016)) claims, "Our entire agricultural industry is based on chemical inputs, but soil is not a chemistry set; it is a biological system. We have treated it like a chemistry set because chemistry is easier to measure and study than soil biology." Not applying chemicals is the key to sustainable food production (Ohlson, 2016). Reducing soil health to a few chemical inputs was the worst of reductionist science (Shapin, 2006). Nature has always had the science right.

Africa has persistently hungry people and difficult weed problems, and three-fourth of arable land is severely degraded (Cummings, 2006). The vast majority—99.7% of human food comes from cropland—is shrinking by more than 10 million hectares per year due to soil erosion. If current trends of soil degradation continue, the continent may be able to feed only one-fourth of its population by 2025. Cummings (2006) estimated that the annual cost of soil erosion in the United States from 1933 to 2010 was $44 billion, similar to the 24 European Union countries. As a result of erosion over the past 40 years, 30% of the world's arable land has become unproductive. Weed management has a role in solving this problem. It may be possible, as unlikely as it may seem, to farm land without creating a weed problem.

Tillage research has shown that much is unnecessary and, in fact, complicates weed management. Soil tillage routinely buries weed seeds for future recovery and growth, which does not seem to be good weed management. Tillage exposes weed seeds to sunlight and encourages germination. No-tillage leaves seeds buried and prevents or inhibits germination. The best tillage for weed management may be none. That does not mean that weeds will not be present. No-tillage creates a niche for weeds that do well without tillage. It will change, but not eliminate, the need for weed management. Research is required to determine the effect of cultivation timing with different types of implements on specific weeds.

2.5 *Aquatic Weeds*

The difficulty of controlling aquatic weeds is related to their habitat, not their life cycle. Seventy percent of the earth's surface is covered with an interconnected water system. Ultimately, all water flows to the sea, and all contaminants can be carried along as dissolved solutes or adsorbed on eroded soil. Almost all water contamination is, in the minds of most people, unacceptable. Because of heightened, enlightened concern about environmental quality, the unacceptability of contaminating water has resulted in legislation to prevent control and, if necessary, punish water pollution. Only a few herbicides can be used in aquatic systems. Present and future herbicides must be compatible with all other actual and potential uses for water. This is a very difficult requirement, and extensive, expensive research will be required if acceptable weed control with herbicides is to be developed. In the aquatic environment, proof of safety will be demanded by an anxious public and a multitude of stakeholders. Reasonable assurance or reasonable doubt will not suffice. Long-term expensive research will be required if non-chemical techniques for aquatic weed management are developed. Because of these concerns, the aquatic environment is likely to be a site for the development of nonchemical control techniques.

2.6 *Parasitic Weeds*

Parasitic weeds are present in the United States, but unless one has them, they are not regarded as major problems. They are major problems in several of the world's developing countries (Gressel et al., 2004). Few selective herbicides or other control techniques are available.

Research has not focused on them because they are very difficult problems and because herbicide developers have concentrated on large acreage crops in the developed world. It takes only a little experience with parasitic weeds to recognize how devastating they can be and how large crop yield losses are. Farmers in developing countries where parasitic weeds are common stop growing susceptible crops, abandon fields, and are often defeated by them. Reliable, affordable management techniques for parasitic weeds would be a significant scientific achievement and a major contribution to feeding people in the world's poor countries. One approach to the Striga problem in Africa involves applying very low doses of an appropriate herbicide to maize seed after the maize plant has been rendered resistant to the herbicide through now standard non-transgenic techniques (Gressel et al., 2004; De Groote et al., 2007). ImiR (non-transgenic maize) was developed by CIMMYT. It is available for short-season maize (95 days) in Kenya and neighboring countries from a variety of seed companies. Resistance will take a long time to evolve as the high local rate (in the rhizosphere—not/ha) requires homozygosity. This will not be so in long-season corn (115 + days)—as toward the end of the season, enough herbicide dissipates to allow heterozygous R individuals to survive. Seed treatment with an imidazolinone herbicide creates a weed-free zone in the vicinity of maize and up to 12 weeks free of Striga, followed by a mid-season application of glyphosate, which requires the crop be glyphosate/imidazolinone resistant. Each herbicide would kill surviving Striga and be appropriate for long-season maize (Gressel, 2009; Ransom et al., 2012). A second approach to witchweed control in maize (Li et al., 2023) is based on a modification of strigolactones exuded from maize roots into the rhizosphere. Reduction in activity of this enzyme and two others involved in the pathway can change strigolactone composition and reduce Striga germination and infection. These results offer prospects for breeding Striga-resistant maize.

An essential resource for researchers is the work of Chris Parker (Parker, 2017, 2021; Parker & Riches, 1993) and his now historic periodic compilation of world research on parasitic weeds. To date, there is little evidence that the herbicide industry, governments, or other research agencies have adequately funded and thereby encouraged research to solve a major third-world weed problem.

2.7 Bioeconomic Models

In the early 1980s, two decision-aid, computer-based models were available. By the early 1990s, 21 mostly crop-based, weed-control-management, decision-aid computer-based software models were available to researchers and farmers (Mortensen & Coble, 1991). Available models will be refined, and new models will be developed. Some weed scientists suggest the ideal is a mechanistic weed-crop competition model that considers and responds to changing environments. The essential knowledge of weed biology to construct such models is not yet available (Schweizer et al., 1998; Weaver, 1993). In most models, weed density is the sole variable used to estimate crop yield loss. Future models will incorporate variables for the relative time of crop and weed(s) emergence, crop density, climate variation, method and amount of fertilizer applied, and weed density (O'Donovan, 1996).

Rather than just predicting crop loss, future models will also enable realistic monitoring of weed population development and long-term implications of failure to control, or, to say it another way, the long-term implication of seed production by uncontrolled weeds (O'Donovan, 1996).

As more biological knowledge of weeds becomes available, as weed seed-sampling techniques improve, and as models improve, they will be used by decision-makers. Weed management decisions guided by models will lead away from prophylactic control methods. Such methods have accounted for the clear and repeatable success of ridding a field of weeds for a season. They have not provided long-term control of weeds in several crops in economically efficient, environmentally sound ways. Models and modeling have emphasized herbicides that have been profitable and efficacious. These will continue to be important criteria, but long-term weed management success will become a more important criterion of success. As new knowledge is incorporated, computer-based weed management decision aid models will provide greater assurance of achieving profitability and appropriate long-term weed management. Some of the knowledge required to develop better models is shown in Table 1.1.

Table 1.1 Examples of knowledge required to develop improved weed management systems and decision aid models

Management Goal	Research Need
Management decision aid model	Relationship of the size of the weed seed bank to final weed population Emergent rate of individual weed species Determination of economic optimum thresholds for control Interaction of management practice and weed seed production Effect of weed density on control
Prediction of seedling emergence	Mechanisms of dormancy Determination of interaction of environmental conditions, seed germination, and dormancy
Effect of management	Effect of crop rotation of weed seed bank size Effect of living and dead mulches Rate of seed predation and decay Rate of seed mortality Light requirement for seed germination Role of tillage and cultural practices
New herbicides and biopesticides	Discovery of new modes of action Genetic engineering and new options for manipulating herbicide selectivity Creation of entirely novel approaches to weed management
Artificial intelligence	Computing power and automation Use of machine vision and global positioning systems

Adapted from Buhler et al. (1997)

2.8 Integrated Weed Management (IWM)

In science, as in most human activities, movements occur, directions change, and progress may result (Kuhn, 1970). Some movements are called bandwagons. Each has associated words and phrases that define and identify. Each movement makes its contribution to the parade of ideas and contributes to the general cacophony of competing ideas, which one hopes will yield a harmonious new paradigm. Some ideas assume a position at the head of the line.

IWM has assumed a position of centrality in weed science. If IWM systems are to endure, change will be required. The direction and scope of change may determine the enduring success of IWM. Historians tell us, after the fact, what things have endured and why, and why other ideas were only temporary. Judgments from the present are often flawed because one is so close, and subjectivity dominates. The perspective of time is often a prerequisite to objectivity.

The risk is that weed management may fade, but IWM makes so much sense that it is likely to endure. It is not perfect, but it is better than anything else we have. The evidence is sufficient to demonstrate that weed management systems for crops are incomplete. They are developing, but research gaps exist and preclude defining complete systems. Weed management has become dependent on herbicides and, within the last decade, herbicide-resistant crops. Truly, IWM sustainable systems will have to include, as Young (2012) points out, components of weed control that must be put together rather than taken apart. Weed science, in his view, "has stopped at the field edge in assembling the components into a truly integrated approach." Integration of other weed-control techniques in addition to herbicides and GM crops has not occurred. Even casual observation of capital, energy, and chemical-dependent agriculture demands answers to questions about the system's sustainability. In view of the rapid appearance of herbicide-resistant weeds, increased herbicide use, and emphasis on short-term solutions, sustainability does not appear to be the goal. Sustainability is and should be a proper goal (Mortensen et al., 2012). "The monocrop factory mode that dominates US agriculture today certainly differs from what has characterized farming for most of history and in most of the world" (Peterson, 2000). "Instead of natural ecosystems with their haphazard mix of species…, farms are taut, disciplined communities conceived and dedicated to the maintenance of a single species: us" (Mann, 2012). This kind of agriculture cannot achieve what all regard as a proper goal—sustainability. The foundation of weed science has been control technology rather than understanding weeds, why and how they compete so well, and their role in the agroecosystem (Buhler, 2006). That must change.

Some challenging research questions are:

- What seeds germinate and emerge first from complex soil seed banks? Why?
- What percentage of seeds, from each species, emerge each year?
- What is the precise percent control for different weed control techniques and for different weeds at different growth stages?

- If soil seed bank composition is known, can the weed complex appearing in a crop be predicted?
- How do weed management techniques affect other pests and other pest management techniques?
- How can the advances in automated machinery for real-time detection and control of weeds developed by engineers and computer scientists be integrated into a weed management system (Young, 2012)?

Although IWM is widely accepted within the weed science community as one of the best ways to manage weeds, it is not widely practiced because current systems fail to meet the performance expectations of growers (Young et al., 2017). Why is that true, and how can it be corrected?

It is likely that successful IWM systems will have to be developed rapidly within the opportunities and constraints of agricultural industrialization. Industrialization is a process whereby agricultural production is structured under the pressure of increasing levels of capital and technology in a way that allows management systems to integrate each step in the economic process to maximize the efficiency of capital, labor, and technology (Urban, 1991, Keeney, 1995). A primary question is whether this process is compatible with and capable of achieving sustainable agriculture. Keeney (1995) thinks that a sustainable rural landscape can be achieved if agriculturalists, environmentalists, and the government work together. If they do not, "today's haphazard and divisive times" will continue. Urban believes the industrializing forces of consumer desires and demand, prescription agricultural products, molecular biology, and the changing nature of farming combine to make agricultural industrialization inevitable. Modern industrial agriculture built on and dependent on scientific knowledge is the only way to feed the 9 billion people expected on the planet by 2050. Others, with equal passion and commitment (e.g., Wendell Berry, Wes Jackson, and Fred Kirschenmann), argue if the dominance of energy, chemical, and capital-dependent agriculture is not drastically changed, the 9 billion will not be fed because the system is not sustainable (Mann, 2012). Without change, the demise of the Jeffersonian agricultural heritage of farming is inevitable. The result of the debate will affect weed science. IWM should not limit its focus to weed control. To be successful, the focus must be the total vegetation complex—habitat management rather than weed control in a year in a crop. Perhaps it is most correct to say that industrialization should, although it may not, change the scale of concern. Sustainable IWM systems should maintain environmental quality and future generations. These are large-scale concerns. Small-scale concerns, such as how to control weeds in a crop in a year, have dominated, but future agricultural systems will require major changes.

Fernandez-Quintanilla et al. (2007) claim the most difficult aspect may be the "need to anticipate the future and change faster than the world around us." Environmental concerns demand large-scale thought. Small-scale thought suffices for individual concerns. Large thoughts are needed for large systems. Everything needs to be integrated to have a complete crop-management system. It will not be easy to do. It is necessary.

2.9 *Perennial Crops*

Weed scientists and all engaged in growing crops need to consider the future role of perennial crops. Wes Jackson created the Land Institute in Salina, Kansas, to create high-seed-yielding perennial crops. Initially, his quest was widely dismissed because perennials do not produce a lot of seeds. An advantage of perennial crops is that land does not have to be plowed, and soil erosion is reduced. Jackson's model has been the Prairie, which does not have weed, fertility, water, or plant competition problems and survives well without human help. Yes, prairies do not produce high-yielding crops, but the Land Institute has demonstrated that high-yielding crops are possible.

Kernza, a perennial grain, has sparked the imagination of farmers, food enthusiasts, bakers, brewers, conservationists, and climate activists. The Chinese are developing perennial rice with yields comparable to annual rice. Soils with perennial crops can accumulate almost a ton of organic carbon/ha/year. Farmers use much less labor and spend less on fertilizer and other inputs. Farmer profits have increased. The most recent development is Baki bean, a perennial version of Sainfoin and a forage crop in the northwestern United States. It is a drought-tolerant legume and will not need supplemental nitrogen fertilizer to succeed. It has protein levels similar to soybean with fat content close to chickpeas.

It is likely that high seed-yielding perennial crops will have weed problems and may become weeds. It is also likely that an IWM approach will be required. Much research remains.

3 Herbicide Technology

Herbicides are the most successful weed-control technology ever developed. They are selective, not too expensive, fairly easy to apply, have persistence that can be managed, and formulations and chemical types are available. In spite of their advantages, herbicides are far from perfect, even in the eyes of their staunchest advocates. The ideal herbicide would be:

- Effective on a spectrum of weeds or able to control a single species selectively.
- Very selective in at least one major crop and some minor crops.
- Not toxic to nontarget species or humans.
- Persistent in soil but not beyond the period of intense weed competition.
- Easily and quickly degraded to innocuous breakdown products by soil microorganisms or nonenzymatic soil processes.
- Applied postemergence to avoid the prophylactic nature of preemergence use.
- Active at very low rates (mg or g/ha).
- Not leachable and not volatile.

There are some that meet nearly all of these criteria. None are perfect, and much remains to be done to develop better herbicides. There is no adequate justification for abandoning herbicides for weed management, but there is adequate justification for reducing their use. Across the world, it is clear that herbicides dominate weed science research and agriculture. For example, from 2013 to 2022, the percentage of papers dealing with herbicides in Weed Science was 49.3, and in Weed Technology, it was 75.3.

There are four ways to discover new herbicides (Beyer Jr, 1991; Evans, 1992):

1. Random selection of chemicals submitted to targeted biological screening based on carefully developed, targeted biological screening techniques to detect chemicals with activity and selectivity.
2. Screening of derivatives of herbicides with known activity found in method 1.
3. Development of leads from natural products with biological activity. Nature is an intense arena of complex chemical activity. Allelopathy is a possible source. Natural products are inherently complex, and the chemicals frequently have insufficient potency or the wrong activity or selectivity to provide strong leads for chemical synthesis (Evans, 1992).
4. Rational design based on biochemical principles and knowledge of plant physiology and biochemistry.

The most attractive concept is biorational design. The organism to be controlled is considered to be the target, and an enzyme, which is essential to its survival, is chosen for direct attack. Candidate chemicals are developed to inhibit an essential plant function based on biochemical knowledge of the target site. To date, there has been little success, but the technique is advancing (Evans, 1992).

Method 1 testing (screening) a large number of chemicals for possible activity has been the most common and successful method of herbicide discovery. Method 4, biorational design, has become increasingly feasible and useful. The combination of biological performance, toxicological properties, and environmental behavior with computer analysis makes this a powerful technique for herbicide development. It is certain that future screens will more precisely target specific plant processes as the biochemistry of those processes becomes known. Discovery of new herbicides has continued, but no new sites of action have been discovered in more than 25 years (Duke, 2012). Reasons include company consolidation, rational management decisions, and the availability of generic formulations. The advent of sulfonylurea and imidazolinone herbicides reduced herbicide rates by an order of magnitude, from kilograms to grams/ha, a desirable direction from an economic and environmental perspective. Most herbicides are now broadcast over the entire target area; some may be applied in a band over plant rows. Advances in application technology and the use of artificial intelligence (AI) combined with GPS systems will allow herbicide application only to weeds rather than an entire field. When knowledge of precise thresholds is combined with precise application, significant reductions in the amount applied per hectare will be achievable. An essential question is, what is the minimum amount that can be applied to achieve the desired control?

It is a significant advantage and a major problem that herbicides persist in soil. Excessive persistence may affect succeeding crops, lead to contaminated ground or surface water, or cause undesirable residues in crops. Some persistence is good because it gives weed control over time, avoids repeated herbicide application, and eliminates or reduces the necessity of using other weed control methods in the crop. Regulating or controlling soil life is a desirable goal for future herbicide development. If non-persistent herbicides could be given a few weeks of persistence, and the soil life of herbicides with long persistence but desirable activity and selectivity could be shortened, it would be good. If these things are achieved, it will probably be through controlled-release formulations that have already achieved some success (e.g., encapsulation). The technology for these formulations is a special challenge in the complex soil environment.

Major herbicide manufacturers have developed safe, properly disposable packaging to protect user safety. Systems to minimize hazards of herbicide handling and container disposal and minimize human exposure will become more common. Safety and efficiency demand adaptations of formulations and application methods. Herbicide labels and use instructions in the world's developed countries are explicit, readily available, often lengthy, and contain adequate instructions for approved uses. Sadly, the same is not true in much of the rest of the world. Labels have to be developed that are clear, simple, adequate to the task, and not too complex. When potential users may be illiterate, clear instructions (often pictorial) are imperative. It is a difficult challenge to create instructions that combine the need for clarity and accuracy, given the complexity of herbicide chemistry and use.

Innovation and progress in herbicide development depend on the agrichemical industry. Consolidation has led to domination by a few large, world-oriented European and US companies. There is a large market in developed countries and limited prospects for quick expansion in the large, diffuse market. The trend toward no- and minimum-tillage, evolving soil tillage practices, genetic modification to create herbicide resistance in crops, and the changing weed spectrum in crops will affect the herbicide market.

Mergers among agrichemical companies began in 1986. In 1999, five European and four US companies dominated the world herbicide market. In 2022, the 10 largest agrochemical companies were Chem-China (China), Bayer (Germany), BASF (Germany), Corteva, FMC (United States), Adama (Israel), UPL (India), Sumitomo (Japan), NU Farm (Australia), and HuaPont (China). There has been no decrease in the availability or number of herbicides. The global agrochemicals market in 2013 was US $41.12 billion, a 43% growth compared to $28.8 billion in 2007. In 2020, it was $231 billion, projected to be $315 billion in 2030. The cost of development for a single agrochemical (that must pay for all failures) was estimated to be $286 million in 2014 (McDougall, 2016). The agricultural chemicals business is not one for the timid. It remains for historians to ascertain the effect of these mergers on agriculture and farmers. There is a downward price pressure due to patent expirations, market saturation, and competition. Biorational approaches, including biological efficacy, toxicology, and environmental interactions, and, presumably, reducing cost while maintaining profit, will dominate the design of new herbicides. These approaches will be combined with more specialized chemical, biological, and safety testing to determine the site of action, use, economic benefit, and environmental acceptability.

The green/environmental movement and widespread public concern helped create ever more restrictive herbicide legislation and regulation, which the public thinks are justified. Some claim that there is more emotion and law than science in decisions that affect herbicide use and development and govern environmental decisions. Restrictive regulations, it is claimed, suppress herbicide development and, therefore, agricultural production.

These brief comments are included to give insight into the business side of agriculture, which may have a greater effect on future weed management programs and the direction of weed science than anything that occurs in research laboratories. Herbicides and herbicide-resistant crops have made significant changes in the way weeds are controlled. Swanton and Valente (2018) argue that the success of this technology has led to the development of cropping systems that rely primarily on these developments rather than using them as components of IWM programs.

It is logical to assume that major herbicide companies have research programs to incorporate herbicide resistance into crops (transgenic, GM). Success has been achieved with several herbicides and several major crops: maize, soybean, wheat, rice, cotton, and tobacco. From the introduction in the mid-1980s to 1994, more than 1500 approvals for field testing of a wide range of transgenic organisms were granted, 40% for herbicide tolerance (Hopkins, 1994). Between 1996 and 1999, the world's commercial area (albeit dominantly in the United States) planted with transgenic crops increased from 1.7 to 39.9 million ha (James, 1999) and to 81 million ha in 2004, a steady double-digit growth rate. By 2005, 8.3 million farmers (http://www.isaaa.org) grew transgenic varieties on more than 90 million ha in 21 countries and 983 million ha in 2021. Transgenic crops (GMs) have been the most rapidly adopted technology in agricultural history. Soybeans are the most common GM crop, and maize is second.

In 2005, more than 7 million of the 8.5 million adopters were small-holder farmers in the world's developing countries (Chassy et al., 2005). For example, transgenic cotton was first planted in China in 1998 and quickly grew to 50% of the cotton acreage. Yields improved 10–30%, pesticide use declined 50–80%, and farmer profit increased (Chassy et al., 2005). Insecticide use in agriculture has declined due to transgenic crops (Chassy et al., www.ers.usda.gov/publications/aer810/). But Lang (2006) warns that all is not rosy. Cotton seed resistant to bollworm was planted widely, worked well, and failed. After 7 years, bollworms resistant to Bt cotton have not appeared, but abundant secondary pests have. Farmers are now spraying cotton up to 20 times a season to control mirids previously controlled by insecticides used to control bollworm. Nevertheless, the great majority of small-scale, resource-poor farmers in developing countries are trying GM crops.

Transgenic crops have been available for some time, and one might ask why they are discussed in a chapter on the future. The reason is that although we know what has been done with agricultural biotechnology and herbicide-resistant crops, the technology is still new and changing rapidly. Scientists do not know, perhaps cannot know, what might be done. The direction of research is clear, but the final destination is not.

Modernity in agriculture cannot be identified with a particular technological or social breakthrough (Kirsch, 2016). It is often a subjective condition, a feeling, or intuition because 'things' are in some profound sense different. We cannot be sure what new possibilities will be discovered as the technology continues to develop. The adoption of molecular-based research will bring a new dimension and can have "far-reaching benefits in agriculture and biotechnology" (Marshall, 2001). One potential, presently unachieved, benefit of genomics research is the discovery of new targets for herbicide action (Hess et al., 2001). Other benefits may include the identification and use of genes to improve a crop's competitive ability (e.g., early shoot emergence, rapid early growth, fast canopy closure, and production of allelochemicals). Genomics may also discover genes that contribute to weediness, a plant's perennial growth habit, seed dormancy, or allelopathy (Weller et al., 2001).

Criticism of herbicide-resistant crops is common and is usually related to all or some of four perceived risks: public health, environmental, social, and weed control (see Zimdahl, 2022 for a discussion on ethical concerns). An important concern is the possibility of environmental, water, or food contamination from increased herbicide use. As biotechnology and herbicide-resistant crops develop, it is important to remember that no technology has ever proved to be perfectly safe. Scientists can measure risk, whereas safety is a subjective matter. Science deals with the probability of risk, which effectively prohibits absolute verification of safety. In addition, scientists cannot prove a negative—that something will not happen. However, it is also logical to conclude that the absence of evidence is not an evidence of the absence of possible harm. This technology has, as all technologies have (e.g., herbicides, cell phones, and computers), good and bad uses. We must be cautious about demonizing the potential but unknown bad effects of legitimate uses by good people and weigh them carefully against illegitimate uses by bad people.

Hubbell and Welsh (1998) suggest that transgenic crops have the potential to create a more sustainable and environmentally favorable agricultural system than chemically based systems but will fail "in enabling a fully sustainable agriculture." They claim genetic traits with a higher potential of enabling truly sustainable agricultural systems have not been developed for three reasons:

1. Lack of US Environmental Protection Agency and USDA regulatory policies that specifically promote sustainable traits.
2. An agricultural biotechnology industry dominated by companies.
3. Patent law and industry policies that prevent farmers from saving transgenic seed and thus tailoring transgenic crops to local ecological conditions.

Social concern is related to a fear that technology favors large farms and will lead to the loss of small farms, small-scale farmers, and rural communities. It is estimated that 1.4 billion small-scale farmers grow 15–20% of the world's food from seed saved from the previous year's crops grown by women are not sold, eaten, or bartered locally and are not included in national production figures. Loss of the ability to save seed may drive these farmers out of farming (Sexton et al., 1998). A secondary concern is the cost of food production, and food costs to the consumer may rise.

The following are three concerns related to weed control:

1. More weeds will become resistant to one or more herbicides because of the continued use of a herbicide to which a crop is resistant (Sandermann, 2006).
2. Resistant gene flow to sexually compatible plants is acknowledged as a potential risk of introducing any genetically engineered (GM) crop. The risk is especially great when a crop and weed are closely related and can interbreed.
3. Resistant crop plants may become hard-to-control volunteer weeds.

The legitimate concerns of epistasis and pleiotropy must be recognized. Epistasis is the suppression of gene expression by one or more other genes, and pleiotropy is when a single gene exerts simultaneous effects on more than one character. In short, an ecological rule applies: It is impossible to do just one thing. When scientists manipulate a genome, specific outcomes are intended, and even when these are achieved, other unplanned things may also occur. Genetic engineering, with the best intentions to do a good thing, may do unexpected things that could be good or bad.

A common critique of herbicide-resistant crops is the technology will promote, not decrease, herbicide use while promoting an unsustainable, chemical, energy, and capital-intensive monocultural agriculture. Herbicide-resistant crops may reinforce farmers' dependence on outside, petroleum-based, potentially polluting technology. An associated concern is that there is no technical reason to prevent a company from choosing to develop a crop resistant to a profitable herbicide that has undesirable environmental qualities such as persistence, leachability, and harm to nontarget species.

Development of herbicide-resistant crops will continue. There are important advantages justifying their development. The technology will provide lower-cost herbicides and better weed control. These powerful, favorable arguments can, but may not, lead to lower food costs for the consumer. Herbicide-resistant crops are providing solutions to intractable weed problems in some crops. Glyphosate resistance has been created in several crops. It is an environmentally favorable herbicide, and therefore, Hubbell and Welsh (1998) argue that it is better to use it in lieu of non-environmentally favorable herbicides. An important argument in favor of the technology is that it has the potential to shift herbicide development away from initial screening for activity and selectivity and later determination of environmental acceptability to the latter occurring first. Resistance to environmentally favorable herbicides that lack adequate selectivity in any crops or in a major crop so their development will be profitable could be engineered, and the herbicide's usefulness could expand. This has important implications for minor crops (e.g., vegetables and fruits) where few herbicides are available because the market is too small to warrant the cost of development. If resistance to a herbicide, successful in a major crop (e.g., cotton), could be engineered into a minor crop, manufacturers and users would benefit.

Biotechnology was discussed by Christianson (1991), a self-acknowledged outsider. His view is quoted here: "I think it would be a pity if the power of the use of mutants and mutation to uncover and describe physiology and development were limited, in the hands of weed scientists, to the isolation and description in yet another species of yet more genes that confer resistance to yet another herbicide." Transgenic crops have developed more rapidly than any preceding agricultural technology (see Hileman, 1995). This brief chapter does not analyze the controversy in depth. Others have done this (Duke, 1996, 2006; Zimdahl, 2022). Scientific journals and popular press articles, too numerous to mention, are available. Much more work will be done and discussed, but it is important to realize that GM technology is widely promoted, accepted, and used. Specter (2016) reminds us that the real question is not whether genetic modification will occur. It is how it will be used.

4 Organic Agriculture

Organic agriculture, with some exceptions, has not been tried and found wanting; it has been found difficult and left untried. Evans (2002, p. 14) offered the same message: Because weeds are inextricably both products of psychology and ecology, weed problems are best addressed by considering not only the agroecosystems that produce them but also the culture that informs how we think and farm. Recognition of this point is potentially threatening and subversive, for it challenges the very social structure of farming upon which the employment of weed scientists depends. They are, in effect, servants of large-scale, single-crop, commercial agriculture, and if that disappears, so too will a large proportion of their jobs.

Weed scientists correctly take pride in achieving agricultural production where no weeds exist, only the crop. That may be the wrong goal. The goals of organic agriculture encourage us to think about what Evans (2002 p. 51) calls a "blindly oppositional attitude" that led scientists to frantically search for immediate solutions to an ever-worsening weed problem. The search often obscured the need for different approaches to weeds and to changes in developed country farming systems. Herbicides helped weed scientists define weeds as the enemy, and only in the late 1990s did this begin to change. Dependence on herbicides led weed scientists and farmers to neglect other weed management strategies such as crop rotation, tillage, fertility management, elimination of herbicides, and a change of attitude about what a weed is and does (Harker, 2004), which organic agriculture has encouraged. The demands and popularity of organic agriculture may compel weed scientists to broaden their horizons and consider other weed management tools more carefully. Proceedings of volume 1 of the second scientific conference of the International Society of Organic Agriculture Research (ISOFAR) will be helpful to those involved in organic agriculture (Neuhoff et al., 2008).

5 Engineering Research

No matter how selective or active herbicides are, they have to be applied in the environment to fulfill their purpose. Over 90% of all herbicides annually applied in the United States are sprayed. Technology has evolved, and application equipment is becoming more precise. Is there a better way? Some of the sprayed herbicides never hit the target. It seems unlikely that the spray not hitting the target could be recovered and reused or at least handled so it does not reside in the environment without fulfilling its intended purpose. Low-volume and ultra-low-volume application techniques are available but not widely used. Reducing the volume of spray required may be possible.

The first robots for agricultural use were developed in the 1920s to incorporate automatic vehicle guidance. The development of the computer in the 1980s made machine vision guidance possible. As with any technology, there are advantages and disadvantages. Robots have higher speeds and closer tolerances, can operate with fewer errors, can or will soon be able to distinguish crops and weeds, sense diseases and other stresses, and can be adapted for weed control and fertilization. Robots are well adapted for time-consuming activities, including weed control, cultivation, and mowing. However, research and development costs are high, and robots are expensive. They require maintenance and repair from people who understand them. It is unlikely they will have much market in developing countries for a few decades.

Robotic sprayers are being developed with remote-sensing technology to detect weeds and spray only where weeds are. Other techniques sense variations in soil type and adjust herbicide concentration to account for differences. Automated detection and identification of weeds and crops are required for the development of practical, affordable site-specific weed management and crop harvesting (e.g., strawberries and lettuce) (Brown & Noble, 2005). However, before site-specific systems can be perfected, accurate, affordable techniques to sample weed populations must be developed (Wiles, 2005). Sampling to determine the location and identity of the weeds is a fundamental principle of IWM. Broadcast herbicide application reduces or eliminates the need to know specific weed identity, but the robot must distinguish weeds from crop plants because weeds are not uniformly distributed in the field. Sampling is a decision-making tool not fully exploited by weed science primarily because there are no simple, affordable methods for weed sampling. Machine vision technology with appropriate data processors can recognize crop growth patterns, use automated systems (steam) to remove intra-row weeds, and will continue to become more effective (Fennimore et al., 2016). These are particularly applicable in specialty crops such as flowers and vegetables where appropriate herbicides are often not available.

In addition to affordable sampling techniques, research necessary to perfect remote sensing technology includes the need for artificial lighting, definition of spectral band requirements, techniques to recognize different weeds and distinguish them from the crop, rapid image processing, multiple spatial resolution systems, and techniques for dealing with multiperspective images (Brown & Noble, 2005). A less sanguine appraisal of the profitability and economic feasibility of site-specific weed management was based on the advantages of reduced herbicide use, the additional costs of scouting for weeds, preparing treatment maps, and applying herbicides only to identified patches (Swinton, 2005).

Herbicide application is not the only area in which engineers can contribute to improved weed management. The effects of tillage over time on recurrent weed populations are not known. There is room for improvement in methods of mechanical control of annual and perennial weeds. Cultivators are available to weed the entire area between crop rows and leave only a narrow band over the row. This is a great improvement over a person with a hoe or an animal-drawn cultivator, but the person with a hoe is selective. Further research will reveal even more specialized tillage and cultivation methods for weeds in crops. Organic farmers use cultivation and several tillage implements effectively.

6 Other Challenges

6.1 *Scientific*

Several other research areas should be considered when planning weed science's future. They include:

- The value, advantages, and disadvantages of monoculture agriculture.
- What is the role of companion cropping and regular inclusion of cover crops in weed management? Can weeds be cover crops? (See Young, 2020).
- What are the long-term effects of soil erosion after regular plowing and cultivation? One effect is all too apparent in the brown color of rivers (Logan, 1995; Montgomery, 2007).
- The future and influence of perennial crops.

Weed scientists were not too concerned with long-term effects when the science was developing. Weeds decreased crop yield—a detrimental long-term effect. The vision did not extend much further because solving the weed problem was a sufficient challenge. Any technology used for enough time has demonstrable environmental and social effects. A longer-term view will help reveal these effects and compel their consideration before widespread use is achieved.

Weed scientists must begin to work more closely with economists who ask what it costs and what it is worth. What is it worth to do the work to develop a more competitive cultivar, deplete the soil seed bank, and achieve assurance of 80–100% weed control? What will it be worth to be able to predict weed problems? No one knows, but the answers are important to IWM systems.

How will nanotechnology affect weed science? Nano integrates biological material with synthetic materials to build new molecular structures. Synthetic biology goes beyond moving existing genes to creating new ones programmed to perform specific tasks. It operates at the nanoscale (one billionth of a meter = 10 – 9 m) of living and nonliving parts. It has enormous potential for good and harm (Shand & Wetter, 2006).

Weed scientists are aware of the scientific research opportunities and challenges. There are equally important, although less discussed social and moral challenges. The primary goal of agricultural scientists has been to develop technologies that enable achieving the maximum yield of a few crops in the world's developed (rich) countries. It is a good goal, but one must ask if it is the right goal (see Kirschenmann, 2012; Zimdahl, 2022). A few relevant questions follow.[1] Liu et al. (2015) found that cultural practices have negative effects on global food production. "It is crucial for agricultural sustainability to increase crop yields and simultaneously decrease environmental impacts of agricultural intensification" (Liu et al., 2015).

6.2 Sustainability

Achieving sustainable agriculture is a goal all agricultural scientists share. Even a cursory review of current writing on agriculture reveals that achieving sustainability has obtained the generally revered status of motherhood with one important difference. There is little debate about what motherhood is or its worth and goodness. The difference is that in spite of the nearly universal adulation of agricultural sustainability; there is little agreement on its nature, what is to be sustained, or how it is to be accomplished (Zimdahl, 2022, p. 135; Gliessman, 2001).

Production is and always will be important, but it is not possible to create sustainable agriculture without a sustainable culture. The reverse is also true (LeVasseur, 2010). It is impossible to have a serious, comprehensive discussion of sustainable agriculture without including community and culture (Holthaus, 2009). Within the agricultural community achieving sustainability is viewed as mainly or wholly technical in nature. It requires different farming methods and the adoption of alternative technologies (Morgan & Peters, 2006), which will be significantly aided by advances in biotechnology. This view ignores the moral, educational, and political tasks that must be considered. In Morgan and Peters' view, it requires a commitment to "philosophical principles that depart from the utilitarian premises of industrial agriculture." It is a demanding task that requires new thinking and a change in attitude toward the earth. It requires us to cease attempting to achieve dominion over the earth and achieve humility and reverence before the world (Berry, 2002). The majority of the mainstream agricultural community does not agree with Liu et al. (2015). The dominant view has been that crop intensification is the best route to feed

[1] Is it more important than enabling the poor of the world to feed themselves? Can the seemingly unending task of discovering new technologies to maximize yields lead to a sustainable agricultural system to feed 9 billion or more people? Is maintaining rural communities a proper goal for agricultural science, or is that someone else's task? Should achieving maximum yield and profit always take precedence over preserving the environment?

9 billion people and protect the environment. But there is no room for complacency, especially since it is invoked by some advocates of biotechnology (Fischer et al., 2014).

Finally, a caution. Those engaged in agriculture and its sub-discipline, weed science, possess a definite but unexamined moral confidence or certainty about the correctness of what they do. The basis of moral confidence is not obvious to those who have it or to the public. Agriculture's unexamined moral confidence is potentially harmful. It is necessary for all engaged in agriculture to analyze what it is about their science and their society that inhibits or limits their science. All should strive to nourish and strengthen the beneficial aspects and change those that are not. To do this, agricultural people must be confident to study themselves, their science, and its institutions, and be dedicated to the task of modifying the goals of both (Zimdahl, 2022).

7 Agriculture's Human Dimension

Doohan et al. (2010) claim that "the human dimension of weed management is most evident when farmers make decisions contrary to science-based recommendations." Agricultural scientists and administrators may be aware that their recommendations are often ignored but usually do not ask why because such questions are beyond their area of expertise. Scientists do science, leading to science-based recommendations. When recommendations are ignored, the reasons could be economic (too expensive), stubbornness, lack of trust, and different perceptions of risk and benefit. Doohan et al. (2010) argue that farmers exhibit an inverse relationship between perceived risk and benefit. If any technology is regarded as beneficial, it is automatically perceived as low risk, which, of course, is not true. Ignoring farmers' reasons is perilous for agriculture's future.

Agricultural scientists have contributed to increasing crop production over several decades. Pesticides have been the primary control technique (Fernandez-Cornejo et al., 2014). Because of their efficacy and ease of use, there has been over-reliance on them at the expense of other control methods (Blackshaw et al., 2008). If the only or primary goal of weed science is to increase production, the quest for better herbicides must continue. If the goal is sustainable weed management in a sustainable environment and society, other control techniques must be investigated and integrated. Research on non-herbicide weed management must show low or no risk of crop failure and reduced profit. The goal should be the development of successful weed management systems with minimal or no effect on the flora and fauna of soil, water, or air and no adverse effects on people or other creatures.

Scientists and others engaged in agriculture are, not by nature or choice, politicians. Failure or inability to consider that we live in a political world and are affected by it is a prescription for disappointment or disaster. Political considerations affect our daily life. A major political accomplishment in some countries is cheap food, especially in urban areas. It affects the way we practice agriculture and manage weeds. If the government removed itself from agricultural policy-making and markets, cheap food might disappear.

Given the agricultural and environmental history, concern about environmental pollution from agriculture is a fairly recent political development. It was not too long ago that pesticide use in agriculture meant prosperity and progress rather than human harm, environmental pollution, and lack of corporate responsibility. For example, a study commissioned by the American Farm Bureau, an organization noted for its defense of agriculture (King, 1991), showed that only 15% of the American public was in favor of abolishing pesticide use in agriculture. However, 66% of those surveyed thought pesticide use should be limited in the future, and 38% thought farmers were using more pesticides than they had in the past. Such information and concern have political meaning and consequences. About 70% of US agricultural produce harbors some trace of pesticides (Gross, 2019). Such challenges are often dismissed by the agricultural community because they are regarded as biased, irrelevant, and lacking in supporting scientific evidence. An example is in the Consumer Reports article by Roberts (2024). The findings are ignored or dismissed by those who willfully ignore the effects of criticism on political action. Political acts change things, and agriculture has to recognize and work in a political milieu or suffer the consequences of regulation by those who do.

8 Conclusions

The American author and farmer Wendell Berry (1981b) has often written eloquently about problems facing American agriculture and their solutions. He advocates solving for patterns. "To the problems of farming, then, as to other problems of our time, there appear to be three kinds of solutions." The first solution causes a ramifying series of new problems. The only limitation of the new problems is that they "arise beyond the purview of the expertise that produced the solution." Those who are encumbered by the new problems are not those who devise solutions for the old problems. This kind of solution shifts the burden away from those who created the problem.

The second solution is one that immediately worsens the problem it is intended to solve. These are often quick-fix solutions that, within weed science, take the form of questions such as: what herbicide will kill the weed? Adopting this kind of problem-solving leads to the need for more quick-fix solutions. Everyone who has tried to fix something is familiar with this kind of solution. What was tried first did not work, and some studies (but perhaps little knowledge) revealed that loosening another bolt or screw would do it. Alas, loosening that screw was the wrong thing to do because it loosened other things, and suddenly parts were everywhere, and neither the source of each part nor a way to fit them back together was known. The third, most desirable solution creates a ramifying series of solutions. These solutions make and keep things whole. For Berry (1981b), a good solution is one that acts constructively on the larger pattern of which it is a part. It is not destructive of the immediate pattern or the whole. Good solutions solve for the whole system, not for a single goal or purpose.

Those who create the next generation of integrated, sustainable agricultural production systems for simple and complex problems will do well to remember Berry's admonition as they search for solutions. One must know the whole system and devise solutions that create more solutions to maintain the pattern and improve the system. Agriculture's inevitable problems should be viewed as a good family physician views patients—in family, not just individual terms. It is the entire system, not just the current problem, that must be managed.

Contributing to the elimination of hunger in the world is a proper goal for weed science. It is a goal consistent with the Millennium goals of the UN (Sachs, 2005, pp. 211–212). Two of the goals are relevant to agriculture and worthy of attention. These large-scale goals include eradicating extreme poverty and reducing hunger by half, and ensuring environmental sustainability.

Although progress has been made, neither goal has been achieved. In his Recollected Essays, Berry (1981a, p. 98) writes eloquently about a vision of the future shared by those who want to create alternative futures, including alternative, improved, and sustainable agricultural systems. His words are a good place to end thoughts about the future. Readers may determine if I have reached beyond my knowledge and ability.

"We have lived by the assumption that what was good for us would be good for the world. We have been wrong. We must change our lives so that it will be possible to live by the contrary assumption that what is good for the world will be good for us. That requires we make the effort to know the world and to learn what is good for it. We must learn to cooperate in its processes and to yield to its limits. Even more important, we must learn to acknowledge that creation is full of mystery; we will never clearly understand it. We must abandon arrogance and stand in awe. We must recover the sense of the majesty of the creation and the ability to be worshipful in its presence. For it is only on the condition of humility and reverence before the world that our species will be able to remain in it." Berry's challenge is clear—change requires more than the contemplation of fixed verities. It must move beyond reproducing the qualities of the science to which we have devoted our careers.

References

Bellah, R. N., Madsen, R., Sullivan, W. M., Swidler, A., & Tipton, S. M. (1991). *The good society* (p. 97). A.A. Knopf.

Berry, W. (1981a). A native hill. In *Recollected essays. 1975–1980* (pp. 73–113). North Point Press.

Berry, W. (1981b). Solving for pattern. In *The gift of good land* (pp. 134–145). North Point Press.

Berry, W. (2002). *The art of the commonplace* (p. 330). Shoemaker and Hoard.

Beyer, E. M., Jr. (1991). Crop protection—Meeting the challenge. *Proceedings of the British Crop Protection Conference—Weeds, 1*, 3–22.

Blackshaw, R. E., Harker, K. N., O'Donovan, J. T., Beckie, H. J., & Smith, E. G. (2008). Ongoing development of integrated weed management systems on the Canadian Prairies. *Weed Science, 56*, 146–150.

Brown, R. B., & Noble, S. D. (2005). Site-specific weed management: Sensing requirements—What do we need to see? *Weed Science, 53*, 252–258.

Buhler, D. D. (2006). Approaches to integrated weed management. In H. P. Singh, D. R. Batish, & R. K. Kohli (Eds.), *Handbook of sustainable weed management* (pp. 813–824). Food Products Press.

Buhler, D. D. (2017). 50th anniversary—Invited article: Challenges and opportunities for integrated weed management. *Weed Science, 50*, 273–280.

Buhler, D. D., Hartzler, R. G., & Forcella, F. (1997). Implications of weed seed bank dynamics to weed management. *Weed Science, 45*, 329–336.

Chandrasena, N. R. (2023). *The virtuous weed: Weedy by name only not weedy by nature* (474 pp). Vivid Publishing.

Chassy, B. M., Parrott, W. A., & Roush, R. (2005, October). Crop biotechnology and the future of food: A scientific assessment. *CAST Commentary*, p. 6. QTA 2005-2.

Chauhan, B. S., Matloob, A., Mahajan, G., Aslam, F., Florentine, S. K., & Jha, P. (2017). Emerging challenges and opportunities for education and research in weed science. *Frontiers in Plant Science, 8*, 1537.

Christianson, M. L. (1991). Fun with mutants: Applying genetic methods to problems of weed physiology. *Weed Science, 39*, 489–495.

Cummings, C. H. (2006). Ripe for change: Agriculture's tipping point. *World Watch, 19*(4), 38–39.

Davis, A. S., Hall, J. C., Jasieniuk, M., Locke, M. A., Luschei, E. C., Mortenson, D. A., Riechers, D. E., et al. (2009). Weed science research and funding: A call to action. *Weed Science, 57*, 442–451.

Dawson, J. H. (1965). Competition between irrigated sugar beets and annual weeds. *Weeds, 13*, 245–249.

De Groote, H., Wangare, L., & Kanampiu, F. (2007). Evaluating the use of herbicide-coated imidazolinone-resistant (IR) maize seeds to control Striga in farmers' fields in Kenya. *Crop Protection, 26*, 1496–1506.

Doohan, D., Wilson, R., Canales, E., & Parker, J. (2010). Investigating the human dimension of weed management: New tools of the trade. *Weed Science, 58*, 503–510.

Duke, S. O. (Ed.). (1996). *Herbicide-resistant crops: Agricultural, environmental, economic, regulatory, and technical aspects* (p. 420). CRC-Lewis Pub.

Duke, S. O. (2006). Taking stock of herbicide-resistant crops ten years after introduction. *Pest Management Science, 61*, 211–218.

Duke, S. O. (2012). Why have no new herbicide modes of action appeared in recent years. *Pest Management Science, 68*, 505–512.

Economist. (2022, November 5). Room at the bottom—Special report climate adaptions, 8–9.

Evans, D. A. (1992). Designing more efficient herbicides. *Proceedings of First International Weed Control Congress, 1*, 34–41.

Evans, C. L. (2002). *The war on weeds in the prairie west: An environmental history* (p. 309). University of Calgary Press.

Fennimore, S. A., Slaughter, D. C., Siemens, M. C., Leon, R. G., & Saber, M. N. (2016). Technology for automation of weed control in specialty crops. *Weed Technology, 30*, 823–837.

Fernandez-Cornejo, J., Nehring, A., Osteen, C., Wincester, S., Martin, A., & Vialou, A. (2014). *USDA–Economic research service*. Economic information bulletin no. 124. 86 pp. www.ers.usda.gov.

Fernandez-Quintanilla, C., Quadranti, M., Kudsk, P., & Barberi, P. (2007). Which future for weed science? *Weed Research, 48*, 297–301.

Fischer, T., Byerlee, D., & Edmeades, G. (2014). Crop yields and global food security—Will yield increase continue to feed the world. *Australian Centre for International Agricultural Research*, 634. Monograph #158.

Gliessman, S. R. (2001). *Agroecosystem sustainability—Developing practical strategies* (p. 210). CRC Press.
Gould, S. J. (2002, January). On the origin of specious critics. *Discover*, 34–42.
Gressel, J. (2009). Crops with target-site herbicide resistance for Orobanche and Striga control. *Pest Management Science, 65*, 560–565.
Gressel, J., Hanafi, A., Head, G., Marasas, W., Obilana, A. B., Ochanda, J., Souissi, T., & Tzotzos, G. (2004). Major heretofore intractable biotic constraints to African food security that may be amenable to novel biotechnological solutions. *Crop Protection, 23*, 661–689.
Gross, L. (2019). More than 90% of Americans have pesticides or their byproducts in their bodies—the real risk from chemicals in our food—for farmers and children, in particular are being ignored. *The Nation*. https://www.thenation.com/article/archive/pesticides-farmworkers-agriculture. Accessed November 2022.
Hall, J. C., Van Errd, L. L., Miller, S. D., Owen, M. D. K., Prather, T. S., Shaner, D. L., et al. (2000). Future research directions for weed science. *Weed Technology, 14*, 647–658.
Hamill, A. S., Holt, J. S., & Mallory-Smith, C. A. (2004). Contributions of weed science to weed control and management. *Weed Technology, 18*, 1563–1565.
Harker, K. N. (2004). My view. *Weed Science, 52*, 183–184.
Harker, K. N., O'Donovan, J. T., Blackshaw, R. E., Beckie, H. J., Mallory-Smith, C., & Maxwell, B. D. (2012). Our view. *Weed Science, 60*, 143–144.
Hess, F. D., Anderson, R. J., & Reagan, J. D. (2001). High throughput synthesis and screening: The partner of genomics for discovery of new chemicals for agriculture. *Weed Science, 49*, 249–256.
Hileman, B. (1995). Views differ sharply over benefits, risks of agricultural biotechnology. *Chemical and Engineering News, 73*, 8–17.
Holthaus, G. (2009). *From farm to the table: What all Americans need to know about agriculture* (363 pp). University Press of Kentucky.
Hopkins, W. L. (1994). *Global herbicide directory* (1st ed., pp. 157–159). Ag. Chem. Info. Services.
Howard, A. S. (1940). *An agricultural testament* (253 pp). Oxford University Press.
Hubbell, B. J., & Welsh, R. (1998). Transgenic crops: Engineering a more sustainable agriculture. *Agriculture and Human Values, 15*, 43–56.
James, C. (1999). Global review of commercialized transgenic crops 1999. In *International service for the acquisition of agri-biotechnology applications* (Briefs no. 12 preview). International Service for the Acquisition of Agri-biotechnology.
Jordan, N., Schut, M., Graham, S., Barney, J. N., Childs, D. Z., Christensen, S., et al. (2016). Transdisciplinary need research: New leverage on challenging weed problems? *Weed Research, 56*, 345–358.
Julien, M., McFadyen, R., & Cullen, J. (Eds.). (2012). *Biological control of weeds in Australia* (620 pp). CSIRO Publishing—Nature.
Keeney, D. (1995). Can sustainable agriculture landscapes accommodate corporate agriculture? In *Environmental enhancement through agriculture. Proceedings of a conference* (pp. 173–181). Tufts University.
King, J. (1991). A matter of public confidence. *Agricultural Engineering, 72*(4), 16–18.
Kirsch, A. (2016, September 5). What makes you so sure? *The New Yorker*, 71–74.
Kirschenmann, F. (2012). From commodities to communities. *Leopold Letters, 24*(1), 5.
Kuhn, T. S. (1970). *The structure of scientific revolutions* (2nd ed., 210 pp). University of Chicago Press.
Lang, S. (2006). Seven-year glitch: Cornell warns that Chinese GM cotton farmers are losing money due to secondary pests. *Chronicle of Higher Education*. July 25. Online, http://www.newscornell.edu/stories/July06.Bt.cotton.China.ssl.html.
Lemonick, M. D. (2006). Are we losing our edge? *Time, 167*, 22–30.
LeVasseur, T. J. (2010). Gary Holthaus: From farm to the table: What all Americans need to know about agriculture. *Journal of Agricultural and Environmental Ethics, 23*, 301–302.

Li, C. L., Dong, J., Durairaj, J., Guan, J. C., Yoshimura, M., Quniodoz, P., et al. (2023). Maize resistance to witchweed through changes in strigolactone biosynthesis. *Science, 379*(6627), 94–99.

Liu, Y., Pan, X., & Li, J. (2015). Current agricultural practices threaten future global food production. *Journal of Agricultural and Environmental Ethics, 28*, 203–216.

Logan, W. B. (1995). *Dirt—The ecstatic skin of the earth* (202 pp). Riverhead Books.

MacLaren, C., Storkey, J., Menegat, A., Metvalfe, H., & Dehnen-Schmutz, K. (2020). An ecological future for weed science sustained crop production and the environment. A review. *Agronomy for Sustainable Development, 40*(4), 1–29. https://doi.org/10.1007/s13593-020-00631-6

Mann, C. C. (2012, November/December). State of the species. Does success spell doom for Homo sapiens? *Orion*, 16–27.

Marshall, G. (2001). A perspective on molecular-based research: Integration and utility in weed science. *Weed Science, 49*, 273–275.

Marshall, E. J. P. (2010). Weed research reaches volume 50! Looking back and looking forward. *Weed Research, 50*, 1–4.

Marshall, E. J. P. (2019). Reflections on 14 years as Editor-in -Chief. *Weed Research, 59*, 1–4.

Mayer, A., & Mayer, J. (1974). Agriculture, the Island Empire. *Daedalus, 103*, 83–95.

McDougall, P. (2016). The cost of new agrochemical product discovery, development and registration in 1995, 2000, 2005–2008 and 2010 to 2014. R&D expenditure in 2014 and expectations for 2019. *CropLife International*. http://www.croplifeamerica.org/cost-of-crop-protection-innovation-increases-to-286-million-per-product/

Montgomery, D. R. (2007). *Dirt—The erosion of civilizations* (285 pp). University of California Press.

Morgan, P. A., & Peters, S. J. (2006). The foundations of planetary agrarianism. Thomas Berry and Liberty Hyde Bailey. *Journal of Agricultural and Environmental Ethics, 19*, 443–468.

Mortensen, D. A., & Coble, H. D. (1991). Two approaches to weed control decision-aid software. *Weed Technology, 5*, 445–452.

Mortensen, D. A., Egan, J. F., Maxwell, B. D., Ryan, M. R., & Smith, R. G. (2012). Navigating a critical juncture for sustainable weed management. *Bioscience, 62*(1), 75–84.

Moss, S. R. (2008). Weed research: Is at delivering what it should? *Weed Research, 48*(5), 389–393.

Neuhoff, D., Halberg, N., Alföldi, T., Lockeretz, W., Thommen, A., Rasmussen, I. A., et al. (2008). *Proceedings—Second scientific conference of the International Society of Organic Agriculture Research*. Volume 1. Modena, Italy.

Neve, P., Barney, J. N., Buckley, Y., Cousens, R. D., Graham, S., Jordan, N. R., et al. (2018). Reviewing research priorities in weed ecology, evolution and management: A horizon scan. *Weed Research, 58*, 250–258.

O'Donovan, J. T. (1996). Computerized decision support systems: Aids to rational and sustainable weed management. *Journal of Plant Sciences, 76*, 3–7.

Ohlson, K. (2016, March/April). Dirt first—A renegade soil scientist is transforming American agriculture. *Orion*, 14–21.

Parker, C. (2017). Parasitic weeds: A world challenge. *Weed Science, 60*, 269–276. https://doi.org/10.1614/ws-d-11-00068.1

Parker, C. (2021). A personal history in parasitic weeds and their control. *Plants, 10*(11), 2249. https://doi.org/10.3390/plants10112249

Parker, C., & Riches, C. R. (1993). *Parasitic weeds of the world: Biology and control* (332 pp). CAB International. Available in the United States from University of Arizona Press.

Peterson, A. (2000). Alternatives, traditions, and diversity in agriculture. *Agriculture and Human Values, 17*, 95–106.

Ransom, J., Kanampiu, F., Gressel, J., De Groote, H., Burnet, M., & Odhiambo, G. (2012). Herbicide applied to imidazolinone resistant-maize seed as a Striga control option for small-scale African farmers. *Weed Science, 60*, 283–289.

Roberts, C. (2024). *6 fruits and vegetables loaded with pesticides*. Consumer Reports April 18. https://www.theguardian.com/environment/2024/apr/18/what-is-pesticide-safety-organic-fruits-vegetables. Accessed April 2024.

Sachs, J. D. (2005). *The end of poverty: Economic possibilities for our time* (396 pp). The Penguin Press.
Sandermann, H. (2006). Plant biotechnology: Ecological case studies on herbicide resistance. *Trends in Plant Science, 11*, 324–328.
Schweizer, E. E., Lybecker, D. W., & Wiles, L. J. (1998). Important biological information needed for bioeconomic weed management models. In J. L. Hatfield, D. D. Buhler, & B. A. Stewart (Eds.), *Integrated weed and soil management* (pp. 1–24). Ann Arbor Press.
Sexton, S., Hildyard, N., & Lehmann, L. (1998). *Food? Health? Hope? Genetic engineering and world hunger*. Cornerhouse Briefing No. 10.
Shand, H., & Wetter, K. J. (2006). Shrinking science: An introduction to nano technology. In L. Starke (Ed.), *State of the world—2006* (pp. 78–95). W.W. Norton & Co..
Shapin, S. (2006, May 15). Paradise sold. *The New Yorker*, 84–88.
Specter, M. (2016, January 2). Rewriting the code of life. *The New Yorker*, 34–43.
Swanton, C. J. (2022). Weed science and the clock of the long now. *Weed Science, 70*, 369.
Swanton, C. J., & Valente, T. (2018). Chapter 4: Key issues and challenges of integrated weed management. In R. L. Zimdahl (Ed.), *Integrated weed management for sustainable agriculture* (Bureigh-Dodds science publishing series in agricultural science, No. 42) (pp. 69, 453 pp–81). Bureigh-Dodds.
Swinton, S. M. (2005). Economics of site-specific weed management. *Weed Science, 53*, 259–263.
Thompson, P. B. (1995). *The spirit of the soil* (196 pp). Routledge.
Urban, T. N. (1991). Agricultural industrialization: Its inevitable. *Choices, 6*(4), 4–6.
Van Wychen, L. (2016). 2015 survey of the Most common and troublesome weeds in the United States and Canada. *Weed science Society of America National Weed Survey Dataset*. Available: http://wssa.net/wp-content/uploads/2015-Weed-Survey_final.xlsx
Van Wychen, L. (2020). 2020 survey of the Most common and troublesome weeds in grass crops, pasture and turf in the United States and Canada. *Weed Science Society of America National Weed Survey Dataset*. Available: http://wssa.net/wp
Van Wychen, L. (2022). 2022 survey of the most common and troublesome weeds in broadleaf crops, fruits & vegetables in the United States and Canada. *Weed Science Society of America National Weed Survey Dataset*. Available: https://wssa.net/weed/surveys/
Ward, S. M., Cousens, R. D., Bagavathiannan, M. V., Barney, J. N., Beckie, H. J., Busi, R., et al. (2014). Agricultural weed research: A critique and two proposals. *Weed Science, 62*, 672–678.
Weaver, S. (1993). Simulation of crop-weed competition: Models and their application. In *Symposium on weed ecology, expert committee on weeds, Edmonton* (pp. 3–11).
Weller, S. C., Bressan, R. A., Goldsbrough, P. B., Fredenburg, T. B., & Hasegawa, P. M. (2001). The effect of genomics on weed management in the 21st century. *Weed Science, 49*, 282–289.
Westwood, J. R. H., Charudattan, S., Duke, O., Fennimore, S. A., Marrone, P., Slaughter, D. C., Swanton, C., & Zollinger, R. (2018). Weed management in 2050: Perspectives on the future of weed science. *Weed Science, 66*, 275–285.
Wiles, L. J. (2005). Sampling to make maps for site-specific weed management. *Weed Science, 53*, 228–235.
William, R. D., Ogg, A., & Rabb, C. (2001). My view. *Weed Science, 49*, 149.
Wyse, D. L. (1992). Future of weed science research. *Weed Science, 6*, 162–165.
Young, S. L. (2012). True integrated weed management. *Weed Research, 52*, 107–111.
Young, S. L. (2020). A unifying approach for IWM. *Weed Science, 68*, 435–436.
Young, S. L., Pitla, S. K., Van Evert, F. K., Schueller, J. K., & Pierce, F. J. (2017). Moving integrated weed management from low level to a truly integrated and highly specific weed management system using advanced technologies. *Weed Research, 57*, 1–5.
Zimdahl, R. L. (2022). *Agriculture's ethical horizon* (3rd ed., p. 318). Elsevier, Inc.

Chapter 2
Herbicide Residues: Environmental Fate and Impact on Food Safety

Shobha Sondhia

Abstract Globally, herbicides constitute about 47% of the crop protection market. Frequent and non-judicial use leads to residues, persistence, phytotoxicity, and adverse consequences on subsequent crops, non-target organisms, and the environment eventually causing health hazards to human and the animal. The fate of herbicide in soil and other matrix depends on various physiochemical processes. The safety of food and associated health concerns for humans and animals is related to acceptable daily intake (ADI), not to the MRL, which is a legal permissible level of a herbicide. In general, after herbicide application, residues are reported in soil and plants during the crop growth period as well as at maturity. However, there is still lack of sufficient data on long-term ecological impacts, herbicide persistence, and residue interactions with ecosystems. Herbicide residues of 2,4-D, alachlor, acetochlor, atrazine, ametryn, butachlor, bentazone, clomazone, diuron, hexazinone, desethylatrazine, dimethoate, fluroxypyr, glyphosate, imazethapyr, isoproturon, linuron, metolachlor mecoprop, metribuzin, molinate, pendimethalin, pretilachlor, propanil, propazine, simazine, tebuthiuron, trifluralin, terbumeton, terbutryn, terbuthylazine, etc. along with some metabolites were detected in several water samples of Brazil, Burkina Faso, Canada, China, Ethiopia, France, Germany, India, Iran, Lebanon, Luxembourg, Portugal, Slovenia, Spain, the United States, Vietnam, etc. Disturbance of sensitive marine ecosystems due to herbicide residues in the Great Barrier Reef lagoon is also reported. For decontamination of herbicides from various matrix, selection or design of a suitable method requires a full understanding of the physicochemical properties of herbicides. Several herbicide residues can be removed by organic adsorbing materials such as activated carbon, activated charcoal, biochar, nanoorganic materials, carbon nanoparticles organoclays, and bioadsorbents. Since in future, the herbicide use is further likely to increase; therefore, herbicide residue problems will be more frequent and may have increased implications for human health and environment. Therefore, cutting-edge residue monitoring technologies, regional persistence studies, and microbial interventions are needed to optimize herbicide residues and associated adverse consequences.

S. Sondhia (✉)
ICAR–Directorate of Weed Research, Jabalpur, Madhya Pradesh, India

B. S. Chauhan, G. Mahajan (eds.), *Recent Advances in Weed Science*,
https://doi.org/10.1007/978-3-031-91014-2_2

Besides, this, the integrated approach of cultural, mechanical, chemical, and biological weed control practices should be adopted as a long-term weed management strategy to avoid any suspected ill effects of herbicide residues on the environment and other living organisms.

Keywords Herbicide residue · Environment impact · Soil contamination · Human health · Residue monitoring

1 Introduction

The global population reached nearly 8.2 billion by mid-2024 and is expected to reach over 9 billion by 2050 (UN News, 2024). To feed the increasing population, world food production must be increased by approximately 70% (www.fao.org). Weeds are considered one of the most significant biotic constraints in crop production faced in global agriculture, which causes approximately 30% of crop yield losses (Sondhia, 2014a; Sondhia & Mishra, 2024), and the degree of loss varies by crop, cultivar, weed species, weed infestation level, location, and farming practices. Besides competing with crops for sunlight, water, nutrients, and space, weeds harbor insects and pathogens, which attack crop plants.

Currently, chemical weed management practices are widely adopted due to their higher weed control efficiency to minimize the yield losses by the weeds, and this is likely to be adopted more frequently under prevalent climate change conditions. Globally, herbicides constitute about 47% of the crop protection market followed by fungicides (17%), insecticides (29%), and others (7%) (Sondhia, 2014a; Sondhia & Mishra, 2024). Among various classes of pesticides, global herbicide consumption in agriculture has increased from 1.72×10^6 t to 1.83×10^6 in the last 10 years (FAO, 2023) (Table 2.1). Successful chemical weed management relies not only on the chemical properties of a herbicide but also on its interaction with the plant, and the environmental factors such as temperature, moisture, relative humidity, and solar radiation influence a plant's physiologic status and its susceptibility to herbicides. Though herbicides are specifically designed to control weeds, besides their targeted function, their residues remain active in the soil or plants and move through drift

Table 2.1 Consumption of various pesticides in the world and India

	World		India	
Pesticide	2021 (t)	% share	2021(t)	% share
Insecticide	822,857	15.02	31,731	51.42
Herbicides	1,831,139	33.42	9749	15.80
Fungicide	1,702,672	31.08	20,092	32.56
Other	1,121,911	20.48	1130	0.16
Total pesticide	5,478,580	100	61,702	100

Source: FAO 2021, https://www.fao.org/faostat/en/#data/RP

beyond their target areas (Sondhia, 2014b). While the judicial use of herbicides provides selective and economical weed control; however, frequent and non-judicial use may lead to residues, persistence, phytotoxicity, and adverse consequences on subsequent crops, non-target organisms such as honey bees, earthworms, and birds and the environment eventually causing health hazards to human (Correia & Moreira, 2010; Sondhia, 2014a; Brühl & Zaller, 2021; Ruuskanen et al., 2023). Some herbicides are reported to be associated with both acute and delayed health effects in exposed humans, ranging from simple skin and eye irritation to more severe impacts on nervous and reproductive disorders, neurological dysfunction, cancer, and respiratory disorders (Tangamornsuksan et al., 2019; Ahmad et al., 2021; Sondhia & Mishra, 2024).

Furthermore, the development of resistance in weeds is also reported due to extensive herbicide use (Manicardi et al., 2023). The lengths of time herbicides persist in different environmental compartments vary significantly, ranging from those that break down quickly into harmless by-products to those that persist in the environment (Sondhia, 2008b; Sondhia and Singh, 2018; Chaudhari et al., 2023). The longer persistence of herbicide in soil poses a hazard to subsequent land use which is undesirable. Generally, herbicides are chemically and photochemically stable in mild conditions, making it difficult to be degraded by physicochemical and biological processes (Sondhia et al., 2023).

Many herbicides are found as bound residues, which make them not only unavailable to the targets but also pollute the ecosystem (Chandra et al., 2013; Sondhia, 2015; Sondhia, 2016). Using the latest hi-tech analytical devices, the presence of herbicide residues can be detected at ppb level or even low detection limits. Accordingly, in this chapter, herbicide residues in soil and agricultural commodities, their persistence, detection methods, and impact on non-targeted organisms are summarized.

2 Herbicide Use Pattern

Globally, the consumption of herbicides was maximum in soybean, followed by cotton, corn, wheat, and others in 2017, and a similar pattern is predicted in 2025 (Herbicides Market, 2019). In India, herbicide application is more common in wheat (44%), followed by rice (31%), plantation crops (10%), soybean (4%), and other crops (11%) (Sondhia, 2014a). India is currently the fourth largest global producer of agrochemicals after the United States, Japan, and China, and herbicide usage is 30% (Sondhia, 2014a, 2019). In many advanced countries, the average annual herbicide consumption is 675–1350 g/ha as compared to 40 g/ha in India. Compared to other countries, India consumes less herbicides due to the relatively cheap labor required for manual weeding. The herbicide consumption in India stood at 0.4 billion USD in 2015 and is expected to grow at a compound annual growth rate (CAGR) of 15% over the next 5 years to reach ~0.8 USD billion by 2020. The global herbicides market size has reached US$ 33.0 billion in 2023 and is expected

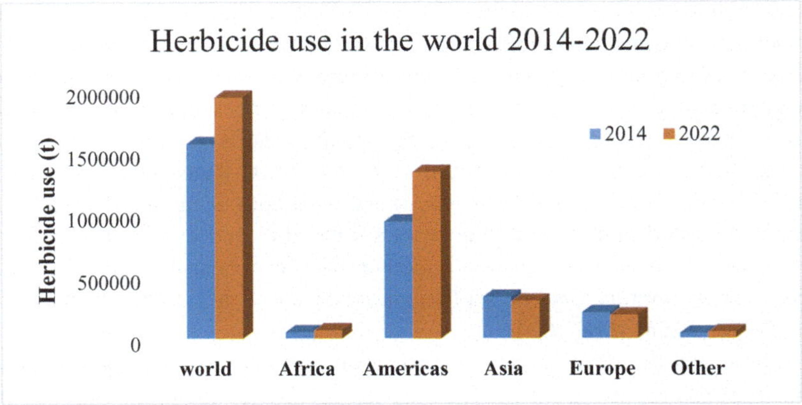

Fig. 2.1 Herbicide use in the world 2014–2022, depicting the major use of herbicides in the Americas' followed by Asia, Europe, and Africa. Source: FAO. 2024. FAOSTAT: Pesticides Use. [Accessed November 11, 2024], Herbicide use graph (Source: Sondhia et al., 2024)

to reach US$ 45.8 billion by 2032, exhibiting a compound annual growth rate (CAGR) of 3.7% during 2024–2032 for effective weed management across the globe (Herbicides Market Size, Share, Trends Forecast 2024–2032). According to Tostado and Bollmohr (2022), the top 10 countries that use pesticides are China, the United States, Argentina, Thailand, Brazil, Italy, France, Canada, Japan, and India. In the world, herbicides are being used in the majority of the Americas, followed by Asia, Europe, and Africa (Fig. 2.1) (FAO stat, 2024; Sondhia et al., 2024).

3 Herbicide Classification

Herbicides can be categorized based on various factors such as their chemical composition, mode of action, activity, selectivity, application method, and timing. They are typically classified into selective and non-selective types. Selective herbicides target specific plant species, controlling or inhibiting their growth without harming other plants. This selectivity arises from factors like differential absorption, translocation, or physiological and morphological variations between plant species. In contrast, non-selective herbicides do not distinguish between plant types and effectively control all plants they come in contact with, impacting both weeds and desirable plants alike.

Some non-selective herbicides such as glyphosate, glufosinate, and paraquat are primarily used to manage weeds in non-cropped areas, wastelands, industrial sites, railway tracts/embankments, etc. Some most prominent group of herbicides based on their mode of action are acetyl-CoA carboxylase (ACCase) inhibitors (Group 1), acetolactate synthase (ALS) and acetohydroxy acid synthase (AHAS) inhibitors (Group 2), cell division inhibitor (Group 3), growth regulator (Group 4), fatty acid and lipid biosynthesis inhibitors (Group 8), chloroplast enzyme 5-enolpyruvyl shikimate-3-phosphate synthase (EPSPS) inhibitors (Group 9), an inhibitor of glutamine synthase (Group 10), mitosis inhibitors (Group 15), photosystem I (PS-I) inhibitor (Group 22), and 4-hydroxyphenyl pyruvatedioxygenase (HPPD) enzyme inhibitor (Group 27). Repeated application of herbicides with the same mode of action can lead to the development of resistance in weed populations. Weeds that evolve resistance to a particular herbicide may also become resistant to other herbicides with the same mode of action, as they target the same binding site. This phenomenon is known as cross-resistance. Additionally, weeds can develop multiple resistances, where they become resistant to herbicides from different modes of action, posing a significant challenge in weed management. Such resistance increases the complexity of controlling weed populations and requires more diversified or integrated approaches to effectively manage these resistant species (Manicardi et al., 2023).

4 Herbicide Persistence and Degradation

The fate of herbicides in soil that results in residues depends on a number of factors which include physicochemical properties of soil, crop growth period, herbicide formulations, dose, time and method of application, weather, and interaction of soil with herbicides (Sondhia, 2014a; Singh et al., 2017; Fernández et al., 2017; Sondhia et al., 2023). After application, herbicide may get adsorbed on soil, taken up by the plant, get volatilized into air, leached into the groundwater, or undergo chemical, photochemical, or microbial degradation in plant or within soil surface to form the toxic or nontoxic degradation products/metabolites (Fig. 2.2). Some amount of herbicide residues in ppm or ppb levels may remain in food commodities and the environment for some time.

Under increasing temperatures, herbicides with medium to high volatility can be vulnerable to volatilization, which may result in reduced bioefficacy against weeds due to enhanced metabolism (Sondhia et al., 2024a, b). The dissipation of 2,4-D, bromoxynil and thifensulfuron-methyl increased with increasing temperature (Cessna et al., 2017; Sondhia et al. 2024). A lower herbicide residue, due to quick dissipation in the soil and crops as a result of increasing carbon dioxide levels and temperature, will result in an increased vulnerability to weeds and, in the future, necessitate more spray often during the crop growing season that will further enhance environmental contamination. Under such environments, excess and extensive herbicide use may be required for effective weed management, which may further burden the environmental load and associated residues as well as contribute to the development of resistance in weeds.

Herbicide persistence in the soil is expressed as half-life or time required to degrade 50% of the original molecule (Table 2.2). However, the half-life is not absolute because it depends on the soil type, temperature, and concentration of the herbicide applied (Sondhia, 2016; Fingler et al., 2017; Sharma et al. 2021, 2023; Sondhia

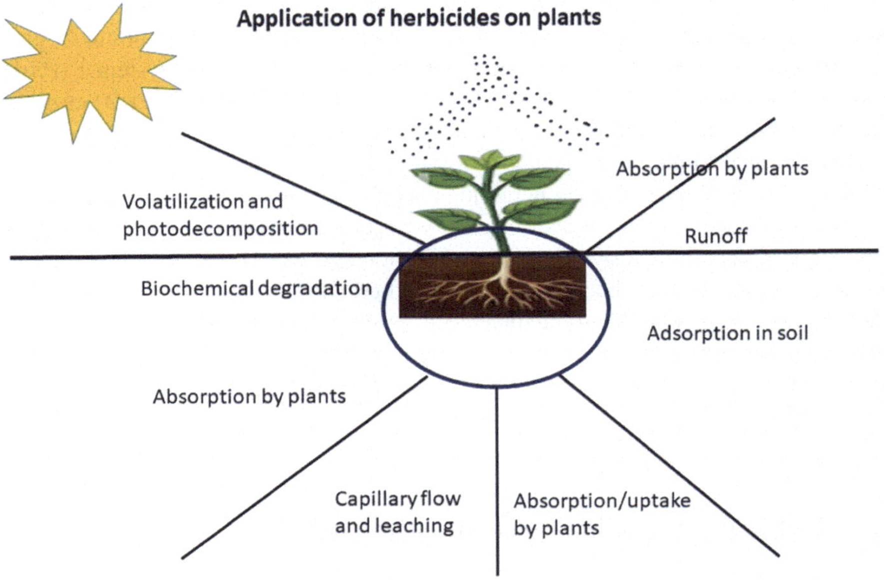

Fig. 2.2 Degradation of herbicide after its pre- and/or post-emergence application in a crop depicting various biochemical degradation pathways of herbicides in the soil

Table 2.2 Half-lives of some herbicides in soil and toxicity based on lethal dose

Herbicides name	Half-lives (days)	Toxicity class based on LD_{50}	Herbicides name	Half-lives (days)	Toxicity class based on LD_{50}
Atrazine	13–58	III	Metribuzin	23–49	III
Butachlor	5–24	III	Metolachlor	8–27	III
Chlorosulfuron	31–93	IV	Metsulfuron-methyl	70–147	IV
Chlorimuron	60	IV	Oxyfluorfen	12–29	III
Dithiopyr	11–25	IV	Pendimethalin	15–77	IV
Diclosulam	5–8.3	IV	Pyrazosulfuron-ethyl	16–21	IV
Haloxyfop-P	3–7	II	Pretilachlor	10–11	IV
Fluazifop-p-ethyl	8–24	III	Pyrozasulfone	46–134	III
Fluchloralin	12–46	IV	2, 4 D	7–22	II-III
Isoproturon	13–21	III	Sulfosulfuron	3–27	IV
Flufenacet	9–22.5	V	Thiobencarb	19–24	III
Imazethapyr	57–71	IV			

Source: Sondhia (2014a, 2019)

et al., 2024a, b). Besides herbicide structure, soil conditions prevailing during and after the application, as well as application methods, influence the fate of the herbicides in the soil (Sondhia & Singh, 2006). Sondhia (2009a) demonstrated that residues of sulfosulfuron in the surface soil (0–15 cm) were 0.229, 0.967, and 1.038 μg/g, which dissipated to 0.003–0.005 μg/g at 25–100 g/ha doses by 100 days. Sulfosulfuron residues were not detected after 200 days in surface and subsurface soils in all the doses.

Variable herbicide bioefficacy is reported in different environments. Foliar herbicides are influenced by several environmental factors, such as surface atmospheric CO_2 levels, light, soil moisture, relative humidity, rainfall, and wind, and can directly affect herbicide bioefficacy by altering the penetration and translocation within the weeds or indirectly by changing the growth and physiological characteristics of the plant, whereas soil-applied herbicides are influenced mainly by soil moisture and temperature (Varanasi et al., 2015). In a recent study, elevated CO_2 or temperature increases the resistance level of *Echinochloa colona* to cyhalofop-butyl (Refatti et al., 2019). The variable herbicide in different environments may also depend on the herbicide's mode of action.

Carbon dioxide and temperature differentially affect the bioefficacy of cyhalofop-butyl in *E. colona* between resistant and susceptible genotypes (Scott et al., 2018). Matzrafi et al. (2019) reported decreased efficacy of ACCase inhibitors under enhanced temperature due to enhanced metabolism. In a study undertaken in open-top carbon chambers (OTC) and free-to-air carbon enrichment (FACE) facilities, cyhalofop-butyl + penoxsulam bioefficacy at 1×, 1.5×, and 2× doses was significantly reduced (Sondhia et al., 2024a, b). However, the efficacy of the soil-applied herbicides (alachlor, ethalfluralin, linuron, and metolachlor) and foliar herbicides (glufosinate-ammonium, bentazone, and mecoprop) was reported to be higher at 30 °C; on the contrary, glyphosate isopropylamine showed similar efficacy regardless of the temperature (Rice et al. 2004; Park et al., 2021).

Soil persistence of imazaquin and imazethapyr has been found to be greater as soil pH decreases as a result of greater adsorption resulting in less herbicide availability for microbial degradation. In soils with greater than 3% organic matter, the herbicide carryover potential increases (Curran, 2001). Soil texture also plays a role in the likelihood of herbicide carryover (Westra et al., 2014). In addition, Kerr et al. (2004) found that herbicide persistence is more likely when soil CEC levels are higher. Herbicide persistence is characterized as long-lasting (>6 months), moderate (3–6 months), low (1–3 months), or non-persistent (<1 month) (Sondhia, 2014a, Kumar et al. 2017; Basu & Rao, 2020) by its half-life (DT_{50}). The repeated application of the same herbicides contributes to their persistence due to slow degradation (Basu & Rao, 2020). Herbicide residues are less persistent on vegetation and plant canopies than in soil (Tiryaki & Temur, 2010). Bipyridiliums, triazines, sulfonylureas, uracils, phenylureas, dinitroanilines, imidazolinones, isoxazolidiones, etc. (Table 2.2) are known for their persistence (Sondhia, 2014a). Goyal and Phogat (2003) reported that intermittent wetting and drying resulted in a very high persistence (90–99%) of trifluralin, whereas with continuous ponding the persistence of trifluralin decreased to 22–40% in alluvial soil. Sulfosulfuron followed first-order dissipation kinetics in the soil at 25–50 g/ha application rates and residues were not detected in the soil at harvest under the wheat cropping system (Ramesh et al., 2007; Sondhia & Singhai, 2008).

Degradation of herbicides by biotic reactions is generally followed by oxidative processes; however, oxidation, reduction, mineralization, hydrolysis, ester formation, photolysis, and polymerization are common by abiotic degradation (Sondhia, 2014a, 2019). The half-lives of some herbicides ranged from 13–147 days showing varying persistence in the soil. Rapid dissipation of metsulfuron-methyl in acidic silty-loam soil as compared to high pH soil is reported (Sondhia, 2008a, 2009d, 2016).

Westra et al. (2014) observed that regardless of sand or clay content, pyroxasulfone and S-metolachlor dissipation rates were positively correlated with the amount of irrigation. Cyhalofop-P-butyl adsorption isotherm indicated that cyhalofop-butyl was highly adsorbed in clayey than sandy-clay loam soil. The adsorption coefficient (Kf) was high in clayey soil (13.39), whereas it was low in sandy-clay soil.

Therefore, more research is required into advanced soil amendments, microbial breakdown processes, and residue monitoring. Need accurate formulations and new approaches, such as biological controls and crop rotation, to minimize persistence and optimize sustainable herbicide application.

5 Maximum Residue Limits and Pre-harvest Interval

Residues are estimated in the harvested produce to determine pre-harvest interval (PHI) in an edible food commodity following application at a recommended dose. Pre-harvest waiting periods describe the safe period for harvesting an edible commodity subsequent to the application of herbicides, especially in vegetables and fruits, for ensuring maximum residue limits (MRL) standards set by FSSAI for India and Codex Alimentations Commission (Codex). The MRL of a herbicide is the largest quantity detected in a commodity, which any regulatory body expects to find in the crop when it has been treated following good agricultural practice. However, the safety of food and associated health concerns for humans and animals is related to acceptable daily intake, not by the MRL. MRL is rather used for trade purposes. If herbicide residues are detected above the MRL, then it may create obstacles in the trading of particular commodities at the international level. The detectable pesticide residues, including herbicides in fruits and vegetables in developed countries like the United States, the United Kingdom, and EU, were found relatively higher as compared to that of India (Sharma et al., 2018). Hence, more research is needed on harmonization of PHI/MRL standards with other countries around the world using sophisticated residue detection technologies and adopting sustainable farming practices to ensure trade compliance and food safety.

6 Herbicide Residues in Soil

6.1 Residues in Soil

World Health Organization (WHO) defines any substance or mixture of substances in food for man or animals resulting from the use of a herbicide that includes degradation and conversion products, metabolites, reaction products, and impurities that have toxicological significance as herbicide residues. Soon after herbicide application, several biochemical processes begin to degrade the herbicide from the original site of application. In soil, chemical degradation by redox reactions is common with anilines, phenols, and dinitroanilines. Hydrolysis, ester formation, oligomerization/polymerization reactions catalyzed by clay surfaces, and photolysis are common with fluchloralin, bentazon, and olefins. Atrazine is reported to be dissipated from the field at the time of crop harvest with the half-life values of 9.38–21.54 days (Sondhia, 2001; Nag & Das, 2009). The 0.051–0.079 μg/g residues of fluazifop-p in soil were reported after its application in soybean fields at 125–500 g/ha (Sondhia, 2007). The 0.093 to 0.081 μg/g clodinafop residues were reported in alluvial, red, and black soil (Sondhia & Mishra, 2005; Roy & Singh, 2006). Certain sensitive crops have also been used to detect the presence of herbicide residues below detectable levels (Paul et al., 2009). Sorghum (*Sorghum bicolor*) and cucumber (*Cucumis sativus*) plants were found to be very sensitive bioassay plants for metribuzin and could detect residues even at 0.010 and 0.046 mg/kg in the post-harvest soil of potato (*Solanum tuberosum*) crop (Sondhia, 2005).

Trifluralin concentrations ranged from ND (not detected) to 66.39 μg/kg dry weight (dw) in the soil samples of cotton fields in China (Li et al., 2021) and showed high risks to wheat (*Triticum aestivum*), barley (*Hordeum vulgare*), and Lucerne (*Medicago sativa*) crop. Atrazine, hexazinone, simazine, terbuthylazine, acetochlor, alachlor, metolachlor, trifluralin, 2,4-D, and MCPA were detected in the majority of samples analyzed with the maximum concentrations of 82.4 ng/g dw. Fish tissue analyses revealed widespread herbicide residues in Lake St Lucia, with total concentrations in the range of 44.3–238 ng/g and 72.2–291 ng/g dw for *Clarias gariepinus* and *Oreochromis mossambicus*, respectively, and demonstrated potential cancer risk to local communities due to exposure to atrazine and simazine bioaccumulation through life-time consumption. Herbicide residues were detected in >95% of coral tissue samples, with total average concentrations across sampling sites ranging between 25.2 and 51.3 ng/g dw (Tyohemba et al., 2020). Rose et al. (2022) frequently reported detection of glyphosate and aminomethylphosphonic acid (AMPA) residues in 87% and 100%, respectively, in topsoil (0–10 cm) with a median concentration of 0.12 to 0.41 mg/kg along with residues of 2,4-dichlorophenoxyacetic acid, trifluralin, and diflufenican in >40% of topsoil samples with median concentrations of <0.05 mg/kg. At harvest, 0.05 mg/kg and 0.01 mg/kg of pendimethalin residue were reported in soils from wheat fields at 125 days after application (DAA), and 0.06 mg/kg and 0.02 mg/kg in the soil of barley fields at 117 DAA (Brar et al., 2023). Atrazine is detected as the most common herbicide residue with the highest mean contamination level of 2.73 mg/kg and 3.84 mg/kg at depths 0–10 cm and 10–20 cm, respectively, along with isoproturon (0.60 mg/kg), diclofop methyl (1.41 mg/kg), bromoxynil (1.25 mg/kg), and pendimethalin (0.37 mg/kg) residues in topsoil (0–10 cm) (Faraj et al., 2024). El Alfy and Faraj (2017) also reported that atrazine was detected in 72.5% of groundwater samples (not detected –1.3 μg/L). Herbicide adsorption and desorption are also influenced by the soil's physicochemical properties. The adsorption–desorption revealed strong adsorption of dithiopyr in alluvial soil with Kd values ranging from 3.97 to 5.78 and a Freundlich capacity factor (K_F) value of 2.41. Sondhia et al. (2006) reported that butachlor residues were not detected after 120 days in clay loam soil applied at 1.0 kg/ha in transplanted rice crops.

In a field experiment, residues of isoproturon were found to be 0.006, 0.041, and 0.022 μg/g in post-harvest soil, wheat grain, and straw, respectively, while 0.021 and 0.096 μg/g residues of clodinafop were detected in soil and grain at a higher level of application (Arora et al., 2013). Residues of clodinafop in soil were found from 0.093 to 0.081 in alluvial, red, and black soil (Sondhia & Mishra, 2005; Roy & Singh, 2006). Glyphosate residues were found to be below detectable levels at 0.5, 1.0, and 2.0 kg/ha application rates (Bandana et al., 2015. Godar et al. 2015). In made tea, the initial concentration of napropamide was found to be in the range of 0.14 to 0.44 μg/g (Biswas et al., 2007). Sondhia (2007b) reported 0.002, 0.006, 0.0075, and 0.010 μg/g residues of imazosulfuron in the soil of transplanted rice field after 60 days at 30–60 g/ha application rates; however, no residues were found after 90 and 120 days. Spinach and lettuce samples were analyzed to determine residues of the four pesticides, viz., dimethoate, 2,4-D, and glyphosate, to ensure food safety. The entire 12 spinach samples analyzed for each pesticide contained residues of the four pesticides. The residues in spinach samples grown in dimethoate-spiked soils at doses D_R and $D_{10\times}$ ranged from 0.10 to 7.65 mg/kg. Eight herbicides were detected in agricultural soil with concentrations ranging between 1.08 and 1724 μg/kg, with the highest concentration (1724.23 μg/kg) of metolachlor, terbuthylazine, and pendimethalin were detected in 35 soil samples. The greatest hazard quotient (HQ) values were identified for terbuthylazine as 2772.48×10^{-7} and 20793.61×10^{-7} for adults and children, respectively. The hazard index (HI) for all herbicides was reported to be 3916.05×10^{-7} for adults and 29370.39×10^{-7} for children (Polat & Tiryaki, 2024).

Soil microbes present at the soil surface and at the rhizosphere are vital for metabolic reactions in which they metabolize herbicides by using them as a source of nutrients, thus leading to herbicide degradation (Singh et al. 2018; Sondhia & Waseem, 2020; Fernandes et al., 2020). Mamy et al. (2005) reported low amounts of metazachlor, sulcotrione, metamitron, trifluralin, glyphosate, and its metabolites residue in the soil after 140 days of application. Chand et al. (2014) demonstrated the adverse impact of halosulfuron methyl (HM) residue on succeeding crops like black gram (*Vigna mumgo*), maize (*Zea mays*), and cucumber (*Cucumis sativus*) after sugarcane (*Saccharum officinarum*).

The above information suggests that cutting-edge residue monitoring technologies, regional persistence studies, and microbial interventions are needed to optimize herbicide breakdown and reduce environmental and health impacts.

6.2 Herbicide Residues in Food Samples

Sondhia (2014a) and Sondhia & Mishra (2024) have recently reviewed the occurrence of some herbicide residues in cereals, pulses, oilseeds, vegetables, plantation crops, cash crops, as well as in soil and aquatic environments. In general, after herbicide application, residues are reported in plants during the crop growth period as well as at maturity. The occurrence of trifluralin in black gram (Aktar et al., 2009), fluazifop-p-butyl in soybean (*Glycine max*, Sondhia, 2007a), alachlor in cotton (*Gossypium hirsutum*, Ramesh & Maheshwari, 2004) and water (Sharma 2002) benthiocarb, imazosulfuron, anilophos, metsulfuron-methyl, pyrazosulfuron, and ethoxysulfuron in rice [(*Oryza sativa*) (Aktar et al., 2007; Sondhia, 2007b, 2009b, 2014a; Sondhia & Dixit, 2012; Singh et al., 2012; Sondhia et al., 2013; Tandon, 2014)], metsulfuron-methyl in wheat (Sondhia, 2008a), napropamide and glyphosate on tea (*Camellia sinensis*, Bandana et al., 2015), and pendimetalin in potato (*Solanum tuberosum*), and cauliflower (*Brassica oleracea*) and radish (*Raphanus sativus*, Sondhia, 2013a) have been reported under Indian tropical conditions and in most cases the residues were reported as below the MRL at harvest. The most predominant pendimethalin was present in 45 (30.2%) samples of 12 commodities [Bitter gourd (*Momordica charantia*), cauliflower (*Brassica oleracea* var. botrytis), cabbage (*Brassica oleracea*), coriander (*Coriandrum sativum*) leaves, cumin (*Cuminum cyminum*), curry (*Murraya Koeingii*) leaves, green chilli (*Capsicum annuum*), red chilli, tomato (***Solanum lycopersicum***), grapes (*Vitis vinifera*), rice (*Oryza sativa*), and wheat. Atrazine was present in 17.6% of the samples of nine commodities, viz. cauliflower, cabbage, okra (*Abelmoschus esculentus*), grapes, curry leaves, coriander (*Coriandrum sativum*) leaves, fennel (***Foeniculum vulgare***), green chilli, and water. At harvest, 0.008, 0.001, and 0.014 μg/g pendimethalin residues were found in tomato (***Solanum lycopersicum***), cauliflower, and radishes (***Raphanus sativus***), respectively, applied as pre-emergence at 1.0 kg/ha in each crop (Sondhia, 2013a, b, c). In cucumber, residues of anilophos (ND-0.042 mg/kg), and in onion (*Allium cepa* L.), fluchloralin (0.012–0.065 mg/kg), and anilophos (ND-0.033 mg/kg) were detected (Srivastava et al., 2011). Residues of chlorpropham and propanil were detected in tomato, capsicum (*Capsicum annuum*), and cucurbits samples (Sharma et al., 2022).

Post-emergence application of imazethapyr in a soybean crop at 100 g/ha resulted in residues in the soybean grains at harvest, and a pre-harvest interval (PHI) of 90–102 days was suggested for the next crop (Sondhia et al., 2015a, b). Sharma et al. (2018) reported 10 herbicides with 0.01–2.45 mg/kg residues in 112 samples at national level in crop, vegetables, spices, and water samples. Butachlor and chlorpropham were detected in 8% of samples each. Other 6 herbicides out of 26 were still lesser in the frequency of detection, and the remaining 16 were not detected. Dimethachlor, fluchloralin, metribuzin, and diuron were found in lesser in frequency. In a 2-year monitoring study, residues of pendimethalin in green chilli were found above MRL (Sharma et al., 2018; Sondhia, 2019) (Table 2.3). Water samples

Table 2.3 Defection of herbicide residues in various food commodities and water during 2014–2016 at the national level in India showing the presence of various herbicide residues

Sample/commodity	Herbicide	Residues (mg/kg)	Mode of detection (GCMS/LCMS/MS)
Apple	Metribuzin	0.01	GCMS
Bitter gourd	Pendimethalin	0.02	GCMS
Brinjal	Chlorpropham	0.01	GCMS
	Dimethachlor	0.02	GCMS
Cauliflower	Atrazine	0.08	GCMS/MS, LCMS/MS
	Butachlor	0.03	GCMS
	Chlorpropham	0.03	GCMS
	Fluchloralin	0.06–0.09	GCMS
	Pendimethalin	0.08	GCMS
Cabbage	Pendimethalin	0.02–1.72	GCMS
	Alachlor	0.10–0.17	GCMS
	Atrazine	0.08–0.56	LCMS/MS, GCMS
	Fluchloralin	0.13	GCMS
	Butachlor	0.09	–
Capsicum	Chlorpropham	0.02–0.04	GCMS
Coriander leaves	Alachlor	0.03–0.11	GCMS
	Butachlor	0.07–0.52	GCMS
	Pendimethalin	0.03–0.54	GCMS
Green chili	Atrazine	0.09	LCMS/MS
	Pendimethalin	0.10–0.43	GCMS/MS, LCMS/MS
Okra	Atrazine	0.09	GCMS/MS, LCMS/MS
	Chlorpropham	0.03	GC–MS
Ridge gourd	Atrazine	0.06	LCMS/MS
	Alachlor	0.15	LCMS/MS
Tomato	Chlorpropham	0.02–0.06	GCMS
	Pendimethalin	0.01	GCMS
	Pretilachlor	0.03	LCMS/MS
	Metribuzin	0.01	GCMS/MS
Grapes	Atrazine	1.19	GCMS/MS, LCMS/MS
	Pendimethalin	0.02	GCMS/MS
	Pretilachlor	0.02	LCMS/MS
Pear	Chlorpropham	0.40	LCMS/MS
Rice	Pendimethalin	0.06	GCMS
Wheat	Pendimethalin	0.22	GCMS
Cumin	Pendimethalin	0.63	GCMS
	Alachlor	0.03–0.035	GCMS
	Butachlor	0.23–0.40	GCMS

(continued)

Table 2.3 (continued)

Sample/commodity	Herbicide	Residues (mg/kg)	Mode of detection (GCMS/LCMS/MS)
Fennel	Atrazine	1.58	GCMS
Curry leave	Atrazine	0.06–0.12	GCMS/MS, LCMS/MS
	Diuron	0.02	LCMS/MS
	Metribuzin	0.02–0.08	GCMS/MS, LCMS/MS
	Pendimethalin	0.02–1.35	GCMS/MS, LCMS/MS
Red chili powder	Pendimethalin	0.01–1.43	GCMS, GCMS/MS
	Fluchloralin	0.19	GCMS
	Diuron	1.0	LCMS/MS
Water	Atrazine	0.03–14.90	GCMS/MS, LCMS/MS
	Diuron	0.85	LCMSMS
	Pendimethalin	2.27–2.45	GCMS/MS, LCMS/MS

Source: Sharma et al. (2018), Sondhia (2019)

were also found to be contaminated with higher quantities of atrazine, pendimethalin, and diuron, requiring immediate attention to decontaminate such water bodies. In a monitoring study of four herbicides, 0.03–1.28 ng/g pendimethalin (36%), 0.02–1.22 ng/g butachlor (61%), and 0.01–0.25 ng/g fluchloralin (3%) residues were detected, and alachlor was not detected in all the locations (Sondhia, 2006; Kumar, 2011, 2012). Argula (***Eruca vesicaria***) (0.27 mg/kg) and tomatoes (***Solanum lycopersicum***) (0.01 mg/kg) had the highest and the lowest concentrations of herbicides, respectively, and South America and Europe had the highest 3.44 mg/kg and the lowest (0.03 mg/kg) of herbicide mean concentrations, respectively (Ahmadi et al., 2024). In another study by Bakırcı et al. (2014), they observed that the mean concentration of herbicide in carrot vegetables of Turkey was 0.012 mg/kg, respectively (Bakırcı et al., 2014).

At harvest no detectable residues of fenoxaprop-ethyl or acid were detected in soil, wheat grain, and straw samples at recommended doses (Sondhia, 2008c; Singh et al., 2013).

In a field experiment, residues of isoproturon were found to be 0.006, 0.041, and 0.022 μg/g in post-harvest soil, wheat grain, and straw, respectively, while 0.021 and 0.096 μg g/g residues of clodinafop were present in soil and grain at higher level of application. The half-life of clodinafop was found to be 29.6 days in the soil of wheat crops (Sondhia & Mishra, 2005; Arora et al., 2013). Sulfosulfuron residues were not found in wheat grains, straw, and subsequent vegetables in the natural ecosystems or in the model ecosystem at the recommended rates for wheat crops (Sondhia & Singhai, 2008). In rice straw, 0.01–0.03 μg/g oxyfluorfen residues were detected at 240–500 g/ha rates. However, in rice grains, 0.018–0.106 μg/g of oxyfluorfen residues were detected in 240–500 g/ha treated plots (Sondhia, 2007b, 2009a). A field dissipation study was conducted to evaluate the persistence of imazosulfuron applied at 30–60 g/ha 4 days after transplanting of rice (Sondhia, 2007). In rice grains 0.002–0.009 μg/g and in rice straw 0.009 and 0.039 μg/g imazosulfuron residues were detected at 40–60 g/ha application rates. Terminal residues of anilofos in soil, rice grain, and rice straw were 0.049–0.070 μg/g and 0.076–0.160 μg/g

(Sondhia, 2014a; Sondhia & Dixit, 2015). In grain and straw, 0.036 and 0.010 µg/g residues of fenoxaprop were detected. Fenoxaprop residues persisted for up to 90 days in the rice field with a half-life of 14.9 days in the soil (Sondhia, 2008c).

The residual effect of atrazine (1.0–2.0 kg/ha) was studied on the succeeding crops of chickpea (*Cicer arietinum*) and Indian mustard (*Brassica juncea*), where fluchloralin was applied at 0.75 kg/ha (Sondhia, 2012a). Atrazine was degraded to undetectable levels at all doses by the time maize (*Zea mays*) crop was harvested (90 days). The half-life of atrazine varied from 23 to 31 days (Sondhia, 2014a). In chickpea and Indian mustard, low levels of fluchloralin residues were detected in soil at 150 days (64–85% and 69–82% losses, respectively). Terminal residues of pendimethalin were monitored in the green field peas (***Pisum sativum***) and chickpeas applied as the pre-emergence herbicide at 750–185 g/ha rates. Terminal pendimethalin residues were found in mature pea grain (0.004-BDL µg/g) and straw (0.007–0.001 µg/g) at 750–185 g/ha treatments, respectively (Sondhia, 2013a, 2013b). In chickpea grains and straw, 0.020 and 0.008 µg/g residues of pendimethalin were found. However, quizalofop-P-ethyl and oxyfluorfen residues were found below <0.001 and 0.003 µg/g, respectively, in grains and straw (Sondhia, 2010). Mandal et al. (2014) demonstrated that at harvest, the residues of quizalofop ethyl on black gram seed, foliage, and soil were found to be below the detection limit of 0.01 mg/kg following a single application of the herbicide at 50–100 g/ha for both the periods.

Flumioxazin residues in the grains and soil were reported below the detection limit (0.001 µg/g) at doses 30–90 g/ha (Sondhia & Dixit, 2008). Application of quizalofop-butyl at 37.5 and 62.5 g/ha in soybean plants resulted in 0.01 µg/g terminal residues in the soybean straw. However, 0.012 µg/g residues were detected at 62.5 g/ha dose (Sondhia, 2014a). The residues of oxyfluorfen in the green onion and mature onion bulbs were 0.041–0.063 and 0.0034–0.0460 µg/g at 150–300/ha rates. Residues of oxyfluorfen applied in mature onion were below the maximum residue limit (0.05 µg/g) (Sondhia & Dixit, 2007a, b; Sondhia, 2009b). A pre-harvest interval of 118 days for the onion crop after the herbicide application was suggested (Sondhia, 2010). Residues of pendimethalin, fluchloralin, and oxadiazon were found below the maximum residue limit in onion bulbs at harvest (125 days after spraying). Sondhia and Dubey (2006) did not find pendimethalin residues in the mature onions. However, 0.007 µg/g pendimethalin residues were detected in green onion at a 1.0 kg/ha application rate. Atrazine residues were found at 0.263 and 0.302 µg/g in the soil of potato crops, which dissipated to 0.119 and 0.138 µg/g, respectively, after 60 days. Application of pendimethalin and trifluralin resulted in below detectable limit residues (0.02 mg/kg) in celery (*Apium graveolens*) seeds (Kaur & Gill, 2012). Pendimethalin applied at 0.6–0.9% to tobacco (***Nicotiana tabacum***) led to 0.198–0.376 mg/kg residues in tobacco leaves and 0.72 mg/kg residues in leaves treated with 0.5% pendimethalin and 0.04–0.079 mg/kg residues treated with 0.25% pendimethalin (Parmar et al., 1998). Residues of fluchloralin and fenoxaprop were also detected in the roots, straw, and fruits of Ashwangandha (*Withania somnifera*) and reduced with anoloides content significantly (Sondhia, 2019).

7 Herbicide Residues in Surface and Groundwater

The presence of herbicide residues in the water samples after its application is mostly due to water irrigation/rainfall and consequential in herbicide leaching into the groundwater (Sondhia, 2014b; Imfeld et al., 2020; Rheinheimer Dos Santos et al., 2020) or transport as colloidal particles in the soil (Terzaghi et al., 2020). A study on groundwater monitoring of 64 EU countries revealed frequent occurrence of herbicide residues (Loos et al., 2010). Rendom et al. (2017) reported glyphosate residues in bottled drinking water, groundwater, and urine of subsistence farmers. Bruggen et al. (2018) showed that glyphosate residues accumulate in the environment, are carcinogenic to humans, and break down into the aminomethyl phosphonic acid (AMPA) in soil and water. Sagratini et al. (2007) reported the detection of atrazine, fluroxypyr, metolachlor, and terbuthylazine residues in irrigation wells and further toxicity to crops. Lewisa et al. (2009) demonstrated that several herbicides were detected in both freshwater and coastal marine waters, and the effects were attributed to specific land uses in the catchment and correlated with sugarcane cultivation in the adjacent catchment. He demonstrated that herbicides reach the Great Barrier Reef lagoon and may disturb sensitive marine ecosystems already affected by other factors such as climate change. The concentration of atrazine ranged from 0.72 to 17.3 μg/L, whereas 0.91–40.97 μg/L simazine concentration was reported in groundwater samples collected from Delhi (Aslam et al., 2013). Occurrence of glyphosate and AMPA in surface waters, storm water wetlands, and streams are reported in suburban and urban areas of Melbourne (Sondhia et al., 2009; Okada et al., 2020) and reported glyphosate and aminomethylphosphonic acid (AMPA) as commonly found herbicides in urban streams and wetlands of Melbourne, Australia. Rheinheimer Dos Santos et al. (2020) have found five herbicides (2,4-D, atrazine, desethyl-atrazine, simazine, and nicosulfuron), along with three fungicides (carbendazim, epoxiconazole, and tebuconazole), and one insecticide (imidacloprid) in epilithic biofilms due to high rainfall (Reichenauer & Germida (2008).

Bento et al. (2019) reported 17.5 times higher glyphosate and aminomethylphosphonic acid (AMPA) residues in sediment than in soil. High concentrations of desethylatrazine, dimethoate, simazine, atrazine, and metolachlor were detected in the Ebro River basin area (Claver et al., 2006). Substantial concentrations of 2,4-D, diazinon, diuron, imidacloprid, and terbutryn were reported in agricultural runoff (García-Gal'an et al., 2020). Herbicide residues of 2,4-D, alachlor, acetochlor, atrazine, ametryn, butachlor, bentazone, clomazone, diuron, hexazinone, imazethapyr, isoproturon, glyphosate, linuron, mecoprop, metribuzin, metolachlor, molinate, pendimethalin, pretilachlor. Propanil, propazine, simazine, tebuthiuron, trifluralin, terbumeton, terbutryn, and terbuthylazine along with some metabolites were detected in 768 water samples of 16 countries [(Brazil, Burkina Faso, Canada, China, Ethiopia, France, Germany, Iran, Lebanon, Luxembourg, Portugal, Slovenia, Spain, United States, Vietnam, and India (West Bengal)] (El-Nahhal & El-Nahhal, 2021; Wan et al., 2021). Occurrences of high concentrations of glyphosate and aminomethylphosphonic acid residues were reported in soil, water, suspended particulate matter, and sediment varied from 82.3% to 99.0% near the Nandu River basin, China, and phenoxy acid and chloroacetanilide herbicides residues were detected at lower frequencies (Tan et al., 2024).

Many herbicides are routinely detected from the surface and groundwater sources in developed countries (Sondhia, 2014a; Bruggen et al., 2018; Tan et al., 2024). The most often detected herbicides are 2, 4-D, atrazine, cyanazine, carbaryl, simazine, bromacil, diuron, diazinon, prometon, metolachlor, dinoseb, picloram, metribuzin, metsulfuron, glyphosate, metolachlor, propanil, butachlor, pendimethalin, oxyfluorfen, etc. Twenty-eight herbicide residues with a concentration range of 0.002–0.27 μg/L were detected in the water samples from Portugal (Palma et al., 2014). Twenty-two herbicide residues with a concentration range of 0.01–4.90 μg/L were detected in the water samples from Brazil (Albuquerque et al., 2016), fourteen herbicide residues with a concentration range of 1.16–32.32 μg/L were detected in the water samples from Spain, and three herbicide residues with a concentration range of 0.001–0.021 μg/L were detected in the water samples of China (Wang et al., 2020a, b). Glyphosate was detected with a concentration of 1.42 μg/L in groundwater (Rendon-von Osten & Dzul-Caamal, 2017). In the groundwater, 0.0023 mg/kg to 0.0154 mg/kg pyrazosulfuron-ethyl residue was detected (Naveen et al., 2012). Persistence and mobility of 2,4-D were found to be dependent on soil water content (Gupta et al., 2012). In the Singoor reservoir, Hyderabad residues of atrazine (0–1.056 μg/L) were detected. The concentration of atrazine residues in Osmansagar water was 0.056 μg/L during post-monsoon (Reddy & Reddy, 2010). El-Nahhal and El-Nahhal (2021) reported HIs of herbicide residues in the range of 0.0001–2.81 and suggested high HI values in the water samples due to high concentrations of herbicide residues that were close to or exceeded the value of the ARfD. There is a lack of data on long-term ecological impacts, regional differences on herbicide persistence and interactions with ecosystems, that can overcome using multi-site longitudinal surveys and advanced ecosystem modeling.

8 Herbicide Leaching and Transport

Herbicides that are less absorbed by the soil are transported after their application through leaching or surface runoff to contaminate groundwater or distant surface water sources, and leaching is high with more water-soluble herbicides (Sondhia et al., 2014; de Medeiros et al., 2024). Heavy rains cause greater leaching and runoff. Sandy soils have a higher leaching potential than clay soils due to larger pore spaces and lower CEC (Sondhia & Yaduraju, 2005; Sondhia, 2009c, 2016). Higher humidity enhances the soil microflora proliferation. Similarly, the persistence of herbicides in dry soil is greater in comparison to wet soil. In a leaching study, metribuzin maximum concentration was recovered from 28 to 32 and 32 to 36 cm depths at 0.80 and 1.20 kg/ha treatments and approximately 0.2% could leach to the depths of 48–52 cm in soil (Fig. 2.3). Leaching results indicated that imazethapyr

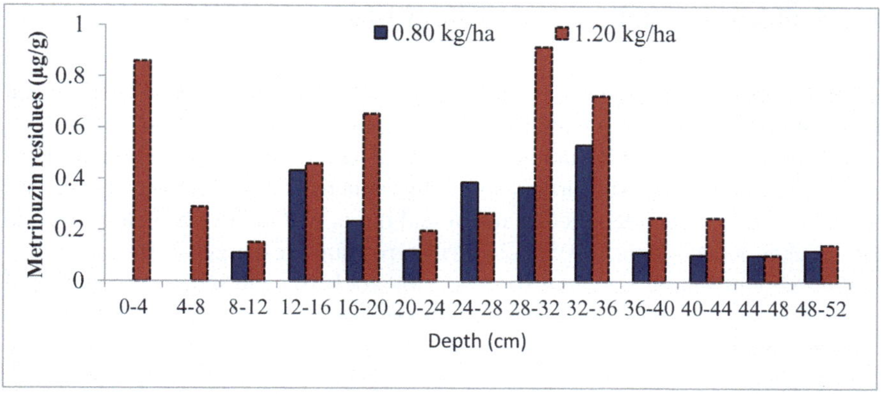

Fig. 2.3 Leaching of metribuzin in soil showing the distribution of residues at various soil depths under laboratory conditions at 0–52 cm soil depth. Imazethapyr recovery was more at 0–10 cm and around 5–8% was detected up to 80 cm

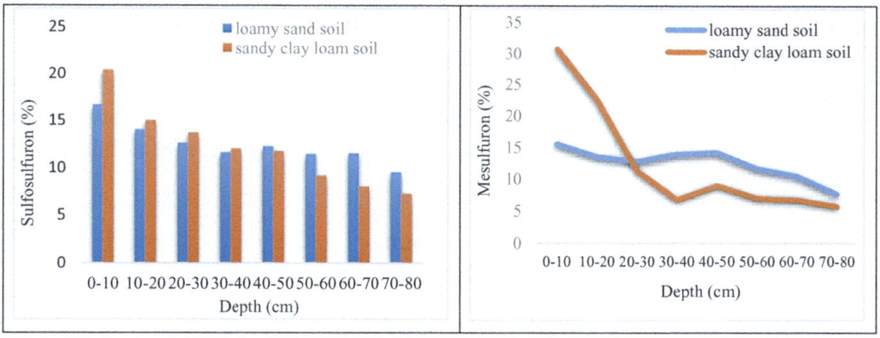

Fig. 2.4 Leaching of sulfosulfuron and metsulfuron residues in a loamy-sand and sandy-clay loam soil column depicting percent distribution of sulfosulfuron and metsulfuron residues at various soil depths

could leach in clay loam soil up to the depth of 70 cm applied at 100 and 200 g/ha (Fig. 2.4) (Sondhia, 2013a). High mobility of metsulfuron-methyl under continuous saturated moisture conditions was reported (Fig. 2.4) (Sondhia, 2009c, 2019).

The concentration of cyhalofop-butyl residues in the range of 1.1–2.2 ng/L, along with major metabolites, was reported from leachates (Sondhia & Khare, 2014; Sondhia & Khare, 2017). However, residues and metabolites of pyrazosulfuron-ethyl were leached up to 100–150 cm depths in sandy clay loam soil (Sondhia, 2019). Residues of pretilachlor were reported to be 87.2–93.0 ng/g up to 50 cm depth. Pendimethalin could leach in clay loam soil up to a depth of 55 cm under 200 mm rainfall conditions (Sondhia, 2007a; 2013a).

9 Herbicide Metabolites and Degradation Products

Herbicide degrades to toxic or nontoxic degradation products due to biochemical degradation in the soil or metabolism in the plants. Desmethyl sulfosulfuron, rearranged amine, sulfonamide, and guanidine were identified as degradation products of sulfosulfuron in the subsurface soil. Chlorophenyl tetrazoline and cyclohexyl ethylamine have been identified as major and minor metabolites of fentazamide in soil (Mukherjee & Gopal, 2005). Clodinafop-propargyl ester generally converts to acid, a major metabolite, and is also responsible for herbicidal action. Dissipation of clodinafop was not affected by specific soil pH and soil type. Major metabolites of cyhalofop-butyl in soil and leachates were detected as (R)-2-4(4-cyanao 2-fluroophenoxy) phenoxypropanoic acid (cyhalofop acid) and (R)-2-4(4-caboxyl-2-flurophenoxy) phenoxypropanoic acid (cyhalofop-diacid), and cyhalofop-phenol (Sondhia & Raj, 2014). Residues of pinoxaden and its metabolites were found below the detectable limit (<0.01 mg/kg) (Dixit et al., 2011). Metabolites of pyrazosulfuron and penoxsulam were detected from soil, pond water, and rice fields (Sondhia et al., 2016; Sondhia, 2019). *Penicillium chrysogenum* and *Aspergillus niger* were found as potent pyrazosulfuron-ethyl degrading fungi (Sondhia et al., 2013). Two metabolites of carfentrazone, viz. rcafentrazone-chloropropanoic acid (I) and 3-desmethylcarfentrazone-chloropropanoic acid (II) were reported in the soil of wheat field. The three degradation products of bispyribac-sodium in the soil, rice plants, and water were reported (Fig. 2.5) (Sondhia, 2019).

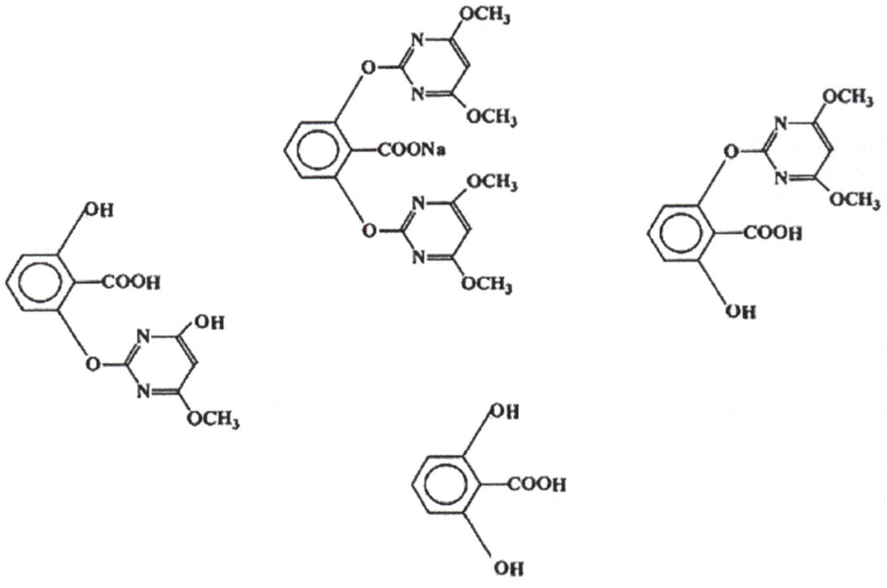

Fig. 2.5 Degradation of bispyribac-sodium in the soil, rice plants, and water, as a result of biochemical degradation

Apart from biochemical degradation of herbicide in the soil, certain fungi have the ability to degrade a wide variety of hazardous chemicals into simpler degradation products (Sondhia et al., 2013, 2016). *Penicillium chrysogenum* and *Aspergillus niger* were reported as potent pyrazosulfuron-ethyl and penoxsulam degrading fungi (Sondhia et al., 2013, 2016). However, there is limited knowledge of how herbicide metabolites in ecosystems are environmentally fated and how harmful they can be in the long term, filled by field research on metabolite persistence, environmental effects, and fungal toxicity in soil types and climates.

10 Impact of Herbicides on Microbe and Plant–Animal Interactions

Herbicide residues may affect plants and their associated microbes (Rainio et al., 2021). However, recent studies indicate a negative correlation between herbicide use and beneficial soil- and root-associated microbes, such as a reduction in the total population of the soil microbial community (Helander et al., 2018; Ramirez-Villacis et al., 2020). However, atrazine is reported to affect the overall microbial community (Yale et al., 2017), while in some environments, a decrease in microbial biomass or an increase in degrading bacteria is reported (Fernandes et al., 2020). Studies also showed a negative impact on root-associated microbes and herbicide-modulated nutrient cycling in soils (Hébert et al., 2019; Ramirez-Villacis et al., 2020). However, less is known about the impacts of herbicides on ectomycorrhizal fungi, although potential sensitivity to glyphosate is reported (Leino et al., 2021).

Changes in the plant's volatile organic compounds attracting pollinators and predatory insects are also reported by the herbicide applications (Fuchs et al., 2022). For example, few volatile organic compounds derived from the shikimate pathway are affected by even low doses of glyphosate (D'Alessandro et al., 2006). Additionally, atrazine is reported to affect animal host gut microbiomes of beetles, drosophila, frogs, and mice (Wang et al., 2020a, b; Brown et al., 2021). Paraquat exposure at low doses is reported to remodel the microbiome of *Drosophila Melanogaster*, simultaneously influencing the adult lifespan (Obata et al., 2018). Likewise, plant rhizosphere bacterial communities are shaped by herbicide exposure (Ramirez-Villacis et al., 2020). Studies show a consistent reduction in the relative abundance of nitrogen-fixing bacteria (Zobiole et al., 2011) and repression in plant-beneficial microbial functions and antifungal enzymes (Lu et al., 2018). In addition to the resistance of weeds toward herbicides, an increased herbicide resistance in free-living soil bacteria is also reported (Rainio et al., 2021) and may severely affect the diversity of the microbial community which may influence soil processes such as nitrogen and carbon flows and may affect non-targeted soil fauna, such as earthworms and other organisms. In plants, sublethal doses of glyphosate can potentially disrupt plants' aboveground (phyllosphere) interactions with other coevolving organisms, such as pathogens, plant-mutualistic microbes, herbivores, and pollinators (Saikkonen et al. 2020; Ramirez-Villacis et al., 2020). Alteration of plant metabolome in *Arabidopsis thaliana* is reported by glyphosate causing a change in the core phyllosphere microbiome (Ke et al., 2021). Therefore, understanding the impact of physiological and behavioral changes by herbicide application on soil, plant/microbial community, and non-targeted organisms will be a future research challenge.

11 Impact of Herbicides Residues on Non-targeted Organisms

Some herbicides exert a negative impact on soil microbes, such as glyphosate, which is reported to negatively affect shikimate pathways present in the majority of microbes, but their genetic resistance to glyphosate varies (Chávez et al., 2022). Furthermore, the use of additives/adjuvants in combination with herbicide-active ingredients can increase their toxicity to the surrounding environment and no target plants and animals (Gandhi et al., 2021). While adjuvants can enhance the performance of herbicides, they can also contribute to the formation of persistent residues in soil and plants, leading to potential adverse effects. For instance, the residue formation of trifluralin in soil and sugar beet roots was found to be 42–49% higher when adjuvants were used (Pacanoski, 2015). The extensive use of herbicides also results in residue accumulation in the aquatic systems near agricultural areas, further contributing to toxicity concerns (Sondhia, 2023).

Herbicides have variable toxicity in addition to acute toxicity from occupational exposure to humans, domestic animals, and wildlife. Intentional/unintentional consumption, inhalation of aerial sprays, direct contact of the herbicide with the applicator, and consumption of food prior to pre-harvest interval are some other ways of herbicide exposure. Their adverse effects on non-target plants, organisms, and human health and their mitigation have been comprehensively reviewed (Sondhia, 2014b; Sondhia et al., 2015a, b). Some phenoxy herbicides are often contaminated with dioxins, which increase cancer risk after their occupational exposure (Kogevinas et al., 1997). Exposure to low concentrations of atrazine has reportedly caused the demasculinization of frogs (Hayes et al., 2002). Incidences of intentional acute poisoning by herbicides like butachlor, fluchloralin, paraquat, 2,4-D, pendimethalin, and glyphosate have also been reported even when the level of contamination of soil, plant, and water was considerably low (Senarathna et al., 2009; Sondhia, 2014b). Apart from residues of parent herbicide molecules, health effects caused by commercial formulations often contain other chemicals, including inactive ingredients, which may also negatively impact human health. Some manufacturers made misleading claims about the safety of their products, viz., practically nontoxic to mammals, birds, and fish, and guaranteed to be even safer than the common salt (Talbot et al., 1991; Samsel & Seneff, 2016).

Several publications on evidence of cancer and other ill effects of some herbicides are reported, but certain scientific community often disagrees on the risk and correlation with higher LD_{50} of herbicides observed against mammals in comparison to common insecticides (Morrison et al., 1992). Triazine exposure has been implicated in an increased risk of breast cancer, although a causal relationship remains unclear (Kettles et al., 1997). Acetochlor increased the chromatid exchange frequency of human lymphocytes and decreased soil microbial community diversity (Luo et al., 2004). The heavy use of certain herbicides has adversely affected wintering migratory birds and other bird populations (Newton, 2004). In a study, Yadav et al. (2013) demonstrated the genotoxic potential of butachlor even at low dose levels (1.0 mg/kg) and suggested that butachlor interferes with cellular activities in fishes at the genetic level, inducing chromosomal aberrations and suggested serious concern toward the potential danger of butachlor for aquatic organisms. In a mortality study, fish mortality was higher with 2,4-DEE and paraquat than with glyphosate (Muniappa et al., 1995).

In an experiment, cultured human lymphocytes were exposed to three different concentrations (2.5, 5.0, and 10.0 μg/ml) of fluchloralin for 24 and 48 h to assess chromosomal aberrations. A significant dose-dependent increase in chromatid-type aberration was observed in these cells. Multiple aberrations (MA) were scored at all concentrations after 48 h treatment and higher concentrations of fluchloralin (20, 40, and 50 μg/mL) showed genotoxic effects of fluchloralin in human cells (Panneerselvam et al., 1995). Nair et al. (2005) and Singh et al., (2008) demonstrated that 2,4-D is capable of inducing higher DNA damage as well as chromosomal aberrations in human lymphocytes. Another study on the indirect effect of herbicides on fish mortality was more with butachlor, followed by analogous and oxyfluorfen (Sondhia, 2014b). In a study, the residue of butachlor after 10 days was 0.1255 mg/kg in gills, 0.3515 in liver, 0.3145 mg/kg in kidney, and 0.2350 mg/kg in brain traces muscle of *Channa punctata*. Bioaccumulation of 0.009–0.496 μg/g sulfosulfuron in fish under static conditions at one-tenth of sub-lethal concentration 9 mg/L for a period of 56 days is reported (Ramesh et al., 2007; Sondhia, 2014a; Sondhia et al., 2015a, b). In the fishes, 0.074, 0.0014, and 0.0230 μg/g residues of oxyfluorfen, butachlor, and anilofos were detected at 90 days in the rainy season, and 0.0014–0.020 μg/g residues of sulfosulfuron, clodinafop, and metsulfuron were detected at 60 days in fishes in the winter season (Sondhia, 2009b, 2019). Pyrazosulfuron-ethyl and pretilachlor application did not alter the quality of the pond and soil significantly (Sondhia, 2019). Fish mortality and toxicity symptoms were recorded initially in the pond where pretilachlor, penoxsulam, and pyrazosulfuron-ethyl herbicides entered through runoff water in the rainy season of 2012. At 60 days pyrazosulfuron-ethyl, penoxsulam, and pretilachlor residues in fishes were < 0.001, 0.017, and 0.010 μg/g, respectively, which were found below the detection limit (<0.001 μg/g) at 100 days (Sondhia, 2019).

12 Good Agricultural Practices and Herbicides Risk Assessment

Good agricultural practices (GAP) is defined by FAO, as a "collection of principles to apply for on-farm production and post-production processes, resulting in safe and healthy food products, while taking into account economic, social and environmental sustainability." In India, the Quality Council of India (QCI) launched GAP (INDGAP), which helps in better implementation of the food regulations in India. The main advantage of the adoption of GAP is the production of safe food at the primary production level by eliminating chances of entering of contaminants like pesticide residues, veterinary (antibiotic) drug residues, metallic residues, aflatoxin residues, and microbiological contaminants from entering the food chain (FAO, 2004). Following GAP, the herbicides are applied in pre-emergent, post-plant pre-emergent, and pre-harvest stages of various crops for the management of various types of weeds. However, it has been seen and reported that such GAP is barely practiced by most of the farmers, instead frequent and repeated use of herbicides is observed and reported in case of less or poor control of weeds which further enhances the risk of higher persistence of residues in soil and crop produce and risk of contamination to the environment.

13 Herbicide Residues Analytical Procedure

Several techniques have been used to analyze herbicide residues in crops, soils, water, and other agricultural commodities. The foremost steps in their analysis include sample preparation, extraction-clean-up, and analysis by chromatographic techniques. The sample is homogenized and extracted with organic solvent(s) of high, medium, or low polarity. Some efficient extraction techniques are solid-phase micro-extraction (SPME), supercritical fluid extraction (SFE), microwave-assisted extraction (MAE), matrix solid-phase dispersion (MSPD), accelerated solvent extraction (ASE), etc. A clean-up procedure is required prior to the determination of herbicide residues in the complex samples. Liquid chromatography (LC) is a technique used for the analysis of thermally unstable herbicides which cannot be analyzed directly by GC due to their breakdown in the GC injection port or the hot column. Most LC methods perform analysis by reversed-phase chromatography using C_8 or C_{18} columns and relatively polar solvents, like acetonitrile, water, and methanol, as mobile phase (Sondhia et al. 2018, 2024). Ultraviolet (UV) detection with fixed or variable wavelength and photodiode array have been the most commonly used detection method for the determination of herbicide residues by LC. Recent developments have led to LC–MS methods, which do not require a derivatization step while providing a high degree of structural information that allows their identification. Besides GC and HPLC, other techniques, such as thin-layer chromatography (TLC) and the enzyme-linked immuno-sorbent assay (ELISA), are also employed.

Table 2.4 Use of various amendments for the enhanced degradation of herbicides

Amendments	Target herbicides	References
Activated sludge	Atrazine and simazine	Leoni et al. (1992)
Maize straw	Methabenzthiazuron	Printz et al. (1995)
Dairy manure	Atrazine	Gan et al. (1996)
Plant residues, ground seed, or commercial meal	Alachlor, metolachlor, atrazine, and trifluralin	Felsot and Dzantor (1997)
Cellulose, straw, and compost	Atrazine	Abdelhafid et al. (2000)
Compost, corn stalks, corn fermentation by-product, peat, manure, and sawdust	Atrazine, trifluralin, and metolachlor	Moorman et al. (2001)
Raw olive cake	Chlorsulfuron, prosulfuron, and bensulfuron	Delgado-Moreno and Peña (2007)
Rice straw, farm yard manure, saw dust, and charcoal	Atrazine	Mukherjee (2009)

QuEChERS (Quick, Easy, Cheap, Effective, Rugged, and Safe) method covers a very wide scope of analytes, including polar, semi-polar, and non-polar herbicide residues in various food matrices, and is now accepted as a standard sample preparation method for fruits and vegetables by AOAC International (method 2007.01) and European Committee for Standardization (CEN method 15662.5) due to its simplicity, inexpensiveness, amenability to high throughput, and high efficiency. A summary of the analytical procedure for the determination of selected herbicides from various matrixes is presented in Table 2.4. Sondhia (2023a, b) and Sondhia et al. (2024) developed a multi-residues method in which approximately 6 and 40 herbicide residues are reported to be analyzed simultaneously. Sondhia (2019) summarized the liquid chromatographic analytical procedure for determining herbicides from various matrixes.

14 Herbicide Residue Management

Selection or design of a suitable method for pesticide removal from soil, water, and other matrices requires a full understanding of the physicochemical properties of herbicides, such as their water solubility, chemical structure, ionic status of molecules, and KOW (Sondhia, 2012b). Adsorption in soil and other matrix is responsible for accumulating herbicide molecules, forming a monolayer or bilayers, depending on the physicochemical properties of herbicides and solid matrix (Khare & Sondhia, 2014). Highly hydrophobic residues can be removed by organic adsorbing materials such as activated carbon, activated charcoal, nanoorganic materials, carbon nanoparticles organoclays, and bioadsorbents such as plant residues (Alves et al., 2020). Organoclays are reported to remove many chloroacetamide herbicides (Nir et al., 2000). Porous materials are reported to remove simazine from water (Esposito et al., 2016; Sousa et al. (2018). Activated carbon granules and carbon nanotubes are reported to remove atrazine, simazine, linuron, and diuron (Alves et al., 2020). Plant materials are reported to remove residues of alachlor, atrazine, simazine, and trifluralin from water (Sharma et al., 2008).

Adsorption of the herbicide molecules in the colloidal fractions of farmyard manure makes (FYM) them unavailable for crops and weeds. The addition of organic matter, bioprocessed materials, or compost naturally initiates microbial activity in the soil and could be utilized to treat contaminated soils. The FYM application at 10 t/ha or green manuring with sesbania to the soil was found to mitigate the residual toxicity of atrazine, sulfosulfuron, and dinitroanilines, (pendimethalin, trifluralin, fluchloralin) in sandy loam soil. Tillage operations help in bringing deep present herbicide residues to the soil surface which would aid in decontamination by volatilization (Oliveira et al. 2015; Sondhia et al., 2015a, b). In the case of deep ploughing, the herbicide layer is inverted and buried in deeper layers, and thereby, the residual toxicity is reduced. Crop rotation is another herbicide residue management practice that spreads the planting and herbicide application season, reducing the risk of encountering widespread herbicide runoff during a single runoff event. Use of cover crops can also be used as a potential decontamination method. However, it depends on the herbicide half-life and availability in the soil and the sensitivity of the cover crop species throughout the growing season.

A study conducted by Wagh (2017) showed that the application of charcoal at 5.0 kg/ha along the seed line reduced the residual toxicity of atrazine in the soybean crop. McGinley et al. (2022) have reported the use of many reusable media as adsorbents for herbicides, such as granular activated carbon, peat fiber, bottom ash, fly ash, blast slag, Phoslock©, and zeolite. Agriculture waste and natural biomass-based materials such as coconut wastes, chitosan, and lignin (James & Yadav, 2021, Soares et al., 2021) have been studied for herbicide removal. Sugarcane bagasse composite with polyaniline, polypyrrole, and sodium alginate was used as an adsorbent for 2,4-D removal by varying the pH, temperature, contact time, and composite dose (Khan et al., 2023). Biostimulants involve the modification of the environment to stimulate existing microorganisms, such as bacteria or other suitable microorganisms capable of bioremediation (Sondhia et al., 2013, 2016). Biostimulation can be perceived by the addition of adequate amounts of water, nutrients, and oxygen into the soil, in order to enhance the activity of indigenous microbial degraders or to promote co-metabolism (Table 2.4). Even though the diversity of natural microbial populations apparently displays the potential for contaminant remediation at polluted sites, factors such as lack of electron acceptors or donors, low nitrogen or phosphorus availability, or a lack of stimulation of the metabolic pathways responsible for degradation can inhibit or delay the remediation process.

The lack of adequate organic matter in the soil generally leads to low microbial population and, hence, lowers decomposition of herbicides leaving the herbicides recalcitrant in the soil for years without degradation. In a study, Masotti et al. (2021) reported that *Achromobacter, Agrobacterium*, and *Ochrobactrum* consume glyphosate with degradation percentages ranging from 41% to 56% after 96 h. In addition, *Bacillus subtilis* exhibits a high capability to degrade pendimethalin and nicosulfuron (Avarseji et al., 2021). Additions of inorganic nutrients have been reported to facilitate the breakdown of certain herbicides in the soil like atrazine (Shin, 2006).

The use of biochar is another method for the deactivation of herbicide residues. Biochar prevents the mobilization of herbicide residues in soil due to its sorption property and hence helps the crop to escape toxicity (Atkinson et al., 2010). Soils amended with 1% and 2% biochar showed enhanced sorption, slower desorption, and reduced biodegradation of isoproturon (Sopeña et al., 2012). Essandoh et al. (2017) reported the removal of 2,4-D and 2-methyl-4-chlorophenoxyacetic acid (MCPA) from simulated wastewater using switchgrass *(Panicum virgatum)* biochar. Biochar prevents soybean seeding injury caused by atrazine residues by regulating the concentration of this herbicide in soil pore water (Xie et al., 2024). A number of nanotechnological interventions such as carbon-based nanotubes (CNTs), graphene, and nanocrystalline metal oxides are becoming popular in terms of herbicide residue mitigation (Xie et al., 2021; Zhang et al. 2022).

Another promising method for removing herbicides from water sources is membrane separation, including nanofiltration (NF), reverse osmosis (RO), and forward osmosis (FO). Atrazine (ATZ) removal from the water using polyamide microfiltration (MF) membranes with functional layers of chitosan/polystyrene sulfonate (CHI/PSS) was studied by Chandra and Usha (2020). Recently, nanoparticle incorporation into membrane technology has garnered considerable interest. This is because adding nanoparticles improves the membrane's separation performance by increasing membrane surface area. One example is the use of synthesized zinc nanoparticles (ZnNPs) in cellulose acetate to remove organic herbicides, i.e., metolachlor and acetochlor (Shad et al., 2021).

Carbon nanotubes are generally composed of graphite carbons arranged in one or several concentric tubules (Firozjaee et al., 2018) and have shown tremendous potential in the removal of several types of herbicides. Graphene has great adsorption capacities for pesticides (ranging from 600 to 2000 mg/g). Graphene and related carbon-based nanomaterials can adsorb contaminants with aromatic rings through π–π interactions (Smith et al., 2015). Graphene can combine with other materials to improve pesticide adsorption capacity. Graphene-coated silica (GCS), as a highly efficient sorbent, has also been used for the removal of residual organophosphorus pesticides from water (Zhang et al., 2015). Nanocrystalline metal oxides are used for the removal of a broad range of herbicides due to their higher adsorption capacity, faster kinetics, shorter intra-particle diffusion distance, and larger number of surface reaction sites (Moradi Dehaghi et al., 2014). Photocatalysis is an environmentally friendly process used for the elimination of a number of organic pollutants and is quite suitable pretreatment for hazardous and non-biodegradable contaminants to enhance their biodegradability. Ikehata and El-Din (2005) reported that ozone-based AOPs, such as ozone/hydrogen peroxide, ozone/ultraviolet irradiation, and ozone/hydrogen peroxide/ultraviolet irradiation, possess a high potential for degrading and detoxifying four major groups of herbicides, namely aniline-based compounds, pyridines and pyrimidines, triazines, and substituted urea relative to ozone alone. Complete degradation of atrazine is achieved using TiO under 306 nm ultraviolet irradiation (El-Saeid et al., 2021). Ag_3PO_4/TiO_2 nanoparticles photocatalyst was used under visible light to degrade 2,4-D herbicide. Other inorganic oxidizing agents, such as persulfate and peroxymonosulfate, can also be used to degrade chloroacetanilide group herbicides such as alachlor and butachlor. Based on this principle, a wide range of pesticides has been treated by photocatalytic degradation.

Colloidal biochar is reported to be suitable for removing atrazine from contaminated water (Yang et al., 2018; An et al., 2020). Phytoremediation has been successfully used to remove clomazone residues in water (Escoto et al., 2019), 56% for acetochlor and 51% for alachlor (Elsayed et al., 2014). Fenton reactions are reported to remove 99% of atrazine residues from water within 15–45 min). High concentrations of 2,4-D, glyphosate, trifluralin, and butachlor were removed using chemical oxidation (graphene oxide/TiO) followed by adsorption (Lara-Moreno et al. 2022; Hosseini & Toosi, 2019).

Various organic and biochemical means of residue mitigation measures are thoroughly researched and available with various levels of residue management in soil, water, and the environment. However, current research gaps include optimizing the efficacy of nanomaterials and bioadsorbents on different soil and water conditions, and scaling membrane and photocatalytic technologies to a scale conducive to large-scale herbicide removal. Filling these gaps would demand rigorous research into material interactions, long-term efficacy, and application in the environment.

15 Conclusions

Herbicide application is an efficient method but a large proportion of it ends up in the environment. In many cases, farmers still use high herbicide dosages and multiple types of herbicides and fail to execute environment-friendly herbicide management programs. Soil texture and physiological properties govern the persistence of herbicides in the soil, and as long as the soil system remains healthy, expected adverse effects from the spray herbicides in the environment can probably be minimized. The persistence of herbicides in tropical countries like India is generally less than in other countries with temperate climates. To conclude, the domain of herbicide residues demonstrates various levels of residue information in soil, water, and crop at harvest. Extensive use of herbicides poses far-reaching consequences, and there is an essential need for efficient technologies for the mitigation of residues. While using herbicides, all the prevention and management aspects should be kept in mind for getting huge harvests as well as for quality food production without deteriorating the environment. Hence, the integration of mechanical and cultural management practices with herbicides for managing weeds is the most viable option. As herbicide application is foreseen to remain a standard part of agricultural practice, the problem of management of herbicide residues will always be there in the future. In summary, there are various methods available to mitigate herbicides in cropped and non-cropped areas. The integrated approach of adopting cultural, mechanical, chemical, and biological weed control practices along with selecting herbicides with shorter residual life should be adopted as a long-term weed management strategy to avoid any suspected ill effects of herbicide residues on the environment and other living organisms.

References

Abdelhafid, R., Houot, S., & Barriuso, E. (2000). How increasing availabilities of carbon and nitrogen affect atrazine behavior in soils. *Biology and Fertility of Soils, 30*(4), 333–340.

Ahmad, J., D'Angelo, K., Rivas, M., Mahal, M., Nookala, V., Kulakauskiene, D., & Makaryus, A.N. (2021). Dilated cardiomyopathy associated with paraquat herbicide poisoning. *Clinical Practice, 11,* 679–686.

Ahmadi, S., Khazaei, S., & Mehri, F. (2024). The concentration of pesticide residues in vegetables: A systematic review and meta-analyses. *Journal of Agriculture and Food Research, 15*, 101027.

Aktar, M. W., Gupta, A., & Gade, V. (2007). Bhattacharyya fate and behavior of benthiocarb a herbicide in transplanted paddy under east-Indian climatic condition. *Bulletin of Environmental Contamination and Toxicology, 79*, 646–649.

Aktar, M. W., Gupta, D., Alam, S., & Chowdhury, A. (2009). Degradation dynamics of a di-nitro aniline herbicide (trifluralin) in/on blackgram (*Vigna mungo*) under East-Indian climatic condition. *Electronic Journal of Environmental, Agricultural and Food Chemistry, 8*(11), 1172–1177.

Albuquerque, A. F., Ribeiro, J. S., Kummrow, F., Nogueira, A. J., Montagner, C. C., & Umbuzeiro, G. A. (2016). Pesticides in Brazilian freshwaters: A critical review. *Environmental Science: Processes & Impacts, 18*(7), 779–787.

Alves, A., Ruiz, G., Nonato, T., Pelissari, C., Dervanoski, A., & Sens, M. L. (2020). Combined microfiltration and adsorption process applied to public water supply treatment: Water quality influence on pesticides removal. *Environmental Technology, 41*(18), 2382–2392.

An, Q., Li, D., Wu, Y., & Pan, C. (2020). Deposition and distribution of myclobutanil and tebuconazole in a semidwarf apple orchard by hand-held gun and air-assisted sprayer application. *Pest Management Science, 76*(12), 4123–4130.

Arora, A., Tomar, S. S., & Sondhia, S. (2013). Efficacy of herbicides on wheat and their terminal residues in soil, grain and straw. *Indian Society of Weed Science, 45*(2), 109–112.

Aslam, M., Alam, M., & Rais, S. (2013). Detection of atrazine and simazine in ground water of Delhi using high performance liquid chromatography with ultraviolet detector current. *World Environment, 8*(2), 323–329.

Atkinson, C. J., Fitzgerald, J. D., & Hipps, N. A. (2010). Potential mechanisms for achieving agricultural benefits from biochar application to temperate soils: a review. *Plant and Soil, 337*, 1–18.

Avarseji, Z., Taliei, F., & Alamdari, E. G. (2021). Investigation of the biodegradability of pendimethalin by *Bacillus subtilis. Pseudomonas Fluorescens, Escherichia coli. Advances in Environmental Technology, 7*(4), 221–229.

Bakırcı, G. T., Acay, D. B., Bakırcı, F., & Ötleş, S. (2014). Pesticide residues in fruits and vegetables from the Aegean region, Turkey. *Food Chemistry, 160*, 379–392.

Bandana et al. (2015), Bandana, Sharma, N., Joshi, R., Gulati A., & Sondhia, S. (2015). Dissipation kinetics of glyphosate in tea under North Western mid hills conditions of India. *Journal of Pesticide Science, 40*(3), 82–86.

Basu, S., & Rao, Y. V. (2020). Environmental effects and management strategies of the herbicides. *International Journal of Bio-resource and Stress Management, 11*, 518–535.

Bento, C., van der Hoeven, S., Yang, X., Riksen, M., Mol, H., Ritsema, C. J., & Geissen, V. (2019). Dynamics of glyphosate and AMPA in the soil surface layer of glyphosate-resistant crop cultivations in the loess Pampas of Argentina. *Environmental Pollution, 244*, 323–331.

Biswas, P. K., Kumar, S., Mitra, S. R., & Bhattacharyya, A. (2007). Persistence of napropamide in/on tea under north-east Indian climatic condition. *Bulletin of Environmental Contamination and Toxicology, 79*, 566–569.

Brar, N. S., Mandal, K., Kaur, S., Sandhu, A. K., Bhullar, M. S., & Singh, M. P. (2023). Quantification of pendimethalin residue in green fodder and silage of winter cereals using gas chromatography-tandem mass spectrometry. *Pesticide Science, 48*(4), 225–233.

Brown, J. B., Langley, S. A., Snijders, A. M., Wan, K. H., Morris, S. N. S., Booth, B. W., Fisher, W. W., Hammonds, A. S., Park, S., Weiszmann, R., Yu, C., Kirwan, J. A., Weber, R. J. M., Viant, M. R., Mao, J. H., & Celniker, S. E. (2021). An integrated host-microbiome response to atrazine exposure mediates toxicity in Drosophila. *Communications Biology, 4*(1), 1324.

Brühl, C.A., & Zaller, J.G. (2021). Indirect herbicide effects on biodiversity, ecosystem functions, and interactions with global changes. In R. Mesnage, & J. G. Zaller (Eds.), Herbicides: Chemistry, Efficacy, Toxicology, and Environmental Impacts Emerging Issues in Analytical Chemistry. Elsevier, Amsterdam, pp. 231–272.

Bruggen, V., He, A. H. C., Shin, M. M., Mai, K., Jeong, V., Finckh, K. C., & Morris, M. R. (2018). Enviormantal and health effect of herbicide glyphosate. *Science of the Total Environment, 616–617*, 255–268.

Cessna, A.J., Knight, J.D., Ngombe, D., & Wolf, T.M. (2017). Effect of temperature on the dissipation of seven herbicides in a biobed matrix. *Canadian Journal of Soil Science, 97(4),* 717–731.

Chávez, O.P., Yunuén, T.T., Larsen, J., & Oliva, F. G. (2022). Glyphosate-based herbicides alter soil carbon and phosphorus dynamics and microbial activity, *Applied Soil Ecology, 169,*104256.

Chand, M., Samar, S., Bir, D., Singh, N., & Kumar, V. (2014). Halosulfuron methyl: A new post emergence herbicide in India for effective control of *Cyperus rotundus* in sugarcane and its residual effects on the succeeding crops. *Sugarcane Technology, 16*(1), 67–74.

Chandra, P. N., & Usha, K. (2020). Removal of atrazine herbicide from water by polyelectrolyte multilayer membranes. *Materials Today Proceedings, 41*(3), 622–627.

Chandra, N., Jain, N. K., Sondhia, S., & Srivastava, A. B. (2013). Deltamethrin induced toxicity and ameliorative effect of alpha-tocopherol in broilers. *Bulletin of Environmental Contamination and Toxicology, 90*, 673–678.

Chaudhari, Y. S., Kumar, P., Soni, S., Gacem, A., Kumar, V., Singh, S., Yadav, V. K., Dawane, V., Sa, P., Jeon, B. H., Ibrahium, H. A., Hakami, R. A., Alotaibi, M. T., Abdellattif, M. H., Cabral-Pinto, M. M. S., Yadav, P., & Yadav, K. K. (2023). An inclusive outlook on the fate and persistence of pesticides in the environment and integrated eco-technologies for their degradation. *Toxicology and Applied Pharmacology, 466*, 116449.

Claver, A., Ormad, P., Rodríguez, L., & Ovelleiro, J. L. (2006). Study of the presence of pesticides in surface waters in the Ebro river basin (Spain). *Chemosphere, 64*(9), 1437–1443.

Correia, F. V., & Moreira, J. C. (2010). Effects of glyphosate and 2,4-D on earthworms (*Eiseniafoetida*) in laboratory tests. *Bulletin of Environmental Contamination and Toxicology, 85*, 264–268.

Curran, W.S. (2001). Persistence of herbicides in soils. Penn State Extension. http://extension.psu.edu. Accessed: June 21, 2015.

D'Alessandro, M., Held, M., Triponez, Y., & Turlings, T. C. (2006). The role of indole and other shikimic acid derived maize volatiles in the attraction of two parasitic wasps. *Journal of Chemical Ecology, 32*(12), 2733–2748.

de Medeiros, R. C. A., Silva, T. S., Teófilo, T. M. D. S., da Silva, F. D., Souza, M. F., Passos, A. B. R. J., Fernandes, B. C. C., Lins, H. A., Chagas, P. S. F. D., Souza, C. M. M., Holanda, I. S. A., & Silva, D. V. (2024). Herbicide leaching in soil with different properties: Perspectives from commercial formulations and analytical standards. *Toxics, 12*(3), 199.

Delgado-Moreno, L., & Peña, A. (2007). Organic amendments from olive cake as a strategy to modify the degradation of sulfonylurea herbicides in soil. *Journal of Agricultural and Food Chemistry, 55*, 6213–6218.

Dixit, A., Sondhia, S., & Varshney, J. (2011). Bio-efficacy of pinoxaden in wheat (*triticum aestivum*) and its residual effect in succeeding rice (*Oryza sativa*) crop. *The Indian Journal of Agricultural Sciences, 81*(7), 659–661.

El Alfy, M., & Faraj, M. T. (2017). Spatial distribution and health risk assessment for groundwater contamination from intensive pesticide use in arid areas. *Environmental Geochemistry and Health, 39*, 231–253.

Okada, E., Allinson, M., Barral, M. P., Clarke, B., & Allinson, G. (2020). Glyphosate and aminomethylphosphonic acid (AMPA) are commonly found in urban streams and wetlands of Melbourne, Australia. *Water Research, 168*, 115139.

El-Nahhal, I., & El-Nahhal, Y. (2021). Pesticide residues in drinking water, their potential risk to human health and removal options. *Journal of Environmental Management, 299*, 113611.

El-Saeid, M. H., Alotaibi, M. O., Alshabanat, M., Al-Anazy, M. M., Alharbi, K. R., & Altowyan, A. S. (2021). Impact of photolysis and TiO_2 on pesticides degradation in wastewater. *Water, 13*(5), 655.

Elsayed, O. F., Maillard, E., Vuilleumier, S., Nijenhuis, I., Richnow, H. H., & Imfeld, G. (2014). Using compound-specific isotope analysis to assess the degradation of chloroacetanilide herbicides in lab-scale wetlands. *Chemosphere, 99*, 89–95.

Escoto, D, F., Gayer, M. C., Bianchini, M. C., Geovana, da Cruz P., Roehrs R., Denardin, E. L.G. (2019). Use of Pistia stratiotes for phytoremediation of water resources contaminated by clomazone. *Chemosphere, 227*, 299–304.

Esposito, S., Garrone, E., Marocco, A., Pansini, M., Martinelli, P., & Sannino, F. (2016). Application of highly porous materials for simazine removal from aqueous solutions. *Environmental Technology, 37*, 2428–2434.

Essandoh, M., Wolgemuth, D., Pittman, C. U., Mohan, D., & Mlsna, T. (2017). Phenoxy herbicide removal from aqueous solutions using fast pyrolysis switchgrass biochar. *Chemosphere, 174*, 49–57.

FAO. (2003). Herbicide-resistance management in developing countries (Bernal E. Valverde.) In Labrada R (ed), Weed Management for Developing Countries Addendum 1 FAO plant production and protection paper 120 Add. 1. http://www.fao.org/docrep/006/Y5031E/y5031e0h.htm.

FAO. (2004). *Good agricultural practices—a working concept*. Background paper for the FAO internal workshop on good agricultural practices. Accessed from ftp://ftp.fao.org/docrep/fao/010/ag856e/ag856e00.pdf

FAO. (2021). https://www.fao.org/faostat/en/#data/RP.

FAO Stat. *Food and agriculture data*, accessed on November 11 2024.

Faraj, T. K., MH, E. L.-S., Najim, M. M. M., & Chieb, M. (2024). The impact of pesticide residues on soil health for sustainable vegetable production in arid areas. *Separations, 11*, 46. https://doi.org/10.3390/separations11020046

Felsot, A. S., & Dzantor, E. K. (1997). Potential of biostimulation to enhance dissipation of aged herbicide residues in land farmed waste. In E. L. Kruger, T. A. Anderson, & J. R. Coats (Eds.), *Phytoremediation of soil and water contaminants* (pp. 77–91). ACS.

Fernandes AFT, Wang P, Staley C, Aparecida SMJ, Miguel AL, Chagas CS, Guedes S E, Jay Sadowsky M. (2020). Impact of atrazine exposure onthe microbial community structure in a Brazilian tropical latosolsoil. Microbes and Environments 35: 2, ME19143.

Fernández, P., Alcántara, R., Osuna, M. D., Vila-Aiub, M. M., & Prado, R. D. (2017). Forward selection for multiple resistance across the non-selective glyphosate, glufosinate and oxyfluorfen herbicides in *Lolium* weed species. *Pest Management Science, 73*(5), 936–944.

Fingler, S., Mendaš, G., Dvoršćak, M., Stipičević, S., Vasilić, Ž., & Drevenkar, V. (2017). Herbicide micropollutants in surface, ground and drinking waters within and near the area of Zagreb, Croatia. *Environmental Science and Pollution Research, 24*, 11017–11030.

Firozjaee, T. T., Mehrdadi, N., Baghdadi, M., & Nabi Bidhendi, G. R. (2018). Application of nanotechnology in pesticides removal from aqueous solutions—A review. *International Journal of Nanoscience, 14*, 43–56.

Fuchs, B., Laihonen, M., Muola, A., Saikkonen, K., Dobrev, P. I., Vankova, R., & Helander, M. (2022). A Glyphosate-based herbicide in soil differentially affects hormonal homeostasis and performance of non-target crop plants. *Frontiers in Plant Science, 12*, 787958.

Gan, J., Becker, R.L., Koskinen, W.C., Buhler, D.D. (1996). Degradation of Atrazine in Two Soils as a Function of Concentration. *Journal of Environmental quality, 22,* 368.

Gandhi, K., Khan, S., Patrikar, M., Markad, A., Kumar, N., Choudhari, A., Sagar, P., & Indurkar, S. (2021). Exposure risk and environmental impacts of glyphosate: Highlights on the toxicity of herbicide co-formulants. *Environmental Challenges, 4*, 100149.

García-Gal'an, M. J., Monllor-Alcaraz, L. S., Postigo, C., Uggetti, E., L'opez de Alda, M., Díez-Montero, R., & García, J. (2020). Microalgae-based bioremediation of water contaminated by pesticides in peri-urban agricultural areas. *Environmental Pollution, 265*(Pt B), 114579.

Godar, A. S., Varanasi, V. K., Nakka, S., Prasad, P. V., Thompson, C. R., & Mithila, J. (2015). Physiological and molecular mechanisms of differential sensitivity of *Palmer amaranth* (*Amaranthus palmeri*) to mesotrione at varying growth temperatures. *PLoS One, 10*, e0126731.

Goyal, M. V. K., & Phogat, V. K. (2003). Persistence of trifluralin in soils under intermittent and continuous ponding condition. *Pesticide Research Journal, 15*(2), 181–184.

Gupta, M., Garg, N. K., Joshi, H., & Sharma, M. P. (2012). Persistence and mobility of 2,4-D in unsaturated soil zone under winter wheat crop in sub-tropical region of India. *Agriculture, Ecosystems & Environment, 146*(1), 60–72.

Hayes, T. B., Collins, A., Lee, M., Mendoza, M., Noriega, N., Stuart, A. A., & Vonk, A. (2002). Hermaphroditic, demasculinized frogs after exposure to the herbicide atrazine at low ecologically relevant doses. *Proceedings of the National Academy of Sciences of the United States of America, 99*(8), 5476–5480.

Hébert, M. P., Fugère, V., & Gonzalez, A. (2019). The overlooked impact of rising glyphosateuse on phosphorus loading in agricultural watersheds. *Frontiers in Ecology and the Environment, 17*, 48–56.

Helander, M., Saloniemi, I., Omacini, M., Druille, M., Salminen, J. P., & Saikkonen, K. (2018). Glyphosate decreases mycorrhizal colonization and affects plant-soil feedback. *Science of the Total Environment, 642*, 285–291.

Herbicides Market. (2019). *Herbicides market size, share, competitive landscape and trend analysis report, by type, mode of action and crop type: Global opportunity analysis and industry forecast, 2018–2025*. Report Code: A00426. https://www.alliedmarketresearch.com/ Herbicides Market Size, Share, Industry Forecast & Trends.

Hosseini, N., & Toosi, M. R. (2019). Removal of 2,4-D, glyphosate, trifluralin, and butachlor herbicides from water by polysulfone membranes mixed by graphene oxide/TiO_2 nanocomposite: Study of filtration and batch adsorption. *Journal of Environmental Health Science and Engineering, 17*(1), 247–258.

Ikehata, K., & El-Din, M. G. (2005). Aqueous pesticide degradation by ozonation and ozone-based advanced oxidation processes: a review (part II). *Ozone Science & Engineering the Journal of the International Ozone Association, 27*(3), 173–202.

Imfeld, G., Meite, F., Wiegert, C., Guyot, B., Masbou, J., & Payraudeau, S. (2020). Do rainfall characteristics affect the export of copper, zinc and synthetic pesticides in surface runoff from headwater catchments? *Science of the Total Environment, 741*, 140437.

James, A., & Yadav, D. (2021). Valorization of coconut waste for facile treatment of contaminated water: a comprehensive review (2010–2021. *Environmental Technology Innovation, 24,* 102075.

Kaur, R., & Gill, B. S. (2012). Analysis of herbicide residues in celery seeds. *Indian Journal of Ecology, 39*(2), 258–260.

Ke, M., Ye, Y., Li, Y., Zhou, Z., Xu, N., Feng, L., Zhang, J., Lu, T., Cai, Z., & Qian, H. (2021). Leaf metabolic influence of glyphosate and nanotubes on the Arabidopsis thaliana phyllosphere. *Journal of Environmental Sciences (China), 106*, 66–75.

Kerr, G. W., Stahlman, P. W., & Dille, J. A. (2004). Soil pH and cation exchange capacity affects sunflower tolerance to sulfentrazone. *Weed Technology, 18*, 243–247.

Kettles, M. K., Browning, S. R., Prince, T. S., & Horstman, S. W. (1997). Triazine herbicide exposure and breast cancer incidence: An ecologic study of Kentucky counties. *Environmental Health Perspectives, 105*(11), 1222–1227.

Khan, A., Bhatti, H. N., Tahira, M., Alqahtani, F. O., Foziah, F., Fawzan, A., Alissa, S. A., & Iqbal, M. (2023). Na-alginate, polyaniline and polypyrrole composites with cellulosic biomass for the adsorptive removal of herbicide: Kinetics, equilibrium and thermodynamic studies. *Arabian Journal of Chemistry, 16*(1), 104399.

Khare, R. R., & Sondhia, S. (2014). Cyhalofop-butyl mobility and distribution of residues in soil at various depths. *Journal of Environmental Science and Health, Part B, 46*(6), 391–399.

Kogevinas, M., Becher, H., Benn, T., Bertazzi, P. A., Boffetta, P., Bueno-de-Mesquita, H. B., Coggon, D., Colin, D., Flesch-Janys, D., Fingerhut, M., Green, L., Kauppinen, T., Littorin, M., Lynge, E., Mathews, J. D., Neuberger, M., Pearce, N., & Saracci, R. (1997). Cancer mortality in workers exposed to phenoxy herbicides, chlorophenols, and dioxins. An expanded and updated international cohort study. *American Journal of Epidemiology, 145*(12), 1061–1075.

Kumar, B. (2011). Residues of pesticides and herbicides in soils from agriculture areas of Delhi region, India. *Journal of Environmental & Earth Sciences, 12*(1), 8.

Kumar, B., Sharma, R., & Singh, S. B. (2012). Evaluation of harvest residues of cyhalofop-butyl in paddy soil. *Bulletin of Environmental Contamination and Toxicology, 89*, 344–347.

Kumar, L., Kumar, N., Sondhia, S., Basu, P. S., & Srivastava, G. K. (2017). Residues, persistence and dissipation pattern of imazethapyr in *Typic Usrtochrept* soil of Indo-Gangetic Plains. *Journal of Food Legumes, 30*(4), 267–272.

Lara-Moreno, A., Morillo, E., Merchán, F., Madrid, F., & Villaverde, J. (2022). Bioremediation of a trifluralin contaminated soil using bioaugmentation with novel isolated bacterial strains and cyclodextrin. *Science of the Total Environment, 840*, 156695.

Leino, L., Tall, T., Helander, M., Saloniemi, I., Saikkonen, K., Ruuskanen, S., & Puigbò, P. (2021). Classification of the glyphosate target enzyme (5-enolpyruvylshikimate-3-phosphate synthase) for assessing sensitivity of organisms to the herbicide. *Journal of Hazardous Materials, 408*, 124556.

Leoni, V., Cremisini, C., Giovinazzo, R., Puccetti, G., & Vitali, M. (1992). Activated sludge biodegradation test as a screening method to evaluate persistence of pesticides in soil. *Science of the Total Environment, 123*, 279–289.

Lewisa, S. E., Brodiea, J. E., Bainbridgea Zoe, T., Rohdeb, K. W., Davisa, A. M., Mastersb, B. L., Maughana, M., Devlina, M. J., Muellerc, J. F., & Schaffelked, B. (2009). Herbicides: A new threat to the great barrier reef. *Environmental Pollution, 157*, 2470–2484.

Li, Y., Li, C., Li, B., & Ma, Z. (2021). Trifluralin residues in soils from main cotton fields of China and associated ecological risk. *Chemosphere, 284*, 131300.

Loos, R., Locoro, G., Comero, S., Contini, S., Schwesig, D., Werres, F., Balsaa, P., Gans, O., Weiss, S., Blaha, L., Bolchi, M., & Gawlik, B. M. (2010). Pan-European survey on the occurrence of selected polar organic persistent pollutants in ground water. *Water Research, 44*(14), 4115–4126.

Lu, G. H., Hua, X. M., Liang, L., Wen, Z. L., Du, M. H., Meng, F. F., Pang, Y. J., Qi, J. L., Tang, C. Y., & Yang, Y. H. (2018). Identification of major rhizobacterial taxa affected by a glyphosate-tolerant soybean line via shotgun metagenomic approach. *Genes (Basel), 9*(4), 214.

Luo, H. F., Qi, H. Y., & Zhang, H. X. (2004). The impact of acetochlor the bacterial diversity in soil. *Acta Microbiologia Sinica, 44*(4), 519–522.

Mamy, L., Barriuso, E., & Gabrielle, B. (2005). Environmental fate of herbicides trifluralin, metazachlor, metamitron and sulcotrione compared with that of gyphosate, a substitute broad spectrum herbicide for different glyphosate-resistant crops. *Pest Management Science, 61*, 905–916.

Mandal, K., Sahoo, S. K., Battu, R. S., & Singh, B. (2014). Estimation of quizalofop ethyl residues in black gram (*Vigna mungo* L.) by gas liquid chromatography. *Bulletin of Environmental Contamination and Toxicology, 92*(1), 115–118.

Manicardi, A., Milani, A., Scarabel, L., Mora, G., Recasens, J., Llenes, J. M., Montull, J. M., & Torra, J. (2023). First report of glyphosate resistance in an *Amaranthus palmeri* population from Europe. *Weed Research*. https://doi.org/10.1111/wre.12579

Masotti, F., Garavaglia, B. S., Piazza, A., Burdisso, P., Altabe, S., Gottig, N., & Ottado, J. (2021). Bacterial isolates from Argentine pampas and their ability to degrade glyphosate. *Science of the Total Environment, 2021*(774), 145761.

Matzrafi, M., Brunharo, C., & Tehranchian, P. (2019). Increased temperatures and elevated CO_2 levels reduce the sensitivity of *Conyza canadensis* and *Chenopodium album* to glyphosate. *Scientific Reports, 9*, 2228.

McGinley, J., Healy, M. G., Ryan, P. C., Mellander, P. E., Morrison, L., O'Driscoll, J. H., & Siggins, A. (2022). Batch adsorption of herbicides from aqueous solution onto diverse reusable materials and granulated activated carbon. *Journal of Environmental Management, 323*, 116102.

Moorman, T. B., Cowan, J. K., Arthur, E. L., & Coats, J. R. (2001). Organic amendments to enhance herbicide biodegradation in contaminated soils. *Biology and Fertility of Soils, 33*, 541–545.

Moradi Dehaghi, S., Rahmanifar, B., Moradi, A. M., & Azar, P. A. (2014). Removal of permethrin pesticide from water by chitosan-zinc oxide nanoparticles composite as an adsorbent. *Journal of Saudi Chemical Society, 18*, 348–355.

Morrison, H. I., Wilkins, K., Semenciw, R., Mao, Y., & Wigle, D. (1992). Herbicides and cancer. *Journal of the National Cancer Institute, 84*(24), 1866–1874.

Mukherjee, I. (2009). Effect of organic amendments on degradation of atrazine. *Bulletin of Environmental Contamination and Toxicology, 53*(6), 832–835.

Mukherjee, I., & Gopal, M. (2005). Evaluation of fentrazamide for weed control and estimation of its residues in rice. *Bulletin of Environmental Contamination and Toxicology, 74*, 667–672.

Muniappa, V. T., Manjunatha, V., Babu, V. S., & Shivkumar, H. R. (1995). Efficacy of post emergent herbicides on control of water hyacinth (*Eichhornia crassipes* Mart.) and their effect on fishes. *World Weeds, 2*, 117–121.

Nag, S. K., & Das, S. K. (2009). Persistence of atrazine in soil under fodder sorghum. *Journal of Crop and Weed, 5*(2), 131–135.

Nair, R. S., Paulmurugan, R., & Wilsanand, V. (2005). Genotoxix effects of coomonly used pesticides of South India in human lymphocytes. *Pollution Research, 24*(1), 7–12.

Naveen, D. V., Gowda, R. C., & Mamatha, B. (2012). Field studies on persistence of pyrazosulfuron-ethyl in soil, ground water and residues in transplanted rice. *Asian Journal of Soil Science, 89*(5), 1032–1036.

Newton, I. (2004). The recent declines of farmland bird populations in Britain: An appraisal of causal factors and conservation actions. *International Journal of Avian Science, 146*(4), 579–600.

Nir, S., Undabeytia, T., Yaron, D., El-Nahhal, Y., Polubesova, T., Serban, S., Rytwo, G., Lagaly, G., & Rubin, B. (2000). Optimization of adsorption of hydrophobic herbicides on montmorillonite preadsorbed by monovalent organic cations: Interaction between phenyl rings. *Environmental Science & Technology, 34*, 1269–1274.

Obata, F., Fons, C. O., & Gould, A. P. (2018). Early-life exposure to low-dose oxidants can increase longevity via microbiome remodelling in Drosophila. *Nature Communications, 9*(1), 975.

Oliveira, H. C., Stolf-Moreira, R., Martinez, C. B. R., Grillo, R., de Jesus, M. B., & Fraceto, L. F. (2015). Nanoencapsulation enhances the post emergence herbicidal activity of atrazine against mustard plants. *PLoS One, 10*(7), e0132971.

Pacanoski, Z. (2015). Herbicides and adjuvants. *InTech,* pp. 125–147.

Palma, P., et al. (2014). Risk assessment of pesticides detected in surface water of the Alqueva reservoir (Guadiana basin, southern of Portugal). *Science of the Total Environment, 488–489*, 208–219.

Panneerselvam, N., Sinha, S., & Shanmugam, G. (1995). Genotoxicity of the herbicide fluchloralin on human lymphocytes in vitro, chromosomal aberration and micronucleus tests. *Mutation Research—Genetic Toxicology, 344*(1&2), 69–72.

Park, H. H., Lee, D. J., & Kuk, Y. I. (2021). Effects of various environmental conditions on the growth of *amaranthus patulus bertol.* and Changes of herbicide efficacy caused by increasing temperatures. *Agronomy, 11*, 1773.

Parmar, N. B., Maraviya, G. V., Shah, P. G., Patel, B. K., Ghelani, L. M., & Patel, A. M. (1998). Pendimethalin residues in tobacco plant. *Tobacco Research, 24*, 57–59.

Paul, R., Sharma, R., Kulshrestha, G., & Singh, S. B. (2009). Analysis of metsulfuron-methyl residues in wheat field soil a comparison of HPLC and bioassay technique. *Pest Management Science, 65*(9), 963–968.

Polat, B., & Tiryaki, O. (2024). Herbicide contamination of Batak plain agricultural soils and risk assessment. *Journal of Environmental Science and Health, Part B, 59*(5), 203–208.

Printz, H., Buraue, P., & Fuhr, F. (1995). Effect of organic amendment on degradation and formation of bound residues of methabenzthiazuron in soil under constant climatic conditions. *Journal of Environmental Science and Health, Part B, 30*, 435–456.

Rainio, M. J., Ruuskanen, S., Helander, M., Saikkonen, K., Saloniemi, I., & Puigbò, P. (2021). Adaptation of bacteria to glyphosate: a microevolutionary perspective of the enzyme 5-enolpy ruvylshikimate-3-phosphate synthase. *Environmental Microbiology Reports, 13*(3), 309–316.

Ramesh, A., & Maheswari, S. T. (2004). Dissipation of alachlor in cotton plant soil and water and its bioaccumulation in fish. *Chemosphere, 54,* 547–652.

Ramesh, A., Sathiyanarayanan, S., & Chandran, L. (2007). Dissipation of sulfosulfuron in water - bioaccumulation of residues in fish LC-MS/MS-ESI identification and quantification of metabolites. *Chemosphere, 68*(3), 495–500.

Ramirez-Villacis, D. X., Finkel, O. M., Salas-González, I., Fitzpatrick, C. R., Dangl, J. L., Jones, C. D., & Leon-Reyes, A. (2020). Root microbiome modulatesplant growth promotion induced by low doses of glyphosate. *mSphere, 5*, e00484–e00420.

Reddy, K. N., & Reddy, H. (2010). Pesticide residues in surface water of lakes around Hyderabad, India. *Pesticide Research Journal, 22*(2), 111–115.

Refatti, J. P., Deavila, L. A., Camargo, E. R., Ziska, L. H., Oliveira, C., Salas-Perez, R., & RomaBurgos, N. (2019). Temperature and CO_2 effects on absorption, translocation and efficacy of cyhalofop-butyl on *Echinochloa colona*. *Frontiers in Plant Science*. https://doi.org/10.3389/fpls.2019.00529

Reichenauer, T. G., & Germida, J. J. (2008). Phytoremediation of organic contaminants in soil and groundwater. *ChemSusChem, 1*(8–9), 708–717.

Rendon-von Osten, J., & Dzul-Caamal, R. (2017). Glyphosate residues in groundwater, drinking water and urine of subsistence farmers from intensive agriculture localities: a survey in hopelch'en, Campeche, Mexico. *International Journal of Environmental Research and Public Health, 14*(6), 595.

Rheinheimer Dos Santos, D., de Castro, M., Lima, J. A., Rosa, P., de Vargas, J., Camotti Bastos, M., Santanna Dos Santos, M. A., Mondamert, L., & Labanowski, J. (2020). Pesticide bioaccumulation in epilithic biofilms as a biomarker of agricultural activities in a representative watershed. *Environmental Monitoring and Assessment, 192*(6), 381.

Rice, P. J., Anderson, T. A., & Joel, C. R. (2004). Effect of sediment on the fate of metalachlor and atrazine in surface water. *Environmental Toxicology and Chemistry, 23*, 1145–1155.

Rose, M. T., Zhang, P., Rose, T. J., Scanlan, C. A., McGrath, G., & Van Zwieten, L. (2022). Herbicide residues in Australian grain cropping soils at sowing and their relevance to crop growth. *Science of the Total Environment, 833*, 155105.

Roy, S., & Singh, S. B. (2006). Effect of soil type, soil pH, and microbial activity on persistence of clodinafop herbicide. *Bulletin of Environmental Contamination and Toxicology, 77*, 260–266.

Ruuskanen, S., Fuchs, B., Nissinen, R., Puigbò, P., Rainio, M., Saikkonen, K., & Helander, M. (2023). Ecosystem consequences of herbicides: the role of microbiome. *Trends in Ecological Evolution, 38*(1), 35–43.

Sagratini, G., Ametisti, M., Canella, M., & Volpini, R. (2007). Well water in Italy: Analysis of herbicide residues as potential pollutants of untreated crops. *Fresenius Environmental Bulletin*, 18–18.

Saikkonen, K., Nissinen, R., & Helander, M. (2020). Toward comprehensive plantmicrobiome research. *Frontiers in Ecology and Evolution, 8*, 61.

Samsel, A., & Seneff, S. (2016). Glyphosate pathways to modern diseases V: Amino acid analogue of glycine in diverse proteins. *Journal of Biological Physics and Chemistry, 16*, 9–46.

Scott, R. C., Barber, L. T., Boyd, J. W., Norsworthy, J. K., & Burgos, N. R. (2018). *Recommended chemicals for weed and brush control* (p. 195). University of Arkansas Division of Agriculture.

Senarathna, L., Eddleston, M., Wilks, M. F., Woollen, B. H., Tomenson, J. A., Roberts, D. M., & Buckley, N. A. (2009). Prediction of outcome after paraquat poisoning by measurement of the plasma paraquat concentration. *QJM: An International Journal of Medicine, 102*(4), 251–259.

Shad, S., Bashir, N., Nault, M.-F. B.-D., & Lynch, I. (2021). Incorporation of biogenic zinc nanoparticles into a polymeric membrane: Impact on the capture of organic herbicides. *Cleaner Engineering and Technology, 5*, 100339.

Sharma, K. K. (2002). Degradation of alachlor in water and tropical soil of India. *Bulletin of Environmental Contamination and Toxicology, 68*, 394–399.

Sharma, R. K., Kumar, A., & Joseph, P. E. (2008). Removal of atrazine from water by low cost adsorbents derived from agricultural and industrial wastes. *Bulletin of Environmental Contamination and Toxicology, 80*(5), 461–464.

Sharma, K.K., Tripathy, V., Gopal, M. & Wali, S. (2018). Good Agricultural Practices and Monitoring of Herbicide Residues in India. In: Herbicide residue in India (Editors: Sondhia, S., Chodhury, P. & Sharma, A.R.), pp 438.

Sharma, N., Devi, S., Kaur, P., & Sondhia, S. (2021). Behaviour of bispyribac sodium in soil and its impact on biochemical constituents of rice. *International Journal of Environmental Analytical Chemistry, I*, 232.

Sharma, K.K., Tripathy, V., Gopal, M. & Wali, S. (2018). Good agricultural practices and monitoring of herbicide residues in India. In S. Sondhia, P. Chodhury, & A. R. Sharma (Eds.), *Herbicide residue research in India,* pp. 438.

Sharma, K. K., Tripathy, V., Sharma, K., Gupta, R., Yadav, R., Devi, S., & Walia, S. (2022). Long–term monitoring of 155 multi–class pesticide residues in Indian 474 vegetables and their risk assessment for consumer safety. *Food Chemistry, 373*(131518), 475.

Sharma, N., Sapehia, S., Kaur, P., Katwal, R., & Sondhia, S. (2023). Dissipation of tembotrione in maize and its effect on biochemical attributes of maize under mid-hill sub-humid zone. *Environmental Monitoring and Assessment, 195*, 1184.

Shin, H.S. (2006). Determination of phenoxy acid pesticides in frog and fish tissues by gas chromatography–mass spectrometry. *Chromatographia, 63*(11–12), 579–583.

Singh, D.K., Singh, G., & Srivastava, A. S. (2008). Harvest time residue of isoproturon in soil, wheat grain and straw. *Pantnagar Journal of Research, 1,* 125–127.

Singh, S. B., Sharma, R., & Singh, N. (2012). Persistence of pyrazosulfuron in rice-field and laboratory soil under Indian tropical conditions. *Pest Management Science, 68*(6), 828–833.

Singh, S. B., Das, T. K., & Kulshrestha, G. (2013). Persistence of herbicide fenoxaprop ethyl and its acid metabolite in soil and wheat crop under Indian tropical conditions. *Journal of Environmental Science and Health—Part B, 48*(5), 324–330.

Singh, P., Sondhia, S., Dubey, R. P., Sushilkumar, K. B., Gharde, Y., & Choudhary, V. K. (2017). Adoption and impact assessment of weed management technologies in wheat and greengram under conservation agriculture system in Central India. *Indian Society of Weed Science, 49*(1), 23–28.

Singh, A., Kewat, M. L., & Sondhia, S. (2018). Studies on the effect of day time application of herbicide mesosulfuronmethyl on soil microbial communities of wheat rhizosphere. *Journal of Environmental Biology, 39*, 59–65.

Smith, M.R. Jr., Hedges, S.W., Lacount R., D. Shah Cessna et al. (2017), Cessna, A.J., Knight, J. D., Ngombe, D., & Wolf Tom. M. (2017). Effect of temperature on the dissipation of seven herbicides in a biobed matrix. *Canadian Journal of Soil Science, 97(4),* 717–731.

Soares, S. F., Amorim, C. O., Amaral, J. S., Trindade, T., & Daniel-da-Silva, A. L. (2021). On the efficient removal, regeneration and reuse of quaternary chitosan magnetite nanosorbents for glyphosate herbicide in water. *Journal of Environmental Chemical Engineering, 9*(3), 105189.

Sondhia, S. (2001). Determination and assessment of atrazine residues in potato (*Solanum tuberosum* L.) soil. *Geobios, 28*(2/3), 140–142.
Sondhia, S. (2005). Phytotoxicity and persistence of metribuzin residues in black soil. *Toxicological and Environmental Chemistry, 87*, 387–389.
Sondhia, S. (2007). Evaluation of leaching potential of pendimethalin in clay loam soil. *Pesticide Research Journal, 19*, 119–121.
Sondhia, S. (2007a). Fluazifop-butyl residues in soybean crop and soil. *Pesticide Research Journal, 19*, 248–250.
Sondhia, S. (2007b). Determination of imazosulfuron persistence in rice crop and soil. *Environmental Monitoring and Assessment, 137*(1–3), 205–211.
Sondhia, S. (2008a). Persistence of metsulfuron-methyl in wheat crop and soil. *Environmental Monitoring and Assessment, 147*(1–3), 463–469.
Sondhia, S. (2008b). Determination of terminal residue of fenoxaprop in soil, wheat grains and straw. *Geobios, 35*, 312–314.
Sondhia, S. (2008c). Evaluation of leaching potential of butachlor in clay soil. *Geobios, 36*(4), 249.
Sondhia, S. (2009a). Persistence and leaching of sulfosulfuron under wheat cropping system. *The Indian Journal of Agricultural Sciences, 79*(6), 484–487.
Sondhia, S. (2009b). Persistence of oxyfluorfen in soil and detection of its residues in rice crop. *Toxicological and Environmental Chemistry, 91*(3), 425–433.
Sondhia, S. (2009c). Persistence of metsulfuron-methyl in paddy field and detection of its residues in crop produce. *Bulletin of Environmental Contamination and Toxicology, 83*(6), 799–802.
Sondhia, S. (2009d). Leaching behaviour of metsulfuron in two texturally different soils. *Environmental Monitoring and Assessment, 154*(1–4), 111–115.
Sondhia, S. (2010). Persistence and bioaccumulation of oxyfluorfen residues in onion. *Environmental Monitoring and Assessment, 162*(1&4), 163–168.
Sondhia, S. (2012a). Dissipation of pendimethalin in soil and its residues in chickpea (*Cicer arietinum* L.) under field conditions. *Bulletin of Environmental Contamination and Toxicology, 89*(5), 1032–1036.
Sondhia, S. (2012b). Phytoremediation of herbicide contaminated soil, a step towards green sustainable agricultural chemistry. In: *Abstract of the INDO-US workshop on green chemistry*, 11–13 March, 2012, p. 53.
Sondhia, S. (2013a). Evaluation of imazethapyr leaching in soil under natural rainfall conditions. *Indian Society of Weed Science, 45*(1), 58–61.
Sondhia, S. (2013b). Harvest time residues of pendimethalin in tomato, cauliflower and radish under field conditions. *Toxicological and Environmental Chemistry, 95*(2), 254–259.
Sondhia, S. (2013c). Dissipation of pendimethalin in the soil of field pea (*Pisum sativum* L.) and detection of terminal residues in plants. *Journal of Environmental Science and Health, Part B, 48*(12), 104–1048.
Sondhia, S. (2014a). Herbicides residues in soil, water, plants and non-targeted organisms and human health implications: An Indian perspective. *Indian Society of Weed Science, 46*(1), 66–85.
Sondhia, S. (2014b). Soil adsorption studies of a rice herbicide, cyhalofop-butyl, in two texturally different soils of India. *Environmental Monitoring and Assessment, 186*, 5969–5976.
Sondhia, S. (2015). Residues of imazethapyr in field soil and plant samples following an application to soybean. *Indian Society of Weed Science, 47*(2), 166–169.
Sondhia, S. (2016). Liquid chromatographic determination of imazosulfuron residue in food commodity, water and soil using photo diode array detector. *Asian Journal of Chemistry, 28*(5), 1021–1023.
Sondhia, S. (2019). Environmental fate of herbicide use in Central India. In S. Sondhia, P. Choudhury, & A. Sharma (Eds.), *Herbicide residue research in India. Environmental chemistry for a sustainable world* (Vol. 12). Springer. https://doi.org/10.1007/978-981-13-1038-6_2
Sondhia, S. (2023a). Dissipation of penoxsulam in paddy soil, water and plants, and quantification of its residues and its residual toxicity to fish. *Environmental Processes, 10*, 6. https://doi.org/10.1007/s40710-023-00619-w

Sondhia, S. (2023b). A new multiresidue thin layer chromatography-ultra violet detection method for the rapid determination of six herbicides residue applied as ready mix formulation in rice. *Separation Science Plus, 6*(5), 2300011. https://doi.org/10.1002/sscp.202300011.

Sondhia, S., & Dixit, A. (2007a). Determination of terminal residues of oxyfluorfen in onion. *Annals of Plant Protection Sciences, 15*(1), 232–234.

Sondhia, S., & Dixit, A. (2007b). Persistence of oxyfluorfen residues in the soil of paddy field and detection of its residues in crop produce. *The Indian Journal of Agricultural Sciences, 80*(10), 926–929.

Sondhia, S., & Dixit, A. (2008). Persistence of flumioxazin residues in soybean (*Glycine max*) crop and soil. *The Indian Journal of Agricultural Sciences, 78*(8), 716–718.

Sondhia, S., & Dixit, A. (2012). Bioefficacy and persistence of ethoxysulfuron in rice. *ORYZA-An International Journal of Rice, 49*(3), 178–182.

Sondhia, S., & Dixit, A. (2015). Bioefficacy and determination of terminal residues of a herbicide anilofos in field soil and plants following an application to the transplanted rice crop. *Communications in Soil Science and Plant Analysis, 46*(20), 2576–2584.

Sondhia, S., & Dubey, R. P. (2006). Terminal residues of butachlor and pendimethalin in onion. *Pesticide Research Journal, 18*, 185–186.

Sondhia, S., & Khare, R. R. (2017). Leaching of rice herbicide, cyhalofop-butyl in vertisols following an application in lysimeters under tropical rainfall conditions. *Asian Journal of Chemistry, 29*(4), 829–834.

Sondhia, S., & Mishra, J. S. (2005). Determination of terminal residue of clodinafop propargyl in soil, wheat grains and straw. *Indian Society of Weed Science, 37*(3&4), 296–297.

Sondhia, S., & Mishra, J. S. (2024). *A scientific appraisal on herbicides proposed for Ban/restriction in India*. Technical Bullet Published by ICAR-Directorate of Weed Research, 68 p.

Sondhia, S., & Singh, V. P. (2006). Persistence of isoproturon in soil and detection of its residues in wheat grains and straw. *Geobios, 33*(2&3), 209.

Sondhia, S., & Singh, P. K. (2018). Bioefficacy and fate of pendimethalin residues in soil and mature plants in chickpea field. *Journal of Research in Weed Science, 1*, 28–39.

Sondhia, S., & Singhai, B. (2008). Persistence of sulfosulfuron under wheat cropping system. *Bulletin of Environmental Contamination and Toxicology, 80*(5), 423–427.

Sondhia, S., & Waseem, U. (2020). Residue dynamics and degradation behaviour of pyrazosulfuron-ethyl in the rice field environment. *Indian Journal of Weed Science, 52*(4), 362–365.

Sondhia, S., & Yaduraju, N. T. (2005). Evaluation of leaching of atrazine and metribuzin in tropical soil. *Indian Society of Weed Science, 37*, 298–300.

Sondhia, S., Singh, V. P., & Yaduraju, N. T. (2006). Persistence of butachlor in sandy clay loam soil and its residues in rice grains and straw. *Annals of Plant Protection Sciences, 14*(1), 206–209.

Sondhia, S., Sushilkumar, & Parmar, V. (2009). Persistence of glyphosate in water used to control *Eichornea crassipes* weed. In: *Abstract in national symposium on weed threat to environment, biodiversity and agricultural productivity*, 2–3 August, 2009, p. 134.

Sondhia, S., Waseem, U., & Varma, R. K. (2013). Fungal degradation of an acetolactate synthase (ALS) inhibitor pyrazosulfuron-ethyl in soil. *Chemosphere, 93*(9), 2140–2147.

Sondhia, S., Singh, P., & Parmar, V. (2014). Assessment of effect of metsulfuron-methyl on water quality using water quality index. *The Indian Journal of Agricultural Sciences, 84*(5), 579–584.

Sondhia, S., Sharma, N., Janaki, P., & Parvinder, K. (2015a). Herbicide residue hazards and their mitigation. *Indian Farming, 65*(7), 34–39.

Sondhia, S., Khankhane, P. J., Singh, P. K., & Sharma, A. R. (2015b). Determination of imazethapyr residues in soil and grains after its application to soybeans. *Journal of Pesticide Science, 40*(3), 106–110.

Sondhia, S., Smita, S., & Varma, R. K. (2016). Biodegradation of the herbicide penoxsulam (triazolopyrimidine sulphonamide) by fungal strains of Aspergillus in soil. *Applied Soil Ecology, 105*, 196–205.

Sondhia, S., Dasari, S., & Pawar, D. (2023). Degradation dynamics, correlations, and residues of carfentrazone-ethyl, fenoxaprop-p-ethyl, and pinoxaden under the continuous application in the wheat field. *Environmental Geochemistry and Health, 45*, 8851–8865.

Sondhia, S., Dasari, S., Mishra, S., Sen, A. K., Pawar, D. V., Rajput, N., & Mishra, J. S. (2024a). Implication of crop weed competitions and bio-efficacy of herbicide in changing climate scenario. In T. R. Sathishkumar & P. Ganeshbabu (Eds.), *The 9th women's meet—AWM 2024, conference proceedings* (pp. 139–140). Venus International Foundation. ISBN: 978–81–969897–4–3.

Sondhia, S., Sreekanth, D., Pawar, D.V., Geetha, K.N. & Sen, A.K. (2024). Crop-weed competition and herbicide bioefficacy: Implications in a changing climate. *Indian Journal of Weed Science, 56*, 391–403.

Sondhia, S., Dasari, S., Pawar, D. V., Geetha, K. N., & Sen, A. K. (2024b). Crop-weed competition and herbicide bioefficacy: Implications in a changing climate. *Indian Journal of Weed Science, 56*, 391–403.

Sopeña, F., Semple, K., Sohi, S., & Bending, G. (2012). Assessing the chemical and biological accessibility of the herbicide isoproturon in soil amended with biochar. *Chemosphere, 88*, 77–83.

Sousa, G. F. M., Gomes, D. G., Campos, E. V. R., Oliveira, J. L., Fraceto, L. F., Stolf-Moreira, R., & Oliveira, H. C. (2018). Post-emergence herbicidal activity of nanoatrazine against susceptible weeds. *Frontiers in Environmental Science, 6*, 12.

Srivastava, A., Trivedi, P., Srivastava, M. K., Lohani, M., & Srivastava, L. P. (2011). Monitoring of pesticide residues in market basket samples of vegetable from Lucknow City, India. Quechers method. *Environmental Monitoring and Assessment, 176*, 465–472.

Talbot, A. R., Shiaw, M. H., Huang, J. S., Yang, S. F., Goo, T. S., Wang, S. H., Chen, C. L., & Sanford, T. R. (1991). Acute poisoning with a glyphosate-surfactant herbicide (roundup): A review of 93 cases. *Human & Experimental Toxicology, 10*(1), 1–8.

Tan, H., Xing, Q., Ling Mo, W. C., Zhang, X., He, X., Liang, Y., & Hao, R. (2024). Occurrence, multiphase partitioning, drivers, and ecological risks of current-use herbicides in a river basin dominated by rice–vegetable rotations in tropical China. *Science of the Total Environment, 908*, 168270.

Tandon, S. (2014). Degradation kinetics of anilofos in soil and residues in rice crop at harvest. *Pest Management Science, 70*, 1706–1710.

Tangamornsuksan, W., Lohitnavy, O., Sruamsiri, R., Chaiyakunapruk, N., Norman S.C., Reisfeld, B., & Lohitnavy, M. (2019). Paraquat exposure and Parkinson'sdisease: A systematic review and meta-analysis. *Achieve of Environmental Occupational Health, 74*, 225–238.

Terzaghi, E., Vitale, C. M., & Di Guardo, A. (2020). Modelling peak exposure of pesticides in terrestrial and aquatic ecosystems: Importance of dissolved organic carbon and vertical particle movement in soil. *SAR and QSAR in Environmental Research, 31*(1), 19–32.

Tiryaki, O., & Temur, C. (2010). The fate of pesticide in the environment. *Journal of Biological and Environmental Sciences, 4*, 29–38.

Tostado, L., & Bollmohr S. (2022). Pesticide atlas: facts and figures about toxic chemicals in agriculture. 2nd Edition, Germany: Heinrich-Böll-Stiftung.

Tyohemba, R. L., Pillay, L., & Humphries, M. S. (2020). Herbicide residues in sediments from Lake St Lucia (Simangaliso World Heritage Site, South Africa) and its catchment areas: Occurrence and ecological risk assessment. *Environmental Pollution, 267*, 115566.

UN News. (2024), http://desapublications.un.org/publications/world-population-prospects-2024-summary-results. Accessed on November 2024.

Varanasi, A., Prasad, P. V. V., & Jugulam, M. (2015). Impact of climate change factors on weeds and herbicide efficacy. *Advanced Agrochem, 135*, 107–146.

Wagh, R. M. (2017). Management of herbicide residues in soil. *International Archive of Applied Science & Technology, 8*(3), 84–87.

Wan, Y., Tran, T. M., Nguyen, V. T., Wang, A., Wang, J., & Kannan, K. (2021). Neonicotinoids, fipronil, chlorpyrifos, carbendazim, chlorotriazines, chlorophenoxy herbicides, bentazon, and selected pesticide transformation products in surface water and drinking water from northern Vietnam. *Science of the Total Environment, 750*, 141507.

Wang, G., Han, Y., Li, X., Andaloro, J., Chen, P., Hoffmann, W. C., Han, X., Chen, S., & Lan, Y. (2020a). Field evaluation of spray drift and environmental impact using an agricultural unmanned aerial vehicle (UAV) sprayer. *Science of the Total Environment, 737*, 139793.

Wang, G. H., Berdy, B. M., Velasquez, O., Jovanovic, N., Alkhalifa, S., Minbiole, K. P. C., & Brucker, R. M. (2020b). Changes in microbiome confer multigenerational host resistance after sub-toxic pesticide exposure. *Cell Host & Microbe, 27*(2), 213–224.e7.

Westra, E. P., Shaner, D. L., & Westra, P. H. (2014). Dissipation and leaching of pyroxasulfone and S-metolachlor. *Weed Technology, 28*, 72–81.

Xie, J., Guo, Y., Ma, Y., Jiang, H., Zhang, L., Mao, L., Zhu, L., Wu, C., & Zheng, Y. (2024). Biochar prevents soybean seedling injury caused by atrazine residue by regulating the concentration of this herbicide in soil pore water. *Biochar, 6*, 59.

Xie, J., Lin, L., Sánchez, O.F., Bryan, C., Freeman, J.L., & Yuan, C. (2021). Pre-differentiation exposure to lowdose of atrazine results in persistent phenotypic changes in human neuronal cell lines. *Environmental Pollution, 15,* 271, 116379.

Yadav, A., Bhatnagar, A., & Kaur, M. (2013). Aberrations in the chromosomes of *Cirrhinus mrigala* (Hamilton) upon exposure to Butachlor. *Iranian Journal of Toxicology, 7*(21), 858–865.

Yale, R. L., Sapp, M., Sinclair, C. J., & Moir, J. W. (2017). Microbial changes linked to the accelerated degradation of the herbicide atrazine in a range of temperate soils. *Environmental Science and Pollution Research, 24*(8), 7359–7374.

Yang, F., Gao, Y., Sun, L., Zhang, S., Li, J., & Zhang, Y. (2018). Effective sorption of atrazine by biochar colloids and residues derived from different pyrolysis temperatures. *Environmental Science and Pollution Research, 25*(19), 18528–18539.

Zhang, C., Zhang, R. Z., Ma, Y. Q., Guan, W. B., Wu, X. L., Liu, X., Li, H., Du, Y. L., & Pan, C. P. (2015). Preparation of cellulose/graphene composite and its applications for triazine pesticides adsorption from water. *ACS Sustainable Chemistry & Engineering, 3*, 396–405.

Zhang, W., Li, J., & Zhang, Y. (2022). Characterization of a novel glyphosate-degrading bacterial species, *Chryseobacterium* sp. Y16C, and evaluation of its effects on microbial communities in glyphosate-contaminated soil. *Journal of Hazardous Materials, 432*, 128689.

Zobiole, L. H., Kremer, R. J., Oliveira, R. S., Jr., & Constantin, J. (2011). Glyphosate affects microorganisms in rhizospheres of glyphosate-resistant soybeans. *Journal of Applied Microbiology, 110*(1), 118–127.

Chapter 3
Harvest Weed Seed Control: An Additional Late-Season Nonchemical Weed Management Tool

Het S. Desai and Lovreet S. Shergill

Abstract Agroecosystems, characterized by significant human-mediated disturbances, have fostered the coevolution of weed species alongside crop cultivation. The persistence and adaptation of these weeds, particularly the rapid emergence of herbicide resistance in conventional farming, highlight the urgent need for diversified selection pressures in weed management. Both conventional and organic systems face unique challenges that require the development of robust integrated weed management (IWM) strategies. This chapter synthesizes the latest advancements in harvest weed seed control (HWSC), a nonchemical technology that has emerged as a critical component of IWM strategies. We explore the fundamental principles, required attributes, and various methods of HWSC, demonstrating its potential to prevent weed seed bank replenishment and reduce the spread of herbicide-resistant populations. By integrating HWSC with other management practices, this chapter provides eco-evolutionary insights crucial for sustainable weed control in both organic and conventional systems. Future research directions are proposed to enhance the adoption and efficacy of HWSC across diverse agricultural landscapes, supporting more resilient and sustainable farming practices.

Keywords Agroecosystems · Herbicide resistance · IWM · Weed adaptation · Nonchemical weed control · Sustainable farming

H. S. Desai (✉)
Southern Agricultural Research Center, Montana State University–Bozeman, Huntley, MT, USA
e-mail: het.desai@student.montana.edu

L. S. Shergill
Department of Agricultural Biology, Colorado State University, Fort Collins, CO, USA

B. S. Chauhan, G. Mahajan (eds.), *Recent Advances in Weed Science*,
https://doi.org/10.1007/978-3-031-91014-2_3

1 Introduction

In global agriculture, weed infestation in both organic and conventional cropping systems is a pervasive annual threat to production goals (Bàrberi, 2002; Walsh et al., 2013). Weed communities in these systems are shaped by a variety of abiotic (e.g., environmental conditions and human-mediated management practices) and biotic (e.g., inter- and intra-specific competitive interactions) filters (Booth & Swanton, 2002). This process of community assembly can lead to differences in the types and diversity of weeds found in each system. Organic systems are comprised of higher numbers of weak filters as opposed to a few intensive filters in conventional systems (Esposito et al., 2023), generating higher species richness and functional diversity. Due to their diverse crop rotations and limited use of herbicides, organic systems foster a wider variety of weed species, making weed management more complicated than conventional systems. From an ecological application perspective, it is imperative to manage weed species exhibiting higher functional similarity with crops to reduce the overall interspecific competition (Borgy et al., 2016; Smith et al., 2010). Although tillage is a relatively intensive and ecologically disruptive filter, it remains a predominant approach in organic systems due to limited weed management options, highlighting the need for alternative methods.

In conventional agricultural systems, adopting conservation practices (e.g., no-till) and genetically engineered crops, especially for glyphosate tolerance, initially brought remarkable economic benefits across the globe (Duke, 2015; Green, 2012). While the development of herbicide-tolerant cultivars may appear to be a pragmatically different approach, it promotes the use of chemical herbicides (Gressel et al., 2017). In herbicide-tolerant cultivars, the ease of weed control and the initial higher efficacy of chemical herbicides often led to weed-free fields. Subsequently, these cultivars have gained widespread adoption. However, the over-reliance on herbicides has led to the evolution of multiple herbicide-resistant weed populations, a major threat to global agriculture (Green, 2012; Kniss, 2018). The International Herbicide-Resistant Weed Database reports over 1600 global cases of herbicide-resistant weeds, with more than 500 unique cases of evolutionary adaptations (mechanisms of action) (Heap, 2023). In North America, herbicide-resistant weeds cause winter wheat and spring wheat yield losses of 3–47%, amounting to 17.1 billion kg and a US $3.6 billion revenue loss (Flessner et al., 2021). In Australia, managing these weeds adds an average cost of AU $187 million annually, or AU $55 per hectare (Llewellyn et al., 2016). Similarly, herbicide resistance development in Europe could increase weed management costs by up to €240 ha^{-1} (Peterson et al., 2018).

The augmentation of weed seed bank is a predominant factor contributing to annual weed infestation and the progression of herbicide resistance into subsequent generations (Buhler et al., 1997; Grundy et al., 2003). Preventing weed plants' contribution to soil seed bank replenishment is a fundamental strategy for mitigating herbicide resistance and reducing infestation in subsequent seasons (Norsworthy et al., 2012). The increasing threat posed by multiple herbicide-resistant weed populations in conventional systems and the lack of management options in organic systems have spurred substantial demand for research into developing non-chemical weed management strategies (Menalled & Dyer, 2011). This chapter covers the late-season non-chemical weed management tactic called harvest weed seed control (HWSC) that could prevent weed seed bank replenishment. Also, we attempt to synthesize the biological and mechanical factors necessary for effective HWSC, the different HWSC methods, the eco-evolutionary consequences of HWSC, the integration of HWSC with other weed management strategies, and the commercial adoption and future research directions of HWSC across the globe.

2 Basis of Harvest Weed Seed Control

In conventional and organic systems, weed populations that survive or escape early-season management practices have often been observed to produce seeds (Schwartz-Lazaro, Shergill, Evans, Bagavathiannan, Beam, Bish, Bond, Bradley, Curran, & Davis, 2021). During conventional harvest, the combine harvester captures plant material (chaff and straw) along with retained weed and crop seeds. In a combine harvester, weed seeds are diverted to unwanted chaff and straw fractions and dispersed across the field. This uncontrolled dispersal of weed seeds through the rear of a combine harvester enriches existing soil seed bank. Moreover, harvesting of weed and crop seeds together typically results in some degree of contamination. Subsequent sowing of contaminated crop seeds enhances weed dispersal and increases in-crop competition (Norsworthy et al., 2012).

Fortunately, HWSC technology provides an opportunity to manage the fate of retained weed seeds within the combine harvester, minimizing their dispersal and progression of herbicide resistance (Walsh & Powles, 2012) (Fig. 3.1). Combine harvesters equipped with HWSC technology, originally developed and adopted in Australia, process retained weed seeds at harvest before disseminating them through the rear outlet (Walsh et al., 2012). This technology involves six methods, each differing slightly in functionality, such as chaff lining, chaff tramlining, narrow windrow burning, bale direct, impact mill, and chaff cart (Fig. 3.1) (Schwartz-Lazaro, Shergill, Evans, Bagavathiannan, Beam, Bish, Bond, Bradley, Curran, & Davis, 2021). At harvest, all methods prevent retained weed seeds from entering the seed bank with varying efficacy.

Harvest weed seed control begins with a standard harvesting of crop and retained weed seeds, followed by the separation of chaff and straw fractions within the combine harvester. Chaff fractions, which typically contain weed seeds, are subjected to different processes based on HWSC methods (Fig. 3.1). These processes primarily follow three principles: (1) direct removal of weed seeds from the field (i.e., chaff cart and bale direct systems), (2) destruction of collected weed seeds by physical means (i.e., impact mill systems), and (3) decreasing dispersal through concentration of weed seeds (i.e., chaff lining/tramlining, narrow windrow burning systems) (Fig. 3.1) (Walsh et al., 2013; Walsh & Powles, 2012).

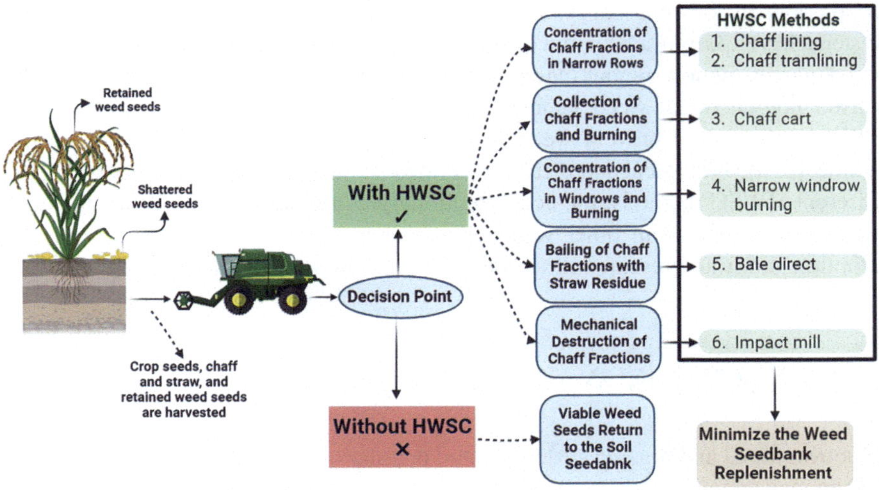

Fig. 3.1 A comprehensive framework demonstrating the positive correlation between harvest weed seed control methods and weed seed retention proportion, indicating that higher seed retention leads to more effective weed control

2.1 Necessary Attributes for Effective Harvest Weed Seed Control

2.1.1 Biological Factors

Seed-shattering, the process by which plants disperse their seeds (Dong & Wang, 2015), is a critical factor influencing the effectiveness of HWSC (Davis, 2008; Walsh et al., 2013). During the domestication of crops, many plants lost adaptive traits vital for their persistence, like seed-shattering, seed dormancy, and non-synchronous flowering (Gregory, 2009). However, arable weed species in mechanically harvested farming systems have adapted by altering the seed-shattering phenology to retain their seeds above crop harvest height, allowing them to be spread by combines at harvest (Walsh & Powles, 2014). This adaptation is driven by natural selection, as weed species with retained seeds are more likely to survive and reproduce in the subsequent season (Maity et al., 2021). The efficacy of HWSC methods is positively correlated with the proportion of weed seed retention above crop harvest height (Walsh & Powles, 2012). Therefore, quantifying weed seed retention at crop harvest is vital to determining the suitability of HWSC methods before area-wide adoption. Walsh et al. (2013) reported varying efficacy (60–90%) of different HWSC methods. However, the success of these methods largely depends on the proportion of weed seed retention at crop harvest because only retained seeds can be effectively processed by HWSC methods (Gill & Holmes, 1997; Walsh, Broster, Schwartz-Lazaro, Norsworthy, Davis, Tidemann, Beckie, Lyon, Soni, & Neve, 2018). Also, the efficacy of HWSC is contingent upon crop species and cultivars due to their differential maturity periods and competitive abilities (Shergill, Schwartz-Lazaro, et al., 2020).

While environmental factors such as humidity, temperature, rainfall, and wind velocity often influence seed-shattering (Liu et al., 2016; Schwartz-Lazaro et al., 2022) indicated that these factors are occasional drivers of weed seed-shattering. High-temperature conditions can accelerate the reproductive transition, resulting in early flowering and early seed-shattering (Pope et al., 2013). Differential weed seed-shattering has been observed in weed populations acclimatized to various environmental conditions and agronomic practices (Schwartz-Lazaro et al., 2017; Soni et al., 2020). Given that numerous factors influence the seed-shattering, it is highly variable among weed populations. Research efforts on weed seed-shattering have been conducted globally, with studies in North America showing high seed retention in dicot species like *Amarnathus plameri* S. Wats., *Amarnathus tuberculatus* (Moq.) J.D.Sauer, *Ambrosia trifida* L., *Bassia scoparia* (L.) A. J. Scott, *Chenopodium album* L, and *Ipomea* spp., while monocots like *Agilops cylendrica* Host, *Avena fatua* L., *Bromus tectorum* L., *Digitaria sanguinalis* (L.) Scop., *Echinochloa crus-galli* (L.) Beauv, and *Setaria* spp. exhibit lower retention (San Martín et al., 2021; Schwartz-Lazaro, Shergill, Evans, Bagavathiannan, Beam, Bish, Bond, Bradley, Curran, Davis, Everman, et al., 2021a, 2021b; Walsh, Broster, Schwartz-Lazaro, Norsworthy, Davis, Tidemann, Beckie, Lyon, Soni, Neve, & Bagavathiannan, 2018). In Europe, Bitarafan and Andreasen (2020) found varying seed retention among dicot species, with monocots like *Alepecurus myosuroides* showing lower retention. Australia, a pioneer in HWSC, has confirmed high seed retention in species like *Lolium regidum* Gaud., *Raphanus raphanistrum* L., *A. fatua*, and *Bromus* spp. (Walsh & Powles, 2014). However, *A. fatua* showed <40% seed retention at wheat harvest in Queensland, Australia (Mahajan & Chauhan, 2021). Despite varying retention levels, the potential of HWSC remains significant, as even moderate seed retention can contribute to effective weed management (Shergill, Schwartz-Lazaro, et al., 2020).

Factors such as weed species, seed size, seed number, chaff load, and chaff type can influence seed destruction efficacy, but the impact of these factors is generally small (Shergill, Bejleri, et al., 2020; Tidemann, Hall, Harker, & Beckie, 2017). The efficacy of HWSC is also influenced by seed bank longevity (i.e., the length of time that weed seeds remain viable in the soil) and fecundity. Weed species with longer seed bank persistence may require multiple years of HWSC implementation to observe a significant reduction in weed populations. For example, in Australia, the utilization of HWSC systems reduced *L. rigidum* density significantly, with reductions ranging widely from 37% to 96% across various studied sites (Walsh, Aves, & Powles, 2017). This variability in effectiveness was attributed to factors such as differential fecundity and the persistence of the residual seed bank. Pending empirical evidence, we hypothesize that the efficacy of HWSC declines for weed species having longer seed bank persistence, requiring long-term and consistent implementation of this technology for managing weed species with prolonged seed bank longevity. Further research is needed to determine the optimal frequency and duration of HWSC implementation for different weed species and environmental conditions.

Despite the widespread adoption across Australia, the research should focus on optimizing the efficacy of this technology by integrating it with other management practices to prevent the evolutionary consequences (discussed later). The following sections summarize the mechanical factors and functionality of each HWSC method.

2.1.2 Mechanical Factors

Harvest weed seed control efficacy hinges upon the mechanical optimization of harvesting equipment. The efficacy of HWSC is influenced by various mechanical factors, including combine harvester setup and speed (Broster et al., 2016). Proper calibration of the harvester can significantly impact weed seed collection, with chaff fraction recovery ranging from 27.9% to 69.6% (Broster et al., 2016). The header is crucial for maximizing weed seed capture. Proper setup involves keeping the cutting bar low, around 6 inches above ground level, which may require slower ground speeds. Additionally, calibrating the reel speed, position, and attack angle is essential to optimize both crop and weed seed collection (Ackroyd et al., 2024).

The type of header used also affects harvest losses and seed control effectiveness (Stefanoni et al., 2022). Currently, HWSC methods are only possible in crops that can be harvested using platform headers, which include soybean, canola, wheat, barley, rice, and sorghum. The utilization of platform headers facilitates weed seed collection by minimizing shattering and maximizing capture. Cotton and corn harvesters utilize forks- and spindle-type headers, which are not suitable for HWSC technology as they increase weed seed-shattering rather than collection. Additionally, using specialty attachments like corn header brushes, air reels, or Duck Foot paddles can improve weed seed capture in certain crops (Ackroyd et al., 2024; Rubione & Unglesbee, 2022). However, weed species of Asteraceae, such as *Conyza* spp. and *Lactuca serriola* L., exhibit a tendency to shatter their seeds upon contact with the combine header, reducing the efficacy of HWSC (Personal observation).

Proper calibration of the combine is also essential for accurate yield monitoring and effective HWSC. This involves calibrating the yield monitor for each specific crop to be harvested, performing multi-point calibrations at different flow rates by varying ground speeds, and recalibrating when crop moisture changes significantly (Powell et al., 2020). To optimize HWSC effectiveness, it is important to test and measure where weed seed and grain losses occur on the combine, particularly at the header. Winans et al. (2023) observed that 22–40% of *A. tuberculatus* seeds dropped to the soil surface due to shatter when the combine reel contacted plants. Adjusting combine settings to minimize these losses and maximize weed seed capture is crucial. By carefully considering these mechanical factors and optimizing combine setup, farmers can significantly improve the effectiveness of HWSC techniques in managing herbicide-resistant weeds and reducing the weed seed bank in their fields.

3 Harvest Weed Seed Control Methods

3.1 Bale Direct

In the bale direct system, a tow-behind square or round baler is directly attached to the combine harvester, which produces bales using the straw and weed seed-containing chaff fractions during harvest (Walsh, Broster, Schwartz-Lazaro, Norsworthy, Davis, Tidemann, Beckie, Lyon, Soni, & Neve, 2018). These bales can be used as a livestock feed. A bale direct system could collect and remove up to 95% of *Lolium rigidum* Gaudin seeds from fields (Walsh & Powles, 2007). Despite the higher efficacy, the worldwide adoption of the bale direct system is relatively limited due to the limited market for bales, the risk of spreading HR weed seeds while transporting bales, and nutrient loss by removing crop residues from the field (Walsh, Broster, Schwartz-Lazaro, Norsworthy, Davis, Tidemann, Beckie, Lyon, Soni, & Neve, 2018).

3.2 Chaff Cart

Chaff carts are a relatively simple HWSC method in which a tow-behind bulk collection bin collects weed seed-containing chaff fractions coming from a rear outlet of a combine harvester (Walsh, Broster, Schwartz-Lazaro, Norsworthy, Davis, Tidemann, Beckie, Lyon, Soni, & Neve, 2018). Chaff carts were shown to have ≥85% efficacy in managing *L. rigidum* populations across 25 sites in Western Australia (Walsh et al., 2014). Although chaff cart systems have shown the potential to remove a greater proportion of weed seed from the field during harvest, global adoption is limited due to difficulties associated with the management of the large volume of collected chaff fractions and nutrient loss due to the removal of crop residues from the fields (Shergill, Schwartz-Lazaro, et al., 2020).

3.3 Chaff Lining and Tramlining

Chaff lining and chaff tramlining, originally developed by the growers of the Esperance region of Western Australia (Ruttledge et al., 2018), are the simplest and the least-cost HWSC methods (Bennett, 2021; Walsh, Broster, Schwartz-Lazaro, Norsworthy, Davis, Tidemann, Beckie, Lyon, Soni, & Neve, 2018). While chaff lining works on confining chaff fractions into narrow rows between crop stubble, chaffs are restricted on the wheel tracks of combine harvesters in controlled traffic systems in chaff tramlining (Walsh, Broster, Schwartz-Lazaro, Norsworthy, Davis, Tidemann, Beckie, Lyon, Soni, & Neve, 2018). These chaff fractions act as a mulch to prevent weed seed germination and/or emergence in a subsequent season. Chaff confinement is achieved by a funnel structure attached to the rear of the combine harvester that places weed seed-containing chaffs into ~20–30 cm rows (Walsh, Broster, Schwartz-Lazaro, Norsworthy, Davis, Tidemann, Beckie, Lyon, Soni, & Neve, 2018). The confined chaff fractions need to remain unaffected by tillage or other agronomic operations throughout the season, as opportunities could arise for weed seeds to emerge by the movement of chaff layers. Studies have shown that these methods could provide up to 90% reduction in weed seedling emergence (Bennett et al., 2023; Walsh et al., 2021); however, the efficacy greatly depends on the chaff type, quantity, and geographical conditions. Due to the low cost and minimal maintenance, these methods are widely adopted in Australia, Europe, and the United States.

3.4 Seed Impact Mill

Mechanical weed seed destruction using an impact mill gained significant attention across the globe due to its greater efficacy and additional benefits compared to other HWSC methods (Walsh et al., 2012). A commercial impact mill was originally developed in Australia and is known as the tow-behind Harrington Seed Destructor (HSD). The impact mill pulverizes weed seed-containing chaff fractions using two spinning cage mills (Walsh et al., 2012). The impact mill allows the on-field retention of the harvest residue, solving the issue of nutrient and moisture losses associated with other HWSC methods. Despite these benefits, the adoption of HSD was limited, partly due to farmers' reluctance to carry an additional unit behind the combine harvester (Walsh et al., 2012). The recent commercialization of integrated seed destructor units such as iHSD™, Redekop™, and Seed Terminator™ aims to overcome the adoption barrier of carrying a large tow-behind unit, thus streamlining the harvesting process (Walsh, Broster, Schwartz-Lazaro, Norsworthy, Davis, Tidemann, Beckie, Lyon, Soni, & Neve, 2018). These impact mills have been shown to be the most effective among other methods, with control rates often exceeding 95%. Winans et al. (2023) concluded that >90% weed seeds were damaged to the extent of losing their viability after exiting from the Seed Terminator™. Similarly, stationary testing of HSD by Tidemann, Hall, Harker, and Beckie (2017) and Shergill, Bejleri, et al. (2020) found >97% and >99% post-processing destruction of weed seeds, respectively.

3.5 Narrow Windrow Burning

Chaff and straw fractions are confined into the narrow windrows (50–60 cm wide) using a chute attached to the rear outlet of a combine harvester. Narrow windrows are subsequently burnt to destroy contained weed seeds. In Australia and the United States, studies have shown that this method could provide up to 99% weed control

(Norsworthy et al., 2020; Walsh & Newman, 2007). Despite the higher efficacy of this method, the adoption is limited at the global level due to the risk of fire outbreaks, fire laws and regulations, high wind velocity, and unpredictable weather patterns (Shergill, Schwartz-Lazaro, et al., 2020). In developing countries, the adoption of HWSC methods is influenced by factors such as agricultural policies, average farm size, and available subsidies. However, methods like chaff lining, tramlining, and narrow windrow burning are relatively low-cost and require minimal modifications to existing equipment. These methods can be easily adopted by small-scale farmers, particularly in organic farming systems, as they align with sustainable practices, support soil health, and are less reliant on costly or environmentally damaging inputs. While narrow windrow burning offers high efficacy in weed control, its fire risks and potential negative impacts on soil health make it less suitable for organic farming, particularly in regions with fire restrictions. In contrast, chaff lining and tramlining are more viable options for marginal farmers, promoting weed control while preserving soil structure and organic matter.

4 Ecological, Evolutionary, and Management Impact

The inclusion of HWSC in integrated weed management (IWM) is a sustainable approach in conventional as well as organic farming systems. A thoughtfully designed IWM protocols, including HWSC as one component, have the potential to prevent the increasing growth rate of weed populations in diverse cropping systems across the globe. In Australia, the utilization of HWSC every 3 years reduced *L. rigidum* density by 98% in <10 years (Walsh, Broster, Schwartz-Lazaro, Norsworthy, Davis, Tidemann, Beckie, Lyon, Soni, & Neve, 2018). The modeling approach by Borger et al. (2016) showed that the overall seed bank replenishment of *Bromus diandrus* Roth could be reduced by 48% with merely 20% HWSC efficacy in 6 years, and the efficacy of 80% could result in >99% reduction. A hypothetical Palmer amaranth management model developed by Shergill, Bejleri, et al. (2020) showed that while 20% efficacy of HWSC was sufficient to maintain the soil seed bank size at stable levels, 80% efficacy could reduce the soil seed bank to <2% within 5 years. Similarly, a model parameterized according to the study conducted in Canada predicted that HWSC efficacy ≥80% is required to maintain the declining growth rate of *A. fatua* (Tidemann, Hall, Harker, Beckie, Johnson, & Stevenson, 2017). An integrated spatial model developed by Somerville et al. (2018) predicted that the efficacies of 72% and 85% are required to maintain low weed densities and prevent the increase in herbicide resistance genes, respectively, when combined with shorter-acting pre-emergence followed by effective post-emergence herbicide application. The same model predicted that when annual HWSC is combined with the longer-acting pre-emergence followed by effective post-emergence herbicide applications, only 45% HWSC efficacy is needed to maintain low weed density and delay the development of herbicide resistance for 25 years.

Although most long-term modeling-based studies found HWSC effective and sustainable practices, weed species could develop adaptive mechanisms to evade the different methods of HWSC (Maity et al., 2021). Empirical evidence suggests that some weed species could shorten their life cycle by adjusting the flowering date. Weed species adapting to early flowering could exhibit a higher seed-shattering rate before crop harvest, potentially evading HWSC. For example, *R. raphanistrum* showed 12 days of early flowering and subsequent maturity due to recurrent selection (Ashworth et al., 2016). Further evaluation confirmed that the early maturing biotypes of *R. raphanistrum* produce and retain more seedpods below the typical wheat harvesting height (Sun et al., 2021). These adaptations suggest that the continuous and standalone utilization of HWSC may result in intense selection pressure, enabling certain weed species to develop adaptive traits to this technology, thus requiring the integration of this technology with other management practices. The following section describes the prospective integration HWSC with management practices.

As weeds adapt to these environmental changes, they may evolve mechanisms such as earlier flowering or altered seed-shattering patterns to evade management practices like HWSC. This evolution of traits could challenge the long-term effectiveness of HWSC and necessitate the integration of additional management strategies to counteract these adaptive responses. Additionally, climate change may influence crop competition dynamics. Changes in seeding rates, row spacing, and crop density—driven by the need to adapt to climate impacts—can alter how crops compete with weeds for light, water, and nutrients. This increased competition could influence weed survival, growth, and seed production, which in turn affects the efficacy of weed control methods like HWSC.

5 Harvest Weed Seed Control: An Important Component in Integrated Weed Management

No technology is a standalone solution in terms of weed management. The continuous utilization of the same tactics (e.g., herbicides or HWSC) could lead to the evolutionary consequences highlighted in previous sections. Consequently, the diversification of selection pressure is a key to developing sustainable weed management programs. One approach for diversifying selection pressure is the development of IWM following the concept of "many little hammers." This concept suggests the applications of multiple weed management tactics at weak links of weeds' life-cycle to achieve overall greater control. Keeping crop rotations at the crux, the concept of "many little hammers" emphasizes the integration of ecologically based weed management tactics whose individual effects are weak, but cumulative effects may prevent weed population growth.

From an ecological standpoint, chemical herbicides and tillage operations are the strongest filters causing significant biodiversity loss which could hamper overall agroecosystem functioning. Given the nonchemical and tillage-free nature of HWSC, there is an opportunity to integrate this technology with other tactics to develop ecologically based and sustainable weed management protocols. For organic growers, the utilization of narrower crop rows could increase the overall efficacy of HWSC for managing weed species that exhibit a high seed-shattering rate. Planting crops in narrower rows is relatively inexpensive and can be achieved without increasing the system complexity (Borger et al., 2016). If the seeding rates remain constant, narrower crop rows do not increase the cropping density, thus preventing the density-dependence mortality of crops by minimally impacting the intensity of intraspecific competition. Additionally, this improves crop resource acquisition, hence increasing overall competitive ability. In terms of interspecific interactions (i.e., crop-weed) in crops planted in narrower rows, weed species are seen to increase their height to capture light following the phenomenon of phototropism. These ecological interactions may alter weed seed-shattering phenology. For example, Walsh (2019) reported that enhancing wheat competitive ability increased overall seed retention above the harvest height by 35% and 31% for *L. rigidum* and *B. diandrus*, respectively, which could positively impact the efficacy of HWSC.

In conventional systems, integration of chemical (i.e., pre-emergence herbicides), cultural (i.e., row width adjustment), and mechanical (i.e., *harvest weed seed control*) weed management tactics could effectively diversify selection pressure. The liberty of using chemical weed management in conventional systems provides opportunities to include effective pre-emergence herbicides, which could delay the weed emergence, hence delaying the flowering and, subsequently, seed-shattering. In the 11 years of experiment, Borger et al. (2016) estimated that the long-term integration of effective pre-emergence herbicides, narrower crop rows, and harvest weed seed control would provide an 80% reduction in *L. rigidum* seed bank. Large-scale field experiments conducted by Norsworthy et al. (2012) concluded that integrating harvest weed seed control methods (e.g., narrow windrow burning and chaff cart) and effective residual herbicide programs drive glyphosate-resistant *A. palmeri* populations near extinction, showcasing >99% reduction in overall seed bank in 3 years. While limited research is available about the inclusion of HWSC in IWM programs, we attempt to highlight some potential prospective research in the future research directions section to sustain this technology.

6 Commercial Adoption Across the Globe

The adoption of HWSC technologies, while initially slow, has gained significant traction in recent years. Australia, the pioneering country for HWSC, has witnessed a notable surge in adoption, spurred by ongoing research, grower awareness, and concerns over herbicide resistance. A systematic survey was conducted by Walsh, Ouzman, et al. (2017) throughout Australian cropping regions, including the participation of 600 growers. Out of these, 43% indicated the routine usage of HWSC; however, interest in different methods varied. The survey concluded that narrow windrow burning had the highest rate at nearly 30%, followed by the chaff tramlining (7%). While this survey was conducted in 2017, 82% of growers showed their interest in the future use of harvest weed seed control, suggesting that adoption rates have likely increased within the past 5 years. In the United States, HWSC adoption is rapidly gaining momentum. While exact figures are difficult to ascertain, anecdotal evidence from organizations like "Getting Rid of Weeds" and research institutions suggests that a growing number of growers are now utilizing HWSC technologies. Notably, Washington State alone has seen significant adoption of impact mills, with an estimated 30 units operating among small grain growers (Drew Lyon, personal communication). This interest is spreading across the country, with growers in regions like the Great Plains and Midwest also exploring and implementing HWSC solutions.

The "Getting Rid of Weeds" initiative, funded by the USDA-ARS, has played a crucial role in increasing HWSC awareness by supporting research, education, and on-farm demonstrations across the United States. Numerous farmers are currently participating in these projects, testing various HWSC technologies (i.e., impact mills, chaff lining) under different conditions and cropping systems. This collaborative effort is generating valuable data and may aid in accelerating the adoption of HWSC nationwide. In Canada, a 2024 survey by Tidemann et al. (2024) documented HWSC adoption among ten growers, primarily in Saskatchewan and Alberta, targeting problematic weeds like *A. fatua* (60%) and *B. scoparia* (50%). While this represents a small sample size, it indicates a growing awareness and interest in HWSC among Canadian farmers.

7 Future Research Direction

Given the aforementioned evolutionary consequences of continuous utilization of HWSC, there is a strong need for research efforts focused on diversifying selection pressure by using this technology in IWM programs. While some efforts documented the usage of this technology in IWM, more research tailoring the requirements of different cropping systems should focus on how different cultural and chemical weed management tactics can complement the efficacy of HWSC based on the concept of "many little hammers."

7.1 Long-Term Evaluation and Participatory Research

To ensure the effectiveness and broader adoption of HWSC, it is essential to conduct long-term studies and evaluate its implementation on commercial farms across North America and elsewhere. Current research in North America has primarily addressed aspects such as the timing of seed-shattering and the effectiveness of impact mills, but extensive field-based studies are limited. The initial promising results and extensive acreage harvested using platform headers suggest a need for continued research and development. Evaluating HWSC through long-term studies and on commercial farms across the continent is crucial. Prioritizing methods like chaff lining and impact mills can help address questions about economic viability, rotational use, regional suitability, and integration with other weed management strategies.

Engaging farmers in conducting large-scale, on-farm research should be prioritized. This participatory approach can help address practical challenges, validate research findings, and promote the adoption of HWSC practices. Collaboration between researchers, farmers, and industry can refine equipment designs suitable for various conditions and provide insights into the long-term impacts on soil seed banks and weed species dynamics. This collaborative effort will identify and correct the limitations of HWSC, supporting the design refinements needed to enhance its effectiveness and sustainability.

7.2 Seed Retention and Shattering Dynamics

Understanding the dynamics of seed retention and shattering is critical for enhancing HWSC efficacy, particularly in managing diverse weed species with varying seed retention traits. Future studies should investigate factors influencing seed shatters, such as weather conditions, planting schedules, crop maturity, and the use of desiccants. Additionally, research must address the challenges posed by weed species with early seed-shattering, which can undermine the effectiveness of HWSC by dispersing seeds before harvest. Considering some previous research efforts, there is an opportunity to promote weed seed retention at crop harvest by increasing the intensity of asymmetric competition and crop-weed niche overlap at early vegetative growth of weed species using cultural practices. However, limited research has documented the concerted impact of cultural (e.g., planting date manipulation, increasing cropping density, row width adjustment, and cover crops) and chemical practices (e.g., residual herbicides and desiccants) on species-specific weed seed retention above harvest height. This gap in the literature highlights the need for studies that evaluate the combined effects of these practices on weed seed retention. Monitoring weed adaptations, such as shifts in species composition or the prevalence of traits like early seed-shattering, is essential for designing resilient weed management strategies. In particular, future work should explore the environmental and genetic factors driving these adaptations to better anticipate and counteract their impacts on HWSC efficacy.

Moreover, the challenge of capturing seeds from diverse weed species, including those with early shattering behavior, necessitates targeted research into the timing and mechanisms of seed shatter. Developing predictive models to assess species-specific shattering dynamics under varying environmental conditions could significantly enhance HWSC systems. Advances in harvest technology should also be leveraged to address the challenge of seed capture, particularly for species prone to early seed-shattering. Innovations in harvesting equipment and seed collection mechanisms that account for variations in seed retention timing and height could further improve HWSC outcomes.

7.3 Technological Innovations and Transdisciplinary Approaches

Despite higher seed retention observed in problematic weed species in crops like corn and cotton, the current commercially available combine headers are not designed to capture retained weed seeds at harvest. This limitation opens the possibility of a transdisciplinary approach involving engineers, agronomists, weed scientists, agroecologists, and growers to design suitable headers that could effectively employ HWSC operations. For example, a collaborative approach by farmers, agronomists, and engineers recently resulted in a vine crusher that seems promising to practice HWSC in potatoes. The potato vine crusher is a customized apparatus consisting of two cylindrical metal rollers that can be mounted on the rear of the potato harvester below the conveyor belts (McKenzie-Gopsill et al., 2023).

7.4 Economic Viability and Farmer Participation

The higher initial investment required for HWSC technology is a significant barrier to its adoption by growers. Therefore, it is crucial to conduct research evaluating the long-term economic impact of HWSC. A comprehensive cost–benefit analysis can demonstrate the economic viability of HWSC in various cropping systems, providing growers with the necessary information to make informed decisions about adopting this technology.

7.5 Expanding Research to New Regions and Species

Numerous studies have identified suitable species for HWSC based on weed seed retention at crop harvest. However, there is still potential to conduct similar research in the regions where this technology has not been introduced. Conducting research in new regions can identify additional candidate species for HWSC and expand the applicability of this technology to different agroecological zones, enhancing its overall effectiveness. Promoting HWSC in developing and underdeveloped countries needs a multi-faceted approach, including farmer workshops and community events to display successful case studies. Affordable and locally adapted HWSC equipment should be developed with engineers and manufacturers tailored to the existing farming practices and resource availability. Support from government and non-government organizations needs to be reinforced, such as subsidies, access to credits, rental services, and customs operations. Farmer participatory approach and feedback loops could identify and resolve on-farm practical challenges unique to the specific regions (Vijayakumar et al., 2022). Finally, expanding HWSC at a global scale may support sustainable agricultural practices that enhance soil health, biodiversity, and carbon sequestration, contributing to a more environment-friendly farming system.

References

Ackroyd, V., Flessner, M., Lazaro, L., Mirsky, S., Rubione, C. & Vangessel, M. (2024). *Harvest weed seed control: A powerful non-herbicide weed control tactic to tackle herbicide resistance*. GrowIWM. Retrieved July 28, 2024, from https://growiwm.org/how-harvest-weed-seed-control/

Ashworth, M. B., Walsh, M. J., Flower, K. C., Vila-Aiub, M. M., & Powles, S. B. (2016). Directional selection for flowering time leads to adaptive evolution in Raphanus raphanistrum (wild radish). *Evolutionary Applications, 9*, 619–629.

Bàrberi, P. (2002). Weed management in organic agriculture: Are we addressing the right issues? *Weed Research, 42*, 177–193.

Bennett, A. J. (2021). *Chaff lining: A harvest weed seed control method to manage herbicide-resistant weeds in US soybean production*. Iowa State University.

Bennett, A. J., Yadav, R., & Jha, P. (2023). Using soybean chaff lining to manage waterhemp (Amaranthus tuberculatus) in a soybean–corn rotation. *Weed Science, 71*, 395–402.

Bitarafan, Z., & Andreasen, C. (2020). Seed production and retention at maturity of blackgrass (*Alopecurus myosuroides*) and silky windgrass (*Apera spica-venti*) at wheat harvest. Weed Science, 68(2), pp.151–156.

Booth, B. D., & Swanton, C. J. (2002). Assembly theory applied to weed communities. *Weed Science, 50*, 2–13.

Borger, C., Riethmuller, G., & D'Antuono, M. (2016). Eleven years of integrated weed management: Long-term impacts of row spacing and harvest weed seed destruction on Lolium rigidum control. *Weed Research, 56*, 359–366.

Borgy, B., Perronne, R., Kohler, C., Grison, A. L., Amiaud, B., & Gaba, S. (2016). Changes in functional diversity and intraspecific trait variability of weeds in response to crop sequences and climate. *Weed Research, 56*, 102–113.

Broster, J. C., Walsh, M. J., & Chambers, A. J. (2016). Harvest weed seed control: The influence of harvester set up and speed on efficacy in South-Eastern Australia wheat crops. In *20th Australasian weeds conference* (pp. 38–41). Weeds Society of Western Australia.

Buhler, D. D., Hartzler, R. G., & Forcella, F. (1997). Implications of weed seedbank dynamics to weed management. *Weed Science, 45*, 329–336.

Davis, A. S. (2008). Weed seed pools concurrent with corn and soybean harvest in Illinois. *Weed Science, 56*, 503–508.

Dong, Y., & Wang, Y.-Z. (2015). Seed shattering: From models to crops. *Frontiers in Plant Science, 6*, 476.

Duke, S. O. (2015). Perspectives on transgenic, herbicide-resistant crops in the United States almost 20 years after introduction. *Pest Management Science, 71*, 652–657.

Esposito, M., Westbrook, A. S., Maggio, A., Cirillo, V., & Ditommaso, A. (2023). Neutral weed communities: The intersection between crop productivity, biodiversity, and weed ecosystem services. *Weed Science, 71*, 301–311.

Flessner, M. L., Burke, I. C., Dille, J. A., Everman, W. J., Vangessel, M. J., Tidemann, B., Manuchehri, M. R., Soltani, N., & Sikkema, P. H. (2021). Potential wheat yield loss due to weeds in the United States and Canada. *Weed Technology, 35*, 916–923.

Gill, G. S., & Holmes, J. E. (1997). Efficacy of cultural control methods for combating herbicide-resistant Lolium rigidum. *Pesticide Science, 51*, 352–358.

Green, J. M. (2012). The benefits of herbicide-resistant crops. *Pest Management Science, 68*, 1323–1331.

Gregory, T. R. (2009). Artificial selection and domestication: Modern lessons from Darwin's enduring analogy. *Evolution: Education and Outreach, 2*, 5–27.

Gressel, J., Gassmann, A. J., & Owen, M. D. (2017). How well will stacked transgenic pest/herbicide resistances delay pests from evolving resistance? *Pest Management Science, 73*, 22–34.

Grundy, A., Mead, A., & Burston, S. (2003). Modelling the emergence response of weed seeds to burial depth: Interactions with seed density, weight and shape. *Journal of Applied Ecology, 40*, 757–770.

Heap, I. (2023). *International herbicide-resistant weed database*. Retrieved August 16, 2023, from https://www.weedscience.org/Pages/filter.aspx

Kniss, A. R. (2018). Genetically engineered herbicide-resistant crops and herbicide-resistant weed evolution in the United States. *Weed Science, 66*, 260–273.

Liu, J., Wang, J., Wang, H., Wang, W., Zhou, R., Mei, D., Cheng, H., Yang, J., Raman, H., & Hu, Q. (2016). Multigenic control of pod shattering resistance in Chinese rapeseed germplasm revealed by genome-wide association and linkage analyses. *Frontiers in Plant Science, 7*, 1058.

Llewellyn, R., Ronning, D., Clarke, M., Mayfield, A., Walker, S., & Ouzman, J. (2016). *Impact of weeds in Australian grain production*. Grains Research and Development Corporation.

Mahajan, G., & Chauhan, B. S. (2021). Interference of wild oat (Avena fatua) and sterile oat (Avena sterilis ssp. ludoviciana) in wheat. *Weed Science, 69*, 485–491.

Maity, A., Lamichaney, A., Joshi, D. C., Bajwa, A., Subramanian, N., Walsh, M., & Bagavathiannan, M. (2021). Seed shattering: A trait of evolutionary importance in plants. *Frontiers in Plant Science, 12*, 657773.

McKenzie-Gopsill, A., MacDonald, A. N., Anderson, L., White, S. N., & Noronha, C. (2023). The potato vine crusher: A new tool for harvest weed seed control. *Canadian Journal of Plant Science, 104*, 115–123.

Menalled, F. D., & Dyer, W. E. (2011). *Preventing and managing herbicide-resistant weeds in Montana*. Montana State University.

Norsworthy, J. K., Green, J. K., Barber, T., Roberts, T. L., & Walsh, M. J. (2020). Seed destruction of weeds in southern US crops using heat and narrow-windrow burning. *Weed Technology, 34*, 589–596.

Norsworthy, J. K., Ward, S. M., Shaw, D. R., Llewellyn, R. S., Nichols, R. L., Webster, T. M., Bradley, K. W., Frisvold, G., Powles, S. B., & Burgos, N. R. (2012). Reducing the risks of herbicide resistance: Best management practices and recommendations. *Weed Science, 60*, 31–62.

Peterson, M. A., Collavo, A., Ovejero, R., Shivrain, V., & Walsh, M. J. (2018). The challenge of herbicide resistance around the world: A current summary. *Pest Management Science, 74*, 2246–2259.

Pope, K. S., Dose, V., Da Silva, D., Brown, P. H., Leslie, C. A., & Dejong, T. M. (2013). Detecting nonlinear response of spring phenology to climate change by B ayesian analysis. *Global Change Biology, 19*, 1518–1525.

Powell, L., Darr, M., & Bergman, R. W. (2020). *Tips for calibrating your combine's yield monitor*. Retrieved July 28, 2024, from https://crops.extension.iastate.edu/blog/levi-powell-matt-darr-ryan-w-bergman/tips-calibrating-your-combines-yield-monitor

Rubione, C., & Unglesbee, E. (2022). *Farmer forum recap: How to set up combine headers for weed seed capture*. GrowIWM. Retrieved July 28, 2024, from https://growiwm.org/farmer-forum-recap-how-to-set-up-combine-headers-for-weed-seed-capture/

Ruttledge, A., Widderick, M., Walsh, M., Broster, J., Bell, K., Rayner, A., Jalaludin, A., Cooray, O., Heuke, L., & Robilliard, S. (2018). *The efficacy of chaff lining and chaff tramlining in controlling problem weeds*. Grain Research and Development Center.

San Martín, C., Thorne, M. E., Gourlie, J. A., Lyon, D. J., & Barroso, J. (2021). Seed retention of grass weeds at wheat harvest in the Pacific northwest. *Weed Science, 69*, 238–246.

Schwartz-Lazaro, L. M., Norsworthy, J. K., Walsh, M. J., & Bagavathiannan, M. V. (2017). Efficacy of the integrated Harrington seed destructor on weeds of soybean and rice production systems in the southern United States. *Crop Science, 57*, 2812–2818.

Schwartz-Lazaro, L. M., Shergill, L. S., Evans, J. A., Bagavathiannan, M. V., Beam, S. C., Bish, M. D., Bond, J. A., Bradley, K. W., Curran, W. S., & Davis, A. S. (2021). Seed-shattering phenology at soybean harvest of economically important weeds in multiple regions of the United States. Part 1: Broadleaf species. *Weed Science, 69*, 95–103.

Schwartz-Lazaro, L. M., Shergill, L. S., Evans, J. A., Bagavathiannan, M. V., Beam, S. C., Bish, M. D., Bond, J. A., Bradley, K. W., Curran, W. S., Davis, A. S., Everman, W. J., Flessner, M. L., Haring, S. C., Jordan, N. R., Korres, N. E., Lindquist, J. L., Norsworthy, J. K., Sanders, T. L., Steckel, L. E., Vangessel, M. J., Young, B., & Mirsky, S. B. (2021a). Seed-shattering phenology at soybean harvest of economically important weeds in multiple regions of the United States. Part 1: Broadleaf species. *Weed Science, 69*, 95–103.

Schwartz-Lazaro, L. M., Shergill, L. S., Evans, J. A., Bagavathiannan, M. V., Beam, S. C., Bish, M. D., Bond, J. A., Bradley, K. W., Curran, W. S., Davis, A. S., Everman, W. J., Flessner, M. L., Haring, S. C., Jordan, N. R., Korres, N. E., Lindquist, J. L., Norsworthy, J. K., Sanders, T. L., Steckel, L. E., Vangessel, M. J., Young, B., & Mirsky, S. B. (2021b). Seed-shattering phenology at soybean harvest of economically important weeds in multiple regions of the United States. Part 2: Grass species. *Weed Science, 69*, 104–110.

Schwartz-Lazaro, L. M., Shergill, L. S., Evans, J. A., Bagavathiannan, M. V., Beam, S. C., Bish, M. D., Bond, J. A., Bradley, K. W., Curran, W. S., Davis, A. S., Everman, W. J., Flessner, M. L., Haring, S. C., Jordan, N. R., Korres, N. E., Lindquist, J. L., Norsworthy, J. K., Sanders, T. L., Steckel, L. E., Vangessel, M. J., Young, B., & Mirsky, S. B. (2022). Seed shattering phenology at soybean harvest of economically important weeds in multiple regions of the United States. Part 3: Drivers of seed shatter. *Weed Science, 70*, 79–86.

Shergill, L. S., Bejleri, K., Davis, A., & Mirsky, S. B. (2020). Fate of weed seeds after impact mill processing in midwestern and mid-Atlantic United States. *Weed Science, 68*, 92–97.

Shergill, L. S., Schwartz-Lazaro, L. M., Leon, R., Ackroyd, V. J., Flessner, M. L., Bagavathiannan, M., Everman, W., Norsworthy, J. K., Vangessel, M. J., & Mirsky, S. B. (2020). Current outlook and future research needs for harvest weed seed control in North American cropping systems. *Pest Management Science, 76*, 3887–3895.

Smith, R. G., Mortensen, D. A., & Ryan, M. R. (2010). A new hypothesis for the functional role of diversity in mediating resource pools and weed–crop competition in agroecosystems. *Weed Research, 50*, 37–48.

Somerville, G.J., Powles, S.B., Walsh, M.J. and Renton, M., 2018. Modeling the impact of harvest weed seed control on herbicide-resistance evolution. Weed Science, 66(3), pp. 395–403.

Soni, N., Nissen, S. J., Westra, P., Norsworthy, J. K., Walsh, M. J., & Gaines, T. A. (2020). Seed retention of winter annual grass weeds at winter wheat harvest maturity shows potential for harvest weed seed control. *Weed Technology, 34*, 266–271.

Stefanoni, W., Latterini, F., Malkogiannidis, V., Salpiggidis, V., Alexopoulou, E., & Pari, L. (2022). Mechanical harvesting of Castor bean (Ricinus communis L.) with a combine harvester equipped with two different headers: A comparison of working performance. *Energies, 15*(9), 2999.

Sun, C., Ashworth, M. B., Flower, K., Vila-Aiub, M. M., Rocha, R. L., & Beckie, H. J. (2021). The adaptive value of flowering time in wild radish (Raphanus raphanistrum). *Weed Science, 69*, 203–209.

Tidemann, B. D., Geddes, C. M., & Sharpe, S. M. (2024). Early adopter insights on physical impact mill technology for harvest weed seed control in Canada. *Canadian Journal of Plant Science*. https://doi.org/10.1139/cjps-2024-0015
Tidemann, B. D., Hall, L. M., Harker, K. N., & Beckie, H. J. (2017). Factors affecting weed seed devitalization with the Harrington seed destructor. *Weed Science, 65*, 650–658.
Tidemann, B. D., Hall, L. M., Harker, K. N., Beckie, H. J., Johnson, E. N., & Stevenson, F. C. (2017). Suitability of wild oat (Avena fatua), false cleavers (Galium spurium), and volunteer canola (Brassica napus) for harvest weed seed control in western Canada. *Weed Science, 65*, 769–777.
Vijayakumar, S., Choudhary, A. K., Deiveegan, M., Subramanian, E., Joshi, E., Goud, B. R., & Kumar, T. S. (2022). The opportunities and challenges for harvest weed seed control (HWSC) in India: An opinion. *Indian Journal of Weed Science, 54*(1), 11–17.
Walsh, M., & Newman, P. (2007). Burning narrow windrows for weed seed destruction. *Field Crops Research, 104*, 24–30.
Walsh, M., Newman, P., & Powles, S. (2013). Targeting weed seeds in-crop: A new weed control paradigm for global agriculture. *Weed Technology, 27*, 431–436.
Walsh, M., Ouzman, J., Newman, P., Powles, S., & Llewellyn, R. (2017). High levels of adoption indicate that harvest weed seed control is now an established weed control practice in Australian cropping. *Weed Technology, 31*, 341–347.
Walsh, M., & Powles, S. (2012). *Harvest weed seed control*. GRDC Agribusiness Crop Updates.
Walsh, M. J., Aves, C., & Powles, S. B. (2014). Evaluation of harvest weed seed control systems. In *19th Australasian weeds conference*. Tasmanian Weed Society.
Walsh, M. J., Aves, C., & Powles, S. B. (2017). Harvest weed seed control systems are similarly effective on rigid ryegrass. *Weed Technology, 31*, 178–183.
Walsh, M. J., Broster, J. C., Schwartz-Lazaro, L. M., Norsworthy, J. K., Davis, A. S., Tidemann, B. D., Beckie, H. J., Lyon, D. J., Soni, N., & Neve, P. (2018). Opportunities and challenges for harvest weed seed control in global cropping systems. *Pest Management Science, 74*, 2235–2245.
Walsh, M. J., Broster, J. C., Schwartz-Lazaro, L. M., Norsworthy, J. K., Davis, A. S., Tidemann, B. D., Beckie, H. J., Lyon, D. J., Soni, N., Neve, P., & Bagavathiannan, M. V. (2018). Opportunities and challenges for harvest weed seed control in global cropping systems. *Pest Management Science, 74*, 2235–2245.
Walsh, M. J., Harrington, R. B., & Powles, S. B. (2012). Harrington seed destructor: A new non-chemical weed control tool for global grain crops. *Crop Science, 52*, 1343–1347.
Walsh, M. J., & Powles, S. B. (2007). Management strategies for herbicide-resistant weed populations in Australian dryland crop production systems. *Weed Technology, 21*, 332–338.
Walsh, M. J., & Powles, S. B. (2014). High seed retention at maturity of annual weeds infesting crop fields highlights the potential for harvest weed seed control. *Weed Technology, 28*, 486–493.
Walsh, M.J. (2019). Enhanced wheat competition effects on the growth, seed production, and seed retention of major weeds of Australian cropping systems. Weed Science, 67(6), pp. 657–665.
Walsh, M. J., Rayner, A. E., Ruttledge, A., & Broster, J. C. (2021). Influence of chaff and chaff lines on weed seed survival and seedling emergence in Australian cropping systems. *Weed Technology, 35*, 515–521.
Winans, T., Massey, R., Schreier, H., Bish, M., & Bradley, K. W. (2023). Harvest weed seed control in soybean with an impact mill. *Weed Technology, 37*, 113–122.

Chapter 4
The Role of Unmanned Aerial Vehicles and Sensor Technology in Site-Specific Weed Management

S. Vijayakumar, Bholuram Gurjar, Muthukumar Bagavathiannan, and Virender Kumar

Abstract Weeds are a persistent foe for farmers, competing for valuable resources and significantly reducing crop yields. Traditional weed control methods often involve applying herbicides uniformly across the entire field without considering weed pressure and distribution, leading to increasing concerns about environmental impact and the risk of evolution of herbicide resistance in weeds. This chapter explores the promising potential of unmanned aerial vehicles (UAVs) equipped with RGB cameras and multispectral/hyperspectral sensors to precisely identify and map weed infestations for site-specific weed management (SSWM). UAVs offer millimeter-level spatial resolution with three-dimensional stereo-vision capabilities, enabling better differentiation of weeds within crop fields. It allows farmers to apply herbicides precisely to affected areas, reducing waste and minimizing environmental impact. UAVs can navigate complex agricultural landscapes, identify weed-infested areas, and apply herbicides with high precision. Despite the advantages of UAVs over other remote sensing platforms, their adoption in crop fields requires significant technological advancements to overcome challenges such as high initial investment costs, complex image acquisition and processing, limited battery endurance, and lower payload capacity. Establishing standardized operating procedures is crucial for effective and environmentally responsible herbicide application using UAVs by reducing the risk of herbicide drift. These protocols should address critical factors like optimal spray height, droplet size, spray volume, wind speed, and suitable weather conditions. Farmers can deploy multiple drones simultaneously, covering larger areas more efficiently and effectively. The future will also witness the integration of UAV data with satellite imagery and ground-based sensors using the

S. Vijayakumar · V. Kumar (✉)
International Rice Research Institute, Los Banos, Laguna, Philippines
e-mail: Vijayakumar.s@icar.gov.in; virender.kumar@irri.org

B. Gurjar · M. Bagavathiannan
Department of Soil and Crop Sciences, Texas A&M University, College Station, TX, USA
e-mail: bholuram.gurjar@ag.tamu.edu; M.Bagavathiannan@ag.tamu.edu

B. S. Chauhan, G. Mahajan (eds.), *Recent Advances in Weed Science*,
https://doi.org/10.1007/978-3-031-91014-2_4

Internet of Things (IoT) and cloud processing to create a more comprehensive and accurate map of weed distribution and crop health.

Keywords Precision agriculture · Herbicide · RGB · Multispectral · Hyperspectral · Weed control · Image processing · Weed mapping

1 Introduction

Weeds pose the most persistent and widespread biotic threat to crop production globally. For example, worldwide, yield losses caused by weeds are around 32%, resulting in an annual economic loss of USD 32 billion (Kubiak et al., 2022). As the agricultural workforce shrinks and labor costs rise, herbicide-based weed control has become a more cost-effective and efficient solution (Shekhawat et al., 2020). Unlike manual weeding, herbicides provide the benefit of early weed targeting, preventing weeds from competing with crops for essential resources. The introduction of herbicide-tolerant crops further increased the dependency on herbicides for weed control.

Although herbicides provide a quick and effective means of weed control, sole dependence on them poses significant environmental, economic, and human health risks. Environmental risks include herbicide-resistance evolution in weeds and negative effects on nontarget organisms (Boutin et al., 2014; Galhano et al., 2011; Heap, 2014). Economically, there are additional costs associated with controlling new weed species that may emerge due to shifts in weed flora, as well as the ongoing expense of herbicides in herbicide-based weed control methods. Human health risks arise from exposure to herbicides during handling and application, especially in the Global South, where farmers often lack adequate personal protective equipment (PPE). These concerns emphasize the importance of identifying strategies that minimize the risks associated with herbicide-based weed control.

These challenges can be addressed through the site-specific application of herbicides, targeting weeds where they are present at the right time and rate while minimizing drift and human exposure as part of the site-specific weed management (SSWM) approach. Unmanned aerial vehicles (UAVs) are used for herbicide application and mapping weed distribution in the field (Bah et al., 2018; Barrero & Perdomo, 2018; Genna et al., 2021). The site-specific application promotes the reduction of herbicide applications by adjusting herbicide doses according to the observed level of weed infestation. The SSWM relies on spatial weed distribution data with infestation level and size to determine where herbicide application is needed most or least or where it can be avoided (Lan et al., 2010). Therefore, accurate field scouting is essential to identify major weed species and their spatial distributions, as weed populations are often unevenly spread across fields. This information will enable farmers to customize their weed control strategies according to variations in weed density, location, and species within the field, thereby optimizing herbicide application (Gerhards et al., 2022).

UAVs (e.g., drones) offer advantages over other methods in field scouting with high-resolution data and precision application of herbicides (Huang et al., 2018). Manual scouting is often challenging, time-consuming, and prone to inaccuracies due to human subjectivity. Ground vehicles equipped with proximal sensors provide an alternative approach to high spatial resolution data for real-time weed assessment and precise herbicide application. However, their effectiveness and large-scale adoption are hampered by adverse field conditions like wet soil conditions and slow operational speeds (Huang et al., 2018). Similarly, satellite remote sensing offers broad-scale spatial information, but its usefulness is restricted by cloud cover and low spatial, spectral, and radiometric resolutions (Vijayakumar, Kumar, et al., 2022). Unlike conventional agricultural tools and implements, which are better suited to large farms due to their size and wide turnaround areas, UAVs offer flexibility for both large and small-scale farmers. Their compact design makes them particularly effective in fragmented agricultural landscapes, where fields are smaller and more dispersed. Thus, UAV technology holds significant potential to transform farming practices in developing countries. By performing multiple operations such as seeding, fertilizer application, pesticide spraying, and transporting farm inputs, UAVs can help address labor shortages and rising wages. Additionally, their adoption can make agriculture more appealing to younger generations, encouraging their participation and retention in farming (Kumar et al., 2020).

UAVs offer a precise and efficient method for applying chemical desiccants to accelerate the drying process of crops before harvest, particularly in crops like cereals, oilseeds, pulses, and potatoes. UAVs can identify crop areas with uneven maturity and apply desiccants precisely, reducing unnecessary chemical use and minimizing environmental impact. UAVs provide uniform coverage over the field, minimizing the risk of uneven drying that could affect grain quality and storage. Unlike ground-based sprayers, UAVs do not cause soil compaction, preserving soil health and structure. UAVs cover large areas quickly, reducing labor requirements and making the agricultural process more efficient (Makam et al., 2024). UAVs operate above the crop canopy, avoiding physical contact with plants and reducing the risk of mechanical damage that can occur with traditional equipment.

This book chapter aims to provide a comprehensive overview of the use of UAVs in weed mapping and control by exploring the different sensors used in UAVs, including RGB, multispectral, and hyperspectral imagery. It discusses the databases and software utilized for processing UAV images. It also examines the application of site-specific precision herbicide spraying with UAVs, highlighting the challenges associated with targeted herbicide application, such as design, sensor resolution, spray dynamics, and environmental factors. Lastly, the chapter outlines future directions and opportunities in UAV technology, focusing on advancements, greater automation, and integration with other technologies.

1.1 Unmanned Aerial Vehicle and Unmanned Aircraft System

Unmanned aerial vehicle (UAV) and unmanned aircraft system (UAS) are closely related terms, but they refer to different aspects of unmanned aerial technology. UAV is the flying aircraft, while a UAS refers to the entire system needed to operate that aircraft, including the drone, control systems, and communication mechanisms. UAVs/UASs are deployed to capture high-resolution images at low altitudes for high-precision and efficient weed scouting over crop canopies. UAV weed mapping facilitates real-time data acquisition and analysis for targeted herbicide applications. The geospatial data management system attached to the UAV platforms handles data acquisition, storage, and transfer through communication protocols. Image preprocessing and information extraction are typically conducted externally. UAV platforms are favored over satellite remote sensing due to their flexibility, ease of use, cost-effectiveness, and rapid technological advancement (Vijayakumar et al., 2021). Two types of UAVs are used in agriculture, viz. rotary-wing UAVs and fixed-wing UAVs (Fig. 4.1). The key features and limitations of both UAVs are given below and summarized in Table 4.1.

Rotary-wing UAVs are most suitable for SSWM. Their unique ability to hover and perform agile maneuvering makes them well-suited for agricultural field inspections. The primary benefit of rotary-wing UAVs in SSWM is their ability to capture ultra-high-resolution imagery, enabling the creation of detailed maps to identify even small individual weeds and patches, an accomplishment not possible with earlier technologies. It also offers the advantage of vertical takeoff and landing (VTOL), eliminating the need for long runways (Vijayakumar et al., 2021). However, rotary-wing UAVs have limitations, including a lower payload capacity and shorter operational range.

Fixed-wing UAVs, while less maneuverable and with lower resolution capabilities, present a strong alternative for large-scale data collection. Their higher payload capacity allows them to carry more advanced equipment, making them ideal for gathering data over large areas in a short period. However, fixed-wing UAVs have their drawbacks. Their high altitude and flight speed result in images with lower

Fig. 4.1 Two major types of UAVs used in agriculture

Table 4.1 The significant difference between fixed-wing and rotary-wing UAVs

UAV parameters	Fixed-wing UAVs	Rotary-wing UAVs
Flight time	~1 h	<25 min
Runway	Required	Not required
Hover	No	Yes
Vertical take-off and landing	No	Yes
Resolution	Relatively coarse	Fine
Image blurring	High	Low
Flight altitude	High	Low
Flight speed	Relatively high	Low
Turning radius	Large	Very small
Area coverage	Large	Small

spatial resolution and limited overlap, which complicates the creation of detailed weed maps. Additionally, high-speed flights at these altitudes can cause image blurring at the edges, making real-time weed detection and herbicide application decisions more computationally demanding (Vijayakumar et al., 2021). Also, the requirement for a long runway or launcher for takeoff and landing makes fixed-wing UAVs less convenient for field deployment than their rotary-wing counterparts.

1.2 Weed Mapping Using UAV

Mapping weeds during the early stages of a crop is essential for successful SSWM. Distinguishing crops and weeds during the seedling stage is challenging due to similarities in their spectral signatures and morphological features (De Castro et al., 2018). For instance, detecting weeds in crops like cotton and rice differs because of variations in crop density and growth patterns. Additionally, the interaction between weeds, crops, and soil further complicates the system. Despite these challenges, sensor-based technologies have proven effective in identifying plants based on optical properties, shape, and texture. Advancements in sensor capabilities have enhanced the accuracy of weed detection and classification. The use of machine learning (ML) or deep learning (DL) algorithms improves the accuracy of classifying crops and weeds in UAV images. Thus, the integration of UAVs, sensors, and artificial intelligence (AI) can optimize weed detection for better weed control.

The widespread adoption of UAVs for weed mapping has been facilitated by advancements in both UAV technology and image processing (photogrammetry). The capture and processing of images into accurate, geo-referenced formats (ortho-rectification) are no longer significant technical hurdles. Studies have shown that a single battery on a small UAV such as the Phantom 4 can map 10 hectares of land in just 20 minutes at a flight altitude of 40 m (Rasmussen et al., 2019). The potential of UAV weed mapping has been demonstrated in row crops such as corn and sunflower (Peña et al., 2013, 2015; Pérez-Ortiz et al., 2015; Torres-Sánchez et al.,

Fig. 4.2 Effect of flight altitude on image resolution and scale

2013). The process of weed mapping in row crops is conducted in three automated phases: crop row identification, distinguishing between crop plants and weeds based on their relative positions, and creating a weed infestation map using a grid system. Weed detection in row crops takes advantage of the wide row spacing, assuming that vegetation coverage between the crop rows is weeds.

High-resolution imagery is critical for accurately assessing weed density, location, and biomass in the field. UAVs provide millimeter-level spatial resolution and detailed information about crops and weeds for accurate weed identification, especially when dealing with species with similar morphological characteristics (Fig. 4.2) (Hao et al., 2023; Huang & Reddy, 2015). Such detailed information is vital for feature extraction, a critical step for ML algorithms that depend on robust datasets for accurate performance (Velumani et al., 2021). However, the use of high-resolution images demands greater processing power and storage capacity, factors that must be considered when planning UAV operations.

1.3 Advantages of UAV-Assisted Weed Mapping and Management

UAV technology has transformed weed management by enabling weed identification, assessing weed pressure, and facilitating targeted herbicide applications (Bah et al., 2019; Li et al., 2016). The benefits of site-specific weed management (SSWM) using UAVs are summarized in Fig. 4.3 and discussed here. Early detection and targeted control improve crop health and can result in higher yields. SSWM reduces herbicide usage, leading to significant cost savings and more sustainable farming practices. This approach allows for swift responses to weed outbreaks, particularly in remote or hard-to-reach areas. The UAV-assisted spot application system accurately delivers the herbicide to the target in the precise amount needed, resulting in herbicide savings of approximately 69–79% in areas with low infestation levels (Nobre et al., 2023). The SSWM using UAVs minimizes herbicide exposure in

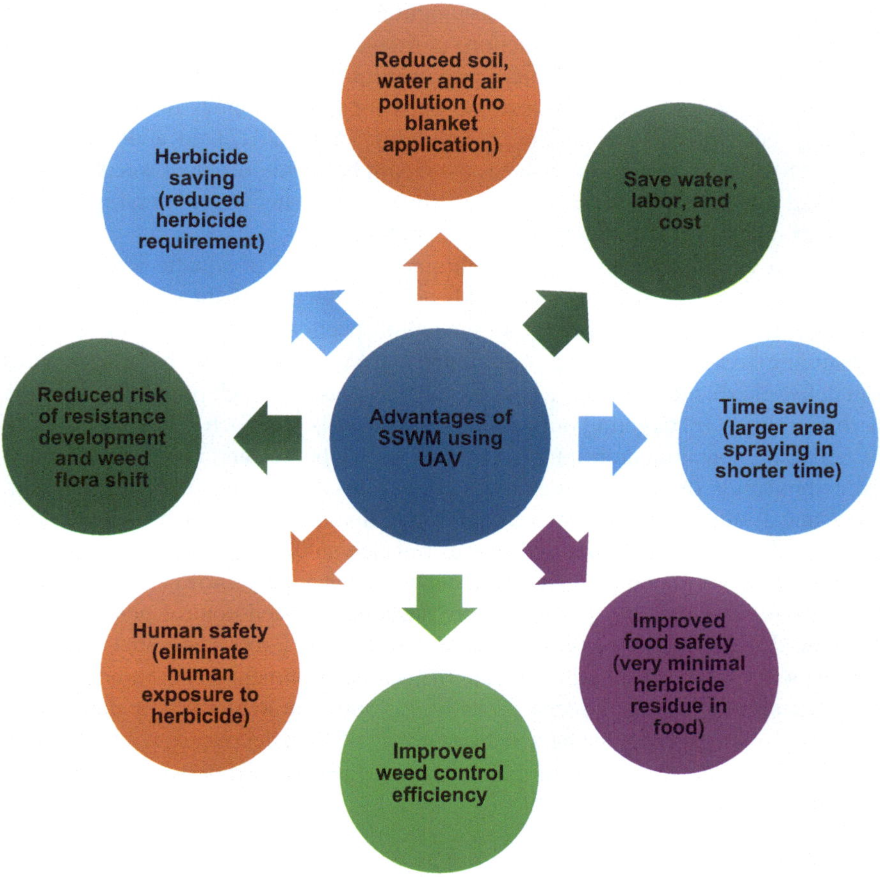

Fig. 4.3 Advantages of SSWM using UAVs

nontargeted areas, thereby reducing its impact on soil, water, and crops (Sapkota et al., 2023). By reducing herbicide consumption by 40–60%, SSWM offers enhanced environmental protection, supports sustainable agricultural practices, improves economic returns, and ensures the long-term sustainability of herbicide use in agriculture (Monteiro & Santos, 2022).

Traditional weed control methods often expose farmers and applicators to hazardous chemicals due to repeated contact during spraying, especially in the global south, where most of the farmers and pesticide applicators are not using personal protective equipment (Kumar et al., 2020). UAV technology offers a significant safety advantage by eliminating the applicator's herbicide exposure. Manual spraying involves considerable water usage, typically between 225 and 450 liters per hectare (L water/ha) (Kumar et al., 2020; Pooja et al., 2021). In comparison, UAV spraying drastically reduces water consumption, often using 5–10 times less water for the same application. This reduction contributes to more sustainable agricultural practices (Paul et al., 2024). Manual spraying requires additional labor to transport water for spraying. UAVs eliminate the need for large volumes of spray liquids, reducing labor costs (Vijayakumar, Madireddy, et al., 2022). In addition, UAV-assisted SSWM approach also reduces the risk of herbicide-resistant weed development and weed flora shifts, saves farmers time by covering larger spraying areas in a shorter time, and reduces herbicide loads in food.

UAVs accurately detect, map, and monitor the extent of aquatic weed infestations in real-time. This capability enables early detection of invasive species and significantly reduces the need for manual inspections, which are often time-consuming and labor-intensive. Additionally, UAVs equipped with spray systems can deliver herbicides directly to weed-infested areas with high precision, reducing herbicide usage by targeting only affected zones. This approach minimizes the risk of herbicide contamination into surrounding water bodies and ensures consistent, even application over large water surfaces. Advanced UAVs can also be fitted with mechanical attachments to physically remove floating weeds such as water hyacinths or duckweed. This method reduces reliance on chemical herbicides, thereby minimizing environmental impact by focusing on physical removal. By integrating these technologies, UAVs promote more sustainable weed control practices with minimal disruption to aquatic ecosystems. Moreover, UAVs can collect valuable data on water quality, weed growth patterns, and environmental conditions, helping authorities make informed decisions on weed control strategies. These data-driven insights provide a comprehensive understanding of the effectiveness of weed management programs over time, further enhancing the precision and sustainability of aquatic weed control efforts.

2 Sensors for Weed Mapping and their Workflow

UAVs use various sensors to create detailed weed maps (Table 4.2). The most common sensor is the RGB camera, which captures color information similar to that of the human eye. Multispectral sensors capture data over a broader range of wavelengths for more advanced weed identification, allowing for calculating vegetation indices that differentiate healthy crops and weeds. Hyperspectral sensors take things a step further, offering an incredibly detailed picture across hundreds of narrow wavelengths, enabling highly precise weed species recognition. Sensors allow for object recognition when combined with adaptive algorithms that can trigger a specific mechanical response. AI algorithms are utilized for image processing, weed infestation map creation, and decision-making.

Table 4.2 Different types of sensors used in UAVs and their specifications[a]

Camera model	Sensor	Spectral range (nm)	Spectral bands	Resolution	GSD
Micasense RedEdge-M	CMOS	475–890	Red, green, blue, red edge, NIR	1280 × 960 (1.2 MP × 5 imagers)	60 m, 4 cm/pixel
Micasense RedEdge-MX	CMOS	475–840	Red, green, blue, red edge, NIR	1280 × 960 (1.2 MP × 5 imagers)	60 m, 4 cm/pixel
Micasense Altum	CMOS	475–842	Red, green, blue, red edge, NIR	2064 × 1544 (3.2 MP × 5 imagers), thermal-160 × 120 (0.01 K)	60 m, 1.25 cm/pixel, thermal sensor-60 m, 17 cm/pixel
DJI Phantom 4 Pro	CMOS	–	Red, green, blue	5472 × 3648	60 m, 1.09 cm/px
DJI Mavic 2 Pro	CMOS	–	Red, green, blue	5472 × 3648 (12 MP)	60 m, 1.53/px
Pika L	CMOS	400–1000	281 channels	–	–
Cubert UHD-185	CMOS	450–995	125 channels (8μm)	1024 × 1024	60 m, 2.06 cm/px
Corning microHSI 410 SHARK	CCD/CMOS hybrid	400–1000	300 channels (2μm)	1364 × 308	60 m, 1.375 m/px
Sony Alpha 7R II	CMOS	–	Red, green, blue	7952 × 5304 (42.4 MP)	60 m, 2.65 cm/px
TetraCam ADC lite	CMOS	520–920	Red, green, NIR	2048 × 1536 (3.2 MP)	60 m, 2.4 cm/px

[a] *CMOS* complementary metal oxide semiconductor, *NIR* near-infrared

2.1 RGB Camera

Red, green, and blue (RGB) cameras are popular for UAV weed mapping due to their affordability, accessibility, size, and weight (Fig. 4.4). RGB cameras are commonly integrated into commercially available UAVs, making them easy to use for a broad range of users. UAVs typically utilize cameras with resolutions between 3 and 12 megapixels (MP), providing a practical balance between cost and image quality (Singh et al., 2020). However, it is important to recognize the limitations of RGB cameras. The broad spectral bandwidths of RGB cameras can make it challenging to distinguish weeds from crops based solely on color information.

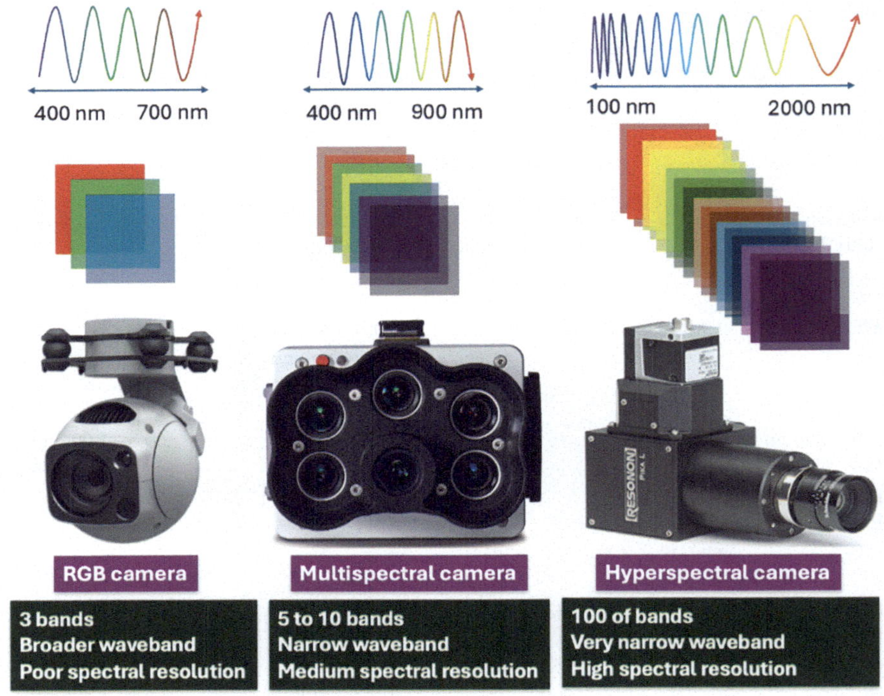

Fig. 4.4 Sensors used in weed mapping

The RGB workflow for UAV weed detection involves three key stages: pre-flight planning, flight and image acquisition, and processing and dataset extrapolation.

(a) *Pre-flight planning:* This step sets the foundation for high-quality data collection. It involves considering the following: defining the study area, flight altitude, image overlap, site topography, weather conditions, and local UAV regulations.
(b) *Flight and image acquisition:* During this stage, ensuring sufficient data storage capacity and platform stability is critical. Inadequate platforms can lead to data flow issues, potentially causing data loss or mission failure.
(c) *Processing and dataset extrapolation:* Unlike multispectral and hyperspectral sensors, RGB data do not require radiometric calibration. It can be used directly or processed to create a georeferenced orthomosaic. This process involves rectifying individual images, georeferencing them using GPS data, and stitching them together. The spatial resolution of the RGB image is the most critical factor for distinguishing different plant species. It depends on the number of pixels in the sensor arrays.
(d) Another critical step is mosaicking, where individual images are stitched together to form a complete picture. Incomplete image overlap or misalignment during flight operations can lead to gaps or inaccuracies in the final mosaic.

Ortho-rectification removes distortions caused by the terrain, creating a flat and accurate representation of the field. The choice of the right mosaicking software can significantly impact the final map quality. Studies have shown that Pix4Dmapper delivers the highest accuracy, while Photoshop CC (spherical mode) offers a cost-effective alternative suitable for many agricultural applications (Singh et al., 2020).

Several RGB-based vegetation indices (VIs), viz., HSV (hue, saturation, and value) and CIE-L*a*b* (L* for illumination, a* for values from red to green, and b* for values from blue to yellow) (Hamuda et al., 2016), excess green (ExG) (Woebbecke et al., 1995), excess red (ExR), the green–red vegetation index (GRVI) (Motohka et al., 2010), and the color index of vegetation extraction (CIVE) are derived from RGB images for crop/weed segmentation. These VIs emphasize the chromatic characteristics of vegetation (Hamuda et al., 2016). For example, ExG enhances plant–soil contrast, as green crops generally have higher green hue values than the soil background. Lambert et al. (2018) used UAV-captured RGB imagery to map black-grass (*Alopecurus myosuroides* Huds.), a problematic grass weed in wheat fields. The random forest (RF) model used in this study to distinguish between areas with (presence) and without (absence) *A. myosuroides* achieved an overall accuracy of 88%. Pahikkala et al. (2015) identified the dicot species dandelion (*Taraxacum officinale* Weber in Wiggers) mixed with oat (*Avena sativa* L.) using UAV-derived RGB images. Statistical models used for weed classification face challenges when leaves overlap within the crop canopy, often resulting in inaccurate classifications. Most existing methods have difficulty handling this issue. In contrast, classification accuracy improves considerably when the leaves are isolated and clearly distinguishable.

The presence of small weeds increases the complexity of detection because they occupy a smaller portion of the image, making detection difficult in UAV images. Furthermore, weed detection is more challenging in dense crop systems compared to row crop systems. For detecting individual grassy weeds, a higher resolution of <0.5 cm/pixel is required (Gao et al., 2018). Detecting individual weed species becomes impractical at a resolution of 2–8 cm/pixel when they are intermixed with crops and other weed seedlings (Singh et al., 2020). At a spatial resolution of 25 cm per pixel, from a flight altitude of 300 m, the system was unable to differentiate crops and weeds due to mixed signatures (Huang et al., 2015). However, weed patches were distinguishable from crops using RGB and color-infrared (CIR) images acquired from UAVs flown at altitudes below 50 m, achieving a spatial resolution of less than 5 cm/pixel (Huang & Reddy, 2015). Similarly, Peña et al. (2015) reported that individual weed plants could be discriminated using RGB and multispectral images with spatial resolutions ranging from 1 to 4 cm acquired by flying UAVs at an altitude of 40 m. The commonly used algorithms and vegetation indices for mapping weeds in RGB imagery, along with their accuracy, are presented in Table 4.3.

Table 4.3 Weed identification accuracy of different algorithms using RGB, multispectral, and hyperspectral imagery

Crop	Algorithm/ classifier	Indices/features	Accuracy	References
(a) RGB camera				
Upland rice	SLIC and RF	Excess green (E × G), excess red (E × R), green–red vegetation index (GRVI), canopy height model (CHM), spatial texture (texture), and color index of vegetation extraction (CIVE)	HSV + texture combination gave 91%	Kawamura et al. (2021)
Canada thistle (*Cirsium arvense*) discrimination in wheat and barley	CNN, TopMaxExG	Green vegetation in preharvest	92–97%	Rasmussen et al. (2019)
Soybean	CNN	Color and texture	99.4%	Haq (2022)
Chilli	RF, SVM, and KNN	Color and texture, E × G, E × R, excess green and red (E × GE × R)	66–96%	Islam et al. (2021)
(b) Multispectral				
Corn	OBIA	Classification of vegetation as either crop or weed based on positions with reference to the crop rows, and creation of a weed infestation map using a grid pattern	86%	Peña et al. (2013)
Sugar beet	DNN	NDVI channel, RedEdge-M (five band), Sequoia (four band)	57% and 78% with 3 and 9 input channels, respectively	Sa et al. (2018)

(continued)

Table 4.3 (continued)

Crop	Algorithm/ classifier	Indices/features	Accuracy	References
Row crops	Unsupervised classification algorithm	Spatial and spectral information, MultiSPEC 4C, AIRINOV, Paris Resolution was 6 mm/pix	75%, 79%, and 89% for spectral, spatial, and spatial and spectral combination methods, respectively	Louargant et al. (2018)
Lettuce	Machine vision identification	Reflectance values from a small spatial area (3 mm diameter) of the leaf, red ratio vegetation index (RVI), and NDVI	Crisphead lettuce: 89.4%, common groundsel: 83.6%, leaf lettuce: 94.0%, Shepherd's-purse: 89.4%, sowthistle: 92.5%, all lettuce: 91.3%, all weeds: 87.8%, overall average: 90.3%	Slaughter et al. (2008)
Rice	Fused RGB-NGRDI	NDVI and NGRDI	70–85%	Barrero and Perdomo (2018)
Soybean	Maximum likelihood, SAM, MED, and FLL	Spectral reflectance	61-83%	Gibson et al. (2004)
Corn	Maximum likelihood algorithm	Spectral information of green, red, and near-infrared	Spectral resolution of 12–16 MP does not distinguish weeds smaller than 9 cm in height	Armstrong et al. (2007)
(c) Hyperspectral				
Sugar beet	SVM	Spectral information	25–79%	Scherrer et al. (2019)
Forest	SVM and ANN	Spectral information	93–99%	Papp et al. (2021)
Sugar beet	ANN	Spectral information	<90%	Koot (2014)
Yellow foxtail (*Setaria pumila*), pheasant's tail grass (*Stipa arundinacea*), meadow buttercup (*Ranunculus acris*), and Canada thistle (*Cirsium arvense*) discrimination in pasture	PLS-DA, SVM, and MLP	Significant spectral regions	70–100%	Li et al. (2021)

(continued)

Table 4.3 (continued)

Crop	Algorithm/ classifier	Indices/features	Accuracy	References
Invasive noxious weed species in the alpine grassland ecosystems	CNN model called 3D&2D-INWS-CNN	Spatial resolution of 2 cm	95%	Xing et al. (2024)
Mile-a-minute weed (*Mikania micrantha* Kunth)	Combination of Savitzky–Golay (SG) smoothing, PCA, and RF	Spectral range of 450–998 nm	88.7%	Huang et al. (2021)

SLIC simple linear iterative clustering, *NDVI* normalized difference vegetation index, *OBIA* object-based image analysis, *NGRDI* normalized green red difference index, *SVM* support vector machine, *ANN* artificial neural network, *DNN* deep neural network, *SAM* spectral angle mapper, *MED* minimum Euclidean distance, *FLL* Fisher's linear likelihood, *RF* random forest, *KNN* k-nearest neighbors, *MLP* multilayer perceptron, *PLS-DA* partial least squares-discriminant analysis, *CNN* convolutional neural network, *PCA* principal component analysis

2.2 Multispectral Imagery

Multispectral sensors, such as the Micasense RedEdge series or the Parrot Sequoia+, offer a wide range of vegetation indices due to their greater number of radiometric bands (4–7 bands) and wider spectral resolution (400–1000 nm). Koger et al. (2003) reported a weed identification accuracy rate of >90% in soybeans using multispectral sensor imagery. However, multispectral sensors come with several trade-offs. They typically capture images more slowly, offer lower spatial resolution, and are more costly compared to RGB cameras. Moreover, extracting valuable insights about weeds from multispectral data requires more sophisticated image processing techniques than working with RGB data. The power of multispectral sensors extends beyond basic weed detection. The additional spectral bands provide valuable data for ML algorithms (Fig. 4.4), which allows for tasks such as early detection of plant diseases and soil water content analysis (Berger et al., 2022; Candiago et al., 2015).

Although multispectral sensors offer advantages over RGB cameras, some factors can limit their effectiveness in weed detection. For example, the way plants reflect light (their spectral signature) can change depending on their growth stage, canopy volume, and exposure to stress factors such as disease or nutrient deficiencies (Lord et al., 1985). Plants lacking nitrogen may reflect more red light due to reduced chlorophyll content, potentially leading to misidentification as weeds (Serrano et al., 2000). This "spectral mixing" can confuse the sensors and lead to inaccurate weed classification (Somers et al., 2016). Studies using multispectral imagery to identify weeds in soybean crops have reported classification errors of up to 39% for both weeds and soybeans, particularly at later growth stages when plant canopies become denser (Gibson et al., 2004). Additionally, multispectral sensors typically have lower image resolution than RGB cameras. This means that they capture less detail and may struggle to detect smaller weeds (Matese & Di Gennaro, 2018). A study using multispectral images to detect weeds in corn found that weeds smaller than 9 cm were entirely undetectable due to the resolution limitations of the sensor (Armstrong et al., 2007). The algorithms used to classify weeds in multispectral imagery along with their accuracy are presented in Table 4.3.

The workflow for using multispectral sensors differs slightly from that for using RGB cameras.

Phase 1: Unlike RGB cameras, multispectral sensors require radiometric calibration to ensure accurate color representation and atmospheric correction to account for weather conditions affecting the data. Due to their lower resolution, multispectral sensors require lower flight altitudes than RGB cameras to capture sufficient detail. Additionally, ensuring adequate image overlap (both front and side) during flight is crucial for creating high-quality mosaics.

Phase 2: The increased data volume from more radiometric bands necessitates careful management to avoid data loss or mission failure.

Phase 3: Multispectral data require specialized processing to address parallax issues (displacement caused by the sensor's multiple lenses). This involves rectification (correcting distortions), georeferencing (adding location data), and stacking the images into a single composite image with calibrated data from each band. Once processed, a multispectral orthomosaic can be generated, allowing for the calculation of desired vegetation indices.

2.3 *Hyperspectral Sensors*

Hyperspectral sensors capture a much wider range of narrow radiometric bands in the visible and infrared regions (100–2000 nm), with bandwidths as precise as 1–15 nm (compared to 50–120 nm for multispectral bands) (Borengasser et al., 2007). This fine-grained spectral resolution allows them to detect specific characteristics in a field based on the unique signature of each band or band combination (Fig. 4.4). However, acquiring hyperspectral imagery over large areas can be time-consuming due to the sheer number of bands captured. Additionally, the vast amount of data generated necessitates significant computational power for analysis, posing a hurdle in creating comprehensive weed maps. Although the cost of hyperspectral sensors has decreased in recent years, they remain pricier than their RGB and multispectral counterparts. Furthermore, their larger size and weight often make them less suitable for integration with UAV systems.

Compared to other sensors, hyperspectral sensors require a more intricate calibration workflow.

Phase 1: Planning must consider both time and space, as hyperspectral sensors have a temporal resolution due to the different acquisition methods.

Phase 2: Hyperspectral images dwarf their multispectral and RGB counterparts in size due to the vast number of spectral bands captured for each pixel. This translates to massive file sizes that can exceed the storage capacity of many UAVs. The payload limitations of UAVs may restrict the transport of adequate file storage systems. Therefore, specialized storage solutions are required to manage this massive data flow.

Phase 3: Ensuring image quality is paramount for hyperspectral data. UAV platform vibrations can introduce issues such as pixel misalignment, requiring meticulous postprocessing techniques.

Once these issues are addressed, researchers can leverage the power of hyperspectral data to calculate specific narrowband indices, such as the chlorophyll absorption ratio index (CARI), greenness index (GI), greenness vegetation index (GVI), modified chlorophyll absorption ratio index (MCARI), modified normalized difference vegetation index (MNDVI), simple ratio (SR), transformed chlorophyll absorption ratio index (TCARI), triangular vegetation index (TVI), modified vegetation stress ratio (MVSR), modified soil-adjusted vegetation index (MSAVI), and photochemical reflectance index (PRI).

With their ability to capture a vast spectrum of light, hyperspectral sensors are powerful tools for distinguishing between crops and weeds. Studies have shown promising results, with accuracies ranging from 25% to 99% depending on the specific application. For example, Scherrer et al. (2019) used hyperspectral data from a UAV to identify kochia weed (*Bassia scoparia*) in sugar beet fields. The support vector machine (SVM) algorithm achieved classification accuracies ranging from 25% to 79%. The lower accuracy observed can be attributed to mixed pixels, background light in uncontrolled field conditions, and UAV vibrations, although higher accuracy was achieved for younger kochia plants, which are inherently easier to distinguish. Similarly, Papp et al. (2021) employed hyperspectral imaging with machine learning algorithms, viz., SVM and artificial neural networks (ANNs), to map common milkweed (*Asclepias syriaca*) in Hungary, achieving impressive accuracies of 93% for SVM and 99% for ANN. Their research suggested that these algorithms perform better at detecting younger plants and in areas with less bare soil. In another study, Koot (2014) integrated RGB and hyperspectral data for volunteer potato plant detection in sugar beet fields using ANN and remote sensing indices, achieving a classification accuracy of 90%, with ANN demonstrating superior performance over remote sensing indices. It is important to note that variations in illumination conditions significantly influence the total recorded reflection. Future research should focus on minimizing the influence of environmental factors and expanding the applicability of these techniques across various agricultural settings. The best-performing algorithms, along with indices and weed classification accuracy for hyperspectral imagery, are presented in Table 4.3.

3 Database and Software for Processing UAV Images

3.1 Database for Weed Mapping

ML algorithms require a substantial amount of training data in the form of images (Bah et al., 2018). These data can be sourced from public databases such as DeepWeeds (Olsen et al., 2019), RoboWeedMap (Teimouri et al., 2022), CottoWeedDet12 (Dang et al., 2023), and ImageWeeds (*Brassica* sp., *Glycine max*, flax, lentils, field peas, *Zea mays*, and *Beta vulgaris*) (Rai et al., 2023). Alternatively, researchers may need to develop their own datasets tailored to specific weed species or local conditions. However, creating a dataset from scratch demands careful planning and precision in image acquisition, selection, and processing. The quality and diversity of the data directly impact the training and application of the ML model, and poorly curated datasets can lead to biased results.

3.2 Commercial Software for Weed Detection

Companies such as DroneDeploy® (San Francisco, CA, USA), Picterra® (Lausanne, Switzerland), Pix4Dfield® (Prilly, Switzerland), Trimble eCognition® (Denver, CO, USA), and Solvi's® (Gothenburg, Sweden) are at the forefront of providing commercial software solutions for weed detection and crop scouting. These companies leverage advanced technologies such as UAVs, ML, and geospatial analysis to identify and manage weeds more effectively. These platforms streamline the process, minimize human error, and enable data-driven decision-making. For example, Picterra and Trimble eCognition leveraged UAV imagery and AI to create detailed field maps. Users first pinpoint a few weed samples, and then the software automatically locates and quantifies the weed presence across the entire field. Similarly, DroneDeploy and Pix4Dfield offer platforms that harness ML to analyze UAV images, effectively distinguishing weeds from crops. These advancements not only expedite the weed detection process but also promote precision in targeting interventions. Additionally, these platforms are compatible with various image formats and can be seamlessly integrated with other agricultural software, making them versatile tools for modern, data-driven farming. A prime example is the Thistle Tool, a semiautomatic program for *C. arvense* detection in preharvest cereals using UAV imagery. It achieved over 90% accuracy in various fields where *C. arvense* was the predominant green vegetation contributor (Rasmussen et al., 2019). The classification accuracy of UAV images is improved by object-based image analysis (OBIA) compared to traditional pixel-based image analysis (Peña et al., 2013). OBIA creates "objects" by grouping adjacent pixels with homogeneous spectral values and then combining them spectrally, topologically, and contextually to improve the precision of image classification (Blaschke, 2010).

4 UAV-Based Site-Specific Precision Herbicide Spray

The targeted application of herbicides via UAVs can be implemented via two different methods. The first method, "spot-on spraying," involves hovering drones directly over individual weed patches to apply chemicals precisely (Fig. 4.5a). This method ensures highly targeted application and is commonly used in orchard spraying (Guo et al., 2022). The second method, i.e., dynamic spot spraying, involves drones navigating over the weed patch while simultaneously dispensing chemicals onto the weed patches (Fig. 4.5b). Although this method is less precise than the hovering technique due to the continuous movement of the drone sprayer and the intermittent activation of the nozzle based on the presence or absence of weed patches, it embodies a form of variable-rate technology. Both methodologies significantly contribute to improved weed control efficacy and the optimized utilization of resources in agricultural and environmental management endeavors.

Numerous studies have explored the efficacy of targeted herbicide applications in agriculture. For instance, Hunter et al. (2020) employed an unmanned aerial system for weed mapping and spot spraying in sod production fields, revealing that unmanned aerial systems exhibited up to threefold greater efficiency in identifying and treating targeted weed patches than ground-based broadcast applications. However, UAV-based site-specific applications overlook 25–30% of targeted weed patches, in contrast to the 2–3% missed with the broadcast method. Campos et al. (2019) noted a 45% reduction in herbicide usage with site-specific drone spray in a vineyard without compromising weed control efficacy. Other UAV-based site-specific applications have demonstrated the potential to limit herbicide use by

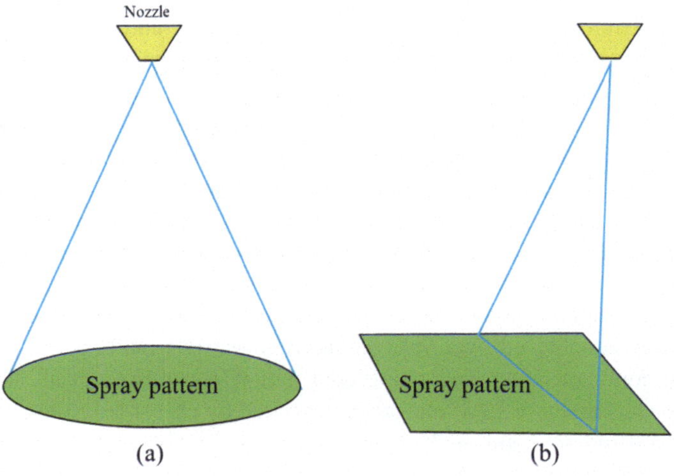

Fig. 4.5 Unmanned aerial system-based targeted herbicide application methods: (**a**) "Spot-on spraying" involves hovering drones directly over individual weed patches to apply chemicals precisely. (**b**) "Dynamic spot spraying" involves drones navigating over the weed patch while simultaneously dispensing chemicals onto the weed patches

45–90% compared to blanket application in the entire field (Berge et al., 2008; Genna et al., 2021; Gerhards & Christensen, 2003; Timmermann et al., 2003). Site-specific spraying of herbicides in corn, sugar beets, cereals, and peas reduced herbicide use by 22–89% without compromising crop yields or increasing weed control costs in subsequent years (Berge et al., 2012; Christensen et al., 2009; Gerhards & Oebel, 2006). It may be possible to reduce herbicide usage even more if these herbicides are species-specific (Wiles, 2009).

Furthermore, a recent economic analysis by Rajmis et al. (2022) revealed a 26–66% reduction in herbicide application costs without impacting crop yield. Additionally, targeted herbicide application has proven advantageous for managing weed escapes in production fields that survive early-season control measures and are responsible for replenishing weed seeds in the soil seed bank (Bagavathiannan & Norsworthy, 2012; Kutugata et al., 2021). As the majority of postemergence herbicides lack labels for application during or past the reproductive stages, UAV-based targeted application has emerged as a potential solution. A study was conducted at Texas A&M University, College Station, TX, aimed at introducing the "no seed, no weed" concept. Using an ML algorithm, weed escapes were identified in rice fields and mapped for targeted herbicide application via a GPS-equipped drone sprayer. Their findings underscored the effectiveness of UAV in controlling escape weeds in rice, with no observed weed seed formation after herbicide application (Fig. 4.6). Similar investigations were conducted in turfgrass and cotton for site-specific weed management. Overall, UAV-based targeted herbicide application represents a promising approach to escape weed management that combines precision, efficiency, and environmental sustainability.

The weed control methods used in precision agriculture are accurate and precise and intended to target specific weeds. UAV spray applications lead to greater efficacy of contact herbicides in controlling weeds compared with backpack spray applications due to the better distribution of spray droplets on leaves, including on lower surfaces (Singh et al., 2019). Gerhards and Oebel (2006), after a 6-year study on the use of herbicides with patch sprayers on different crops and fields, reported

Fig. 4.6 Unmanned aerial system-based targeted herbicide application for controlling escape weeds in rice

that herbicide use in winter cereals was reduced by 60% against broad-leaved weeds and by 90% against grassy weeds using this approach. In the case of sugar beet and corn crops, the herbicide savings for grassy weeds were approximately 36% and 78%, respectively. For broad-leaved weeds, the average savings of herbicides were approximately 11% for corn and 41% for sugar beet. In barley crops, herbicide savings ranged from 20 to 80% against broad-leaved weeds and 18 to 90% against grassy weeds.

Nikolić et al. (2021) integrated site-specific and time-specific weed control methods to reduce herbicide usage, benefiting both the environment and on-farm variable costs. They employed the AlertInf model for weed emergence prediction and utilized a UAV with visible range sensors operating at 3 cm resolution for weed detection. The artificial neural network (ANN) and visible atmospherically resistant index (VARI) classification methods are used to distinguish between weeds, crops, and soil. Both classification methods demonstrated high overall accuracy, with ANN achieving 98.6% and VARI 98.1%. The percentage reduction in the area to be sprayed ranged from 65.3% to 93.4% using VARI and from 42.4% to 87.8% using ANN. Additionally, the potential reduction in herbicide use ranged from 1.32 L/ha to 0.28 L/ha for ANN and from 0.80 L/ha to 0.15 L/ha for VARI.

5 Challenges for UAV-Assisted Targeted Herbicide Application

5.1 UAV Design and Flight Challenges

The stability of UAVs usually decreases with increasing drag and payload, leading to increased power consumption. Improper weight distribution on smaller UAVs further exacerbates this issue. Moreover, frequent payload changes or modifications require constant rebalancing of the aircraft, which consumes significant ground time. Furthermore, weather conditions can significantly impact data collection. High winds, in particular, complicate the process of maintaining optimal aircraft balance. Light conditions also play an important role in image resolution. Improper camera settings (e.g., color imbalance) and light exposure during image acquisition can result in poor image quality. Variations in light levels, overlapping leaves and stems, and shadows make it ineffective for distinguishing individual plants in the field (Smith et al., 2018; Zhang et al., 2016).

High-resolution sensors are essential for accurately identifying and distinguishing weed locations from crop locations, ensuring that herbicides are applied only where needed (Fig. 4.7a). The advancement of technology and newer cameras enabled better detection of individual weeds, but challenges such as the weight of these cameras and compatibility with existing systems remain. The sensors used in UAVs require specific operating systems and software. Such software is often expensive and requires frequent upgrades. The integration of new sensor packages

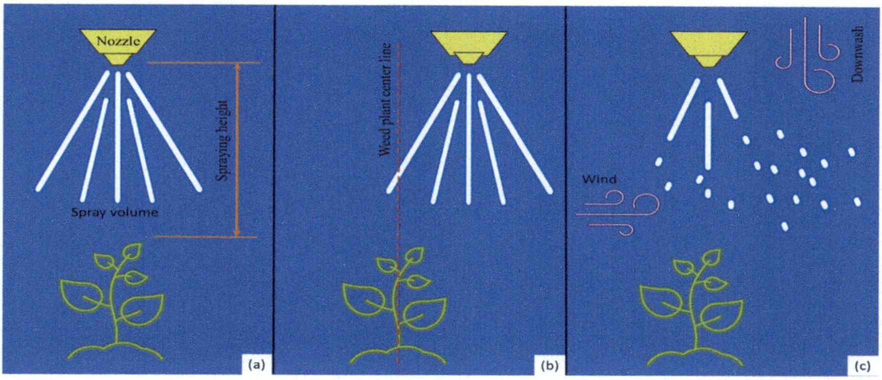

Fig. 4.7 Examples of phenomena affecting targeted herbicide application

into UAVs involves intensive standardization and compatibility issues. In addition, the accuracy of GPS systems plays a crucial role in guaranteeing precise application (Fig. 4.7b) (Czyża et al., 2023). High-precision GPS allows drones to navigate accurately and apply herbicides with minimal drift, further enhancing the precision of the operation. Without these high levels of sensor and GPS accuracy, the effectiveness and efficiency of UAV-assisted herbicide applications would be significantly compromised. Adverse weather conditions, especially strong winds, pose significant challenges to targeting weeds with high precision during aerial spraying (Fig. 4.7c).

Most UAV spray units currently available on the market do not have site-specific application capabilities based on GPS input. These units utilize the waypoint feature for herbicide applications, which cannot auto-trigger spray applications at a prespecified location (Tables 4.4 and 4.5). SSWM operations using existing UAV spray platforms depend solely on the remote pilot's skill set to manually trigger applications at a required location. While this approach may work for small areas, it is not feasible for spot applications over a large farm.

5.2 *Sensor Resolution and Flight Height*

The resolution of an image significantly affects the performance of weed detection algorithms. Commercially available UAV sensors often lack the ideal combination of spatial, spectral, and temporal resolutions for early-stage weed detection. Most sensors excel at classifying larger, established weeds rather than young seedlings. Early weed detection (15–40 days after crop/weed emergence) is crucial for minimizing weed competition and protecting crop yields (Ramesh et al., 2021). A high spatial resolution (1–2 cm/pixel) is ideal for early detection and can be achieved by lowering the flight altitude. UAVs flown between 2.5 and 4 m achieved the highest accuracy (87%) and precision (Pflanz et al., 2018). Studies have shown a clear

Table 4.4 Technical specifications of DJI multipurpose rotary-type UAVs commercially available for agriculture

UAV	Maximum takeoff (kg)	Tank capacity (liters)	Battery type	Battery capacity (mAh)	GNSS	Recharge time and hovering time (min/battery)	Application	Spray droplet size (µm)	Effective spray width	Coverage (ha/day)
AGRAS T50	Spraying: 92 Spreading: 103	Spray: 40 Spread: 50	Intelligent flight battery	30,000	GPS + Galileo + BeiDou	Recharge: 9 to 12 min	Spraying, survey, spreading,	50–500	4–11 at 3 m above the crop	21 ha/h: Farmland 4 ha/h: Orchards
AGRAS T25	Spraying: 52 Spreading: 58	Spray: 20 Spread: 25	Intelligent flight battery	15,000	GPS + Galileo + BeiDou	Recharge: 9 to 12 min	Spraying, survey, spreading,	50–500	4–7 at 3 m above the crop	12 ha/h
AGRAS T20P	Spraying: 52 Spreading: 58	Spray: 20 Spread: 25	Intelligent flight battery	13,000	RTK positioning enabled. GPS L1, GLONASS F1, Galileo E1, BeiDou B1	Recharge: 9 to 12 min spraying: 7 min/battery Spreading: 6 min/battery	Spraying, mapping, spreading	50–300	7 at 2.5 m above the crop	12 ha/h: Farmland 2.6 ha/h: Orchards
AGRAS T40	Spraying: 90 Spreading: 101	Spray: 40 Spread: 50	Intelligent flight battery	30,000	RTK positioning enabled. GPS L1, GLONASS F1, Galileo E1, BeiDou B1	Recharge: 9 to 12 min spraying: 7 min/battery Spreading: 6 min/battery	Spraying, spreading, surveying, mapping	50–300	11 m at 2.5 m above the crop	21 ha/h: Farmland 4 ha/h: Orchards
AGRAS T30	78	Spray: 30 Spread: 35	BAX501	29,000	RTK enabled. GNSS: GPS L1, GLONASS F1, and Galileo E1	Hovering time: 7.8 min/battery	Spraying, spreading	–	–	16 ha/h
AGRAS T10	24.8	Spray: 8	BAX501	9500	RTK enabled. GNSS: GPS L1, GLONASS F1, and Galileo E1	Hovering time: 9 min/battery	Spraying	–	–	6 ha/h

Table 4.5 The Directorate General of Civil Aviation (DGCA)-approved rotary type UAVs in India for agriculture

UAV name	Manufacture	Maximum payload (kg)	Spray tank capacity (liters)	Battery type	Battery cycles	Battery capacity (mAh)	Flying time (min/battery)	Application	Coverage (ha/day)
Krishi Viman	WOW go Green	25	10	LiPo	~220, 50 min full charge	16,000	17 (1 ha/ battery)	Spraying	10
AG-365S	Marut drones	25	10	LiPo	300	16,000	22 (1–1.2 ha/ battery)	Spraying, fertilizer and seed spreading, crop monitoring	30
AG-365	Marut drones	30	11	LiPo	300	22,000	22	Spraying	30
Kisan/ Agri	Garuda	24.5	8	5 L petrol/h			8	Spraying, aerial image	8–12
Agribot A6	IOTECH,	23.3	10	Li-ion	500	21,000	20	Spraying, broadcasting, and mapping	12
Agribot MX	IOTECH,	23.3	10	Li-ion	500	25,200		Spraying, broadcasting	12
Drishti	IOTECH,						45	Aerial surveys, HD video transmission, and field mapping	

LiPo lithium polymer, *mAh* milliampere-hour

decline in weed detection accuracy as altitude increases, dropping from 92% at 10 m (spatial resolution of 5.42 mm/pixel) to 84% at 30 m (spatial resolution of 20.31 mm/pixel) (Che'Ya, 2016). However, this can introduce challenges with image alignment for some sensors. For example, the MicaSense camera requires a minimum altitude of 15 m for automatic image alignment (MicaSense, 2018). In contrast, Rasmussen et al. (2019) reported that flight altitudes below 50 m, corresponding to 15 mm per pixel resolution, did not enhance weed detection.

The optimal flight altitude plays a crucial role in weed mapping. There is always a trade-off between flight altitude and spatial resolution (Curcio et al., 2022; Kubišta & Surový, 2021). At higher altitudes, UAVs can cover larger areas, but the spatial resolution decreases, making it more difficult to distinguish small weeds from crops. Conversely, flying at lower altitudes improves spatial resolution but limits the coverage area. In high-altitude images, the soil pixel signatures are greater than those of vegetation, and it is extremely difficult to differentiate the edges of vegetation from the soil background (Singh et al., 2020). High-altitude image collection also leads to the inclusion of more plants/leaves per pixel.

5.3 *Droplet Deposition and Drift Hazard*

To ensure the maximum effectiveness and minimal environmental impact of herbicide spray using UAVs, several factors need to be carefully considered. A key factor is droplet deposition, where achieving 70–75 droplets/cm^2 (for fine and medium droplets) and 10–15% target area coverage is crucial for optimal product efficacy. The spraying height, droplet size, wind speed, and evaporation all play significant roles in deposition (Liao et al., 2019; Martin et al., 2019). Studies have shown that a flight altitude between 1.5 and 2.0 m offers the best balance. Lower heights (1.5 m) minimize evaporation and drift compared to higher altitudes (4.0 m) (Meesaragandla et al., 2024). Low spraying heights can also generate strong downward airflows that cause plant swaying, potentially hindering deposition on the underside of leaves. In contrast, a higher spraying height increases the drift hazard because even small winds can carry droplets away from the target and increase the effective spray width.

The spraying height also influences the wind dynamics and, in turn, drift potential. Lower flight heights create stronger downward air currents that cause plants to sway, facilitating deposition on the underside of leaves (Qin et al., 2014). This also reduces drift, with studies showing an average drift rate of 7.9% at 1.5 m compared to 20% at 2 m (Lou et al., 2018). This is because lower flight heights maintain a stronger vertical wind field above the crop canopy, preventing droplets from escaping. Alternatively, a higher flight height reduces the downward pressure wind field beneath the rotor (Lou et al., 2018). In the spot-on spraying method, where the UAV sprayer hovers over the crop canopy, the soft vegetative tissues can deflect due to the downwash created by the UAV rotors (Hentschke et al., 2018). This deflection alters the plant's position, resulting in an uneven surface that can lead to inaccuracies in altitude measurement by the sensor (Hodgson & Bresnahan, 2004). However, it

Fig. 4.8 Deflection of plants due to the downwash created by the UAVs

occurs only in late-season weed control, as the escaped weeds are taller at the harvesting stage and are shaken by a slight breeze, which causes them to shift their plant position (Fig. 4.8). It does not occur with early-stage weed control, as the weeds are short.

Droplet size also plays a part. While coarse droplets (>400μm) applied at a low altitude (2 m) effectively concentrate deposition within 1–2 m along the center axis of the UAV, finer droplets sprayed from higher altitudes (>3 m) increase the risk of drift, although they may cover a wider area (Qin et al., 2016). Fine droplets (<50μm) tend to drift easily and lose kinetic energy quickly after release, causing external winds to direct them into the drift zone. In contrast, coarser droplets (>400μm) substantially reduce drift compared to finer droplets but encounter difficulties in achieving penetration (Hussain et al., 2019).

5.4 Spray Volume

The limited UAV endurance caused by shorter battery life leads to restricted payload-carrying capacity. This typically translates to a spray volume of 20–40 L/ha for UAVs, which is 10–20 times lower than the volume used with knapsack sprayers (200–400 L/ha). However, the efficiency of UAV spraying often matches that of knapsack sprayers (Paul et al., 2024). A lower spray volume can sometimes lead to increased phytotoxicity (plant injury), reduced droplet density on target plants, and weaker penetration of the herbicide (Meesaragandla et al., 2024). For example, a study by Chen et al. (2019) showed that a limited spray volume (22.5 L/ha) for a postemergence herbicide resulted in 10–20% wheat injury and reduced weed control compared to knapsack sprayers. This is because the concentration of herbicide used for UAV spraying is 30 times greater than that used for traditional spraying methods. However, there is a trade-off. Compared with UAVs, knapsack sprayers with large volumes of water lead to lower weed control efficiency. This is due to the loss of herbicide solution to the soil, as it is not retained on the weed foliage (Meesaragandla et al., 2024).

5.5 Weather Conditions

Strong winds pose a significant challenge for UAV weed control. Crosswinds exceeding 4 m/s can cause substantial spray drift, leading to uneven herbicide distribution. This translates to missed target areas, reduced overall coverage, and potentially sublethal herbicide doses that fail to control weeds effectively (Richardson et al., 2019). Wind also impacts data acquisition. High wind speeds can blur images captured during flight, making it more difficult to distinguish weeds during postprocessing steps such as mosaicking (creating a complete image from multiple image tiles). Studies have shown that flights conducted under calm conditions with wind speeds less than 5 km/h significantly improve the ability to identify weeds in imagery.

5.6 Shortage of Technical Expertise/Manpower

The lack of technical knowledge and expertise among farmers is another key challenge to the adoption of precision weed control using UAVs. Operating and maintaining the equipment, interpreting the collected data, and formulating targeted weed management plans all require a level of expertise that many farmers may not possess, especially in developing countries (Vijayakumar et al., 2021). This is further compounded by the lack of readily available specialized service providers, limiting access to professional support and training. As a result, the significant benefits of drone-assisted weed control, including improved efficiency and reduced herbicide application, remain unrealized in many agricultural regions.

5.7 High Initial Investment and Lack of Multipurpose UAVs

Currently, the high cost of UAVs makes them unaffordable for small and marginal farmers in developing countries. Moreover, most UAVs have limited payload capacities (<10 kg) and short flight durations (<15 min). While increasing payload and flight time enhances UAV performance, it also significantly raises the overall cost (Kumar et al., 2020). Another limitation is the single-function design of many UAVs, such as those used solely for spraying, which reduces their appeal to farmers. In contrast, farmers prefer multipurpose UAVs capable of performing various tasks, such as seed spreading, fertilizer application, pesticide spraying, and transporting farm inputs and tools, making the investment more worthwhile. Therefore, the development of UAVs with higher payload capacities, extended flight durations, and multifunctional capabilities can increase their adoption in agriculture (Vijayakumar, Madireddy, et al., 2022). Additionally, providing subsidies and tax incentives, particularly for small and marginal farmers, would further encourage them to invest in UAV technology.

5.8 Research Gaps in UAV Technology

Differentiating crop residue from weed species: UAV-assisted remote sensing systems face significant challenges in distinguishing crop residues from emerging weeds due to similarities in color, texture, and reflectance. Current spectral imaging techniques often struggle to make accurate distinctions, particularly during early weed growth stages (Pflanz et al., 2018). Advancing multispectral or hyperspectral imaging combined with ML algorithms is crucial for improving detection accuracy. Weed identification under variable soil conditions: Fallow lands typically have diverse soil types and conditions that impact UAV detection accuracy. Variations in soil color, texture, and organic matter can interfere with the spectral signatures of weeds. Developing robust models that adapt to varying soil backgrounds is essential to improve weed identification in such fields (Singh et al., 2020). Challenges in identifying weeds in moist and dry soil conditions: Soil moisture levels influence reflectance characteristics captured by UAVs. Wet soils can cause glare or dark images, while dry soils may reflect excessive light, leading to inconsistent weed detection. Algorithms that adjust for varying soil moisture conditions are needed to enhance detection accuracy.

Detection under uneven ground cover: Fallow lands often feature irregular ground cover, with exposed soil, crop stubble, and emerging weeds creating complex detection environments. UAV systems must overcome these challenges to accurately classify weeds. Enhancing image processing techniques to handle mixed ground cover is necessary for improving classification precision. Weed emergence timing and growth patterns: Weeds in fallow fields emerge at different times and grow at varying rates based on environmental factors. UAV-based systems must account for these variations to ensure effective detection. Dynamic weed detection models that incorporate growth stages and seasonal changes are essential to address this gap. Integration of soil and crop residue data: Most UAV systems focus on aerial imaging without incorporating soil characteristics or residue data, limiting detection accuracy. Integrating UAV imagery with ground-based soil sensors could significantly improve weed identification, particularly in fields with high crop residue cover (Singh et al., 2020). Addressing these gaps will enhance UAV technology's effectiveness in managing fallow lands, improving weed control strategies, and optimizing resource use in agricultural systems.

6 Future Directions and Opportunities

6.1 Technological Advancements

The development of lightweight, low-cost, high-resolution sensors coupled with advanced imaging technologies will enable more precise identification and differentiation of weed species from crops. This will narrow the trade-off between precision and extent. The development of new AI algorithms, whether ML or DL, which can accurately detect and classify crops and weeds regardless of the crop type or location, will enhance UAV capabilities and their adoption. Future UAV systems will likely incorporate real-time data processing, enabling immediate decision-making and action in the field. This integration of AI will not only improve the precision of weed control but also optimize the overall management of agricultural resources. Overall, rapid advancements in UAV technology and image processing will ensure large-scale adoption of targeted herbicide applications sooner in the farmer's field.

6.2 Increased Automation and Autonomy

The future of UAV-based precision weed control will see a significant increase in automation and autonomy. Autonomous drones equipped with advanced navigation systems and AI-driven algorithms will be capable of independently conducting weed control operations without human intervention. These UAVs will be able to navigate complex agricultural landscapes, identify weed-infested areas, and apply herbicides with high precision. This increased autonomy will not only save time and labor but also enhance the scalability of UAV-based weed control. Farmers will be able to deploy multiple drones simultaneously, covering larger areas more efficiently and effectively. Furthermore, autonomous UAVs can operate in conditions that may be challenging or unsafe for human operators, such as at night or in extreme weather conditions, ensuring consistent and reliable weed control.

6.3 Integration with Other Technologies

The integration of UAVs with other emerging agricultural technologies presents exciting opportunities for precision weed control. Combining UAV data with satellite imagery and ground-based sensors will provide a more comprehensive and accurate picture of weed distribution and crop health. This multisource data fusion will enhance the ability to monitor and manage weeds over time, allowing for proactive and precise interventions. Additionally, the use of Internet of Things (IoT) devices and cloud computing will facilitate seamless data sharing and analysis. UAVs can transmit real-time data to central databases, where they can be processed and analyzed alongside data from other sources (Makam et al., 2024). This interconnected ecosystem will enable more informed decision-making and foster collaborative efforts among farmers, agronomists, and researchers.

Despite these promising future directions, several challenges and considerations must be addressed to fully realize the potential of UAV-based precision weed control. The high initial investment required for advanced UAVs and related technologies can be a significant barrier for small-scale farmers. Developing cost-effective alternatives and offering financial support or subsidies is essential to promote broader adoption. Additionally, establishing standardized regulations and guidelines is crucial to ensure the safe and responsible use of UAVs in agriculture. Privacy concerns, data security, and ethical considerations regarding AI-driven decision-making must be carefully managed to build trust among stakeholders.

7 Conclusions

UAV-based precision weed control offers significant potential to enhance the efficiency and sustainability of weed management practices. Its economic and environmental advantages, along with the prospects of higher crop yields and reduced chemical use, position it as a transformative solution for the future of agriculture. However, despite its promise, UAV applications in weed detection and management remain in the early stages of development. This technology is currently facing challenges, including sensor integration, multiresolution solutions, flight altitude vs. efficiency, payload capacity, battery life, image processing, and information extraction. Technological advancements, increased automation, and integration with other technologies will drive the future of UAV applications in agriculture. By overcoming these hurdles and fostering innovation, UAV technology can help the agricultural sector meet the growing need for sustainable, resilient food production systems.

References

Armstrong, J. J. Q., Dirks, R. D., & Gibson, K. D. (2007). The use of early season multispectral images for weed detection in corn. *Weed Technology, 21*(4), 857–862.

Bagavathiannan, M. V., & Norsworthy, J. K. (2012). Late-season seed production in arable weed communities: Management implications. *Weed Science, 60*(3), 325–334.

Bah, M. D., Dericquebourg, E., Hafiane, A., & Canals, R. (2019). Deep learning based classification system for identifying weeds using high-resolution UAV imagery. In *Intelligent computing: proceedings of the 2018 computing conference* (Vol. 2, pp. 176–187). Springer International Publishing.

Bah, M. D., Hafiane, A., & Canals, R. (2018). Deep learning with unsupervised data labeling for weed detection in line crops in UAV images. *Remote Sensing, 10*(11), 1690.

Barrero, O., & Perdomo, S. A. (2018). RGB and multispectral UAV image fusion for Gramineae weed detection in rice fields. *Precision Agriculture, 19*, 809–822.

Berge, T. W., Goldberg, S., Kaspersen, K., & Netland, J. (2012). Towards machine vision based site-specific weed management in cereals. *Computers and Electronics in Agriculture, 81*, 79–86.

Berge, T. W., Rene Cederkvist, H., Aastveit, A. H., & Fykse, H. (2008). Simulating the effects of mapping and spraying resolution and threshold level on accuracy of patch spraying decisions and herbicide use based on mapped weed data. *Acta Agriculturae Scandinavica Section B Soil and Plant Science, 58*(3), 216–229.

Berger, K., Machwitz, M., Kycko, M., Kefauver, S. C., Van Wittenberghe, S., Gerhards, M., Verrelst, J., Atzberger, C., Van der Tol, C., Damm, A., & Rascher, U. (2022). Multi-sensor spectral synergies for crop stress detection and monitoring in the optical domain: A review. *Remote Sensing of Environment, 280*, 113198.

Blaschke, T. (2010). Object based image analysis for remote sensing. *ISPRS Journal of Photogrammetry and Remote Sensing, 65*(1), 2–16.

Borengasser, M., Hungate, W. S., & Watkins, R. (2007). *Hyperspectral remote sensing: Principles and applications*. CRC Press.

Boutin, C., Strandberg, B., Carpenter, D., Mathiassen, S. K., & Thomas, P. J. (2014). Herbicide impact on non-target plant reproduction: What are the toxicological and ecological implications? *Environmental Pollution, 185*, 295–306.

Campos, J., Llop, J., Gallart, M., García-Ruiz, F., Gras, A., Salcedo, R., & Gil, E. (2019). Development of canopy vigour maps using UAV for site-specific management during vineyard spraying process. *Precision Agriculture, 20*(6), 1136–1156.

Candiago, S., Remondino, F., De Giglio, M., Dubbini, M., & Gattelli, M. (2015). Evaluating multispectral images and vegetation indices for precision farming applications from UAV images. *Remote Sensing, 7*(4), 4026–4047.

Che'Ya, N. N. (2016). *Site-specific weed management using remote sensing*. The University of Queensland.

Chen, Y., Qi, H., Li, G., & Lan, Y. (2019). Weed control effect of unmanned aerial vehicle (UAV) application in wheat field. *International Journal of Precision Agricultural Aviation* 2(2), 25–31. https://doi.org/10.33440/j.ijpaa.20190202.45

Christensen, S., Søgaard, H. T., Kudsk, P., Nørremark, M., Lund, I., Nadimi, E. S., & Jørgensen, R. (2009). Site-specific weed control technologies. *Weed Research, 49*(3), 233–241.

Curcio, A. C., Peralta, G., Aranda, M., & Barbero, L. (2022). Evaluating the performance of high spatial resolution UAV-photogrammetry and UAV-Lidar for salt marshes: The Cádiz Bay study case. *Remote Sensing, 14*(15), 3582.

Czyża, S., Szuniewicz, K., Kowalczyk, K., Dumalski, A., Ogrodniczak, M., & Zieleniewicz, Ł. (2023). Assessment of accuracy in unmanned aerial vehicle (uav) pose estimation with the real-time kinematic (rtk) method on the example of dji matrice 300 rtk. *Sensors, 23*(4), 2092.

Dang, F., Chen, D., Lu, Y., & Li, Z. (2023). YOLOWeeds: A novel benchmark of YOLO object detectors for multi-class weed detection in cotton production systems. *Computers and Electronics in Agriculture, 205*, 107655.

De Castro, A. I., Torres-Sánchez, J., Peña, J. M., Jiménez-Brenes, F. M., Csillik, O., & López-Granados, F. (2018). An automatic random forest-OBIA algorithm for early weed mapping between and within crop rows using UAV imagery. *Remote Sensing, 10*(2), 285.

Galhano, V., Gomes-Laranjo, J., Fernández-Valiente, E., Videira, R., & Peixoto, F. (2011). Impact of herbicides on non-target organisms in sustainable irrigated rice production systems: State of knowledge and future prospects. In *Herbicides and environment* (pp. 45–72). IntechOpen.

Gao, J., Liao, W., Nuyttens, D., Lootens, P., Vangeyte, J., Pižurica, A., He, Y., & Pieters, J. G. (2018). Fusion of pixel and object-based features for weed mapping using unmanned aerial vehicle imagery. *International Journal of Applied Earth Observation and Geoinformation, 67*, 43–53.

Genna, N. G., Gourlie, J. A., & Barroso, J. (2021). Herbicide efficacy of spot spraying systems in fallow and postharvest in the pacific northwest dryland wheat production region. *Plants, 10*(12), 2725.

Gerhards, R., Andujar Sanchez, D., Hamouz, P., Peteinatos, G. G., Christensen, S., & Fernandez-Quintanilla, C. (2022). Advances in site-specific weed management in agriculture—A review. *Weed Research, 62*(2), 123–133.

Gerhards, R., & Christensen, S. (2003). Real-time weed detection, decision making and patch spraying in maize, sugarbeet, winter wheat and winter barley. *Weed Research, 43*(6), 385–392.

Gerhards, R., & Oebel, H. (2006). Practical experiences with a system for site-specific weed control in arable crops using real-time image analysis and GPS-controlled patch spraying. *Weed Research, 46*(3), 185–193.

Gibson, K. D., Dirks, R., Medlin, C. R., & Johnston, L. (2004). Detection of weed species in soybean using multispectral digital images. *Weed Technology, 18*(3), 742–749.

Guo, S., Yao, W., Xu, T., Ma, H., Sun, M., Chen, C., & Lan, Y. (2022). Assessing the application of spot spray in Nanguo pear orchards: Effect of nozzle type, spray volume rate and adjuvant. *Pest Management Science, 78*(8), 3564–3575.

Hamuda, E., Glavin, M., & Jones, E. (2016). A survey of image processing techniques for plant extraction and segmentation in the field. *Computers and Electronics in Agriculture, 125*, 184–199.

Hao, Y., Pei, H., Lyu, Y., Yuan, Z., Rizzo, J. R., Wang, Y., & Fang, Y. (2023). Understanding the impact of image quality and distance of objects to object detection performance. In *2023 IEEE/RSJ international conference on intelligent robots and systems (IROS)* (pp. 11436–11442). IEEE.

Haq, M. A. (2022). CNN based automated weed detection system using UAV imagery. *Computer Systems Science and Engineering, 42*(2), 837–849. https://doi.org/10.32604/csse.2022.023016

Heap, I. (2014). Global perspective of herbicide-resistant weeds. *Pest Management Science, 70*(9), 1306–1315.

Hentschke, M., Pignaton de Freitas, E., Hennig, C. H., & Girardi da Veiga, I. C. (2018). Evaluation of altitude sensors for a crop spraying drone. *Drones, 2*(3), 25.

Hodgson, M. E., & Bresnahan, P. (2004). Accuracy of airborne lidar-derived elevation. *Photogrammetric Engineering & Remote Sensing, 70*(3), 331–339.

Huang, Y., Li, J., Yang, R., Wang, F., Li, Y., Zhang, S., Wan, F., Qiao, X., & Qian, W. (2021). Hyperspectral imaging for identification of an invasive plant Mikania micrantha Kunth. *Frontiers in Plant Science, 12*, 626516. https://doi.org/10.3389/fpls.2021.626516

Huang, Y., & Reddy, K. N. (2015). Unmanned aerial vehicles: A unique platform for low-altitude remote sensing for crop management. In *Weed science for sustainable agriculture, environment and biodiversity* (pp. 185–192). Indian Society of Weed Science.

Huang, Y., Reddy, K. N., Fletcher, R. S., & Pennington, D. (2018). UAV low-altitude remote sensing for precision weed management. *Weed Technology, 32*(1), 2–6.

Huang, Y., Reddy, K. N., Thomson, S. J., & Yao, H. (2015). Assessment of soybean injury from glyphosate using airborne multispectral remote sensing. *Pest Management Science, 71*(4), 545–552.

Hunter, J. E., Gannon, T. W., Richardson, R. J., Yelverton, F. H., & Leon, R. G. (2020). Integration of remote-weed mapping and an autonomous spraying unmanned aerial vehicle for site-specific weed management. *Pest Management Science, 76*(4), 1386–1392.

Hussain, S., Masud Cheema, M. J., Arshad, M., Ahmad, A., Latif, M. A., Ashraf, S., & Ahmad, S. (2019). Spray uniformity testing of unmanned aerial spraying system for precise agro-chemical applications. *Pakistan Journal of Agricultural Sciences, 56*(4), 897–903.

Islam, N., Rashid, M. M., Wibowo, S., Xu, C. Y., Morshed, A., Wasimi, S. A., Moore, S., & Rahman, S. M. (2021). Early weed detection using image processing and machine learning techniques in an Australian chilli farm. *Agriculture, 11*(5), 387.

Kawamura, K., Asai, H., Yasuda, T., Soisouvanh, P., & Phongchanmixay, S. (2021). Discriminating crops/weeds in an upland rice field from UAV images with the SLIC-RF algorithm. *Plant Production Science, 24*(2), 198–215.

Koger, C. H., Shaw, D. R., Watson, C. E., & Reddy, K. N. (2003). Detecting late-season weed infestations in soybean (Glycine max). *Weed Technology, 17*(4), 696–704.

Koot, T. M. (2014). *Weed detection with unmanned aerial vehicles in agricultural systems*. Wageningen University and Research Centre.

Kubiak, A., Wolna-Maruwka, A., Niewiadomska, A., & Pilarska, A. A. (2022). The problem of weed infestation of agricultural plantations vs. the assumptions of the European biodiversity strategy. *Agronomy, 12*(8), 1808. https://doi.org/10.3390/agronomy12081808

Kubišta, J., & Surový, P. (2021). Spatial resolution of unmanned aerial vehicles acquired imagery as a result of different processing conditions. *Central European Forestry Journal, 67*(3), 148–154.

Kumar, V., Nayak, A. K., Poonam, A., Aravindan, S., & Khanam, R. (2020). Unmanned aerial vehicle and its application in Indian agriculture: A perspective. *Indian Farming, 70*(8), 34–37.

Kutugata, M., Hu, C., Sapkota, B., & Bagavathiannan, M. (2021). Seed rain potential in late-season weed escapes can be estimated using remote sensing. *Weed Science, 69*(6), 653–659.

Lambert, J. P. T., Hicks, H. L., Childs, D. Z., & Freckleton, R. P. (2018). Evaluating the potential of unmanned aerial systems for mapping weeds at field scales: A case study with Alopecurus myosuroides. *Weed Research, 58*(1), 35–45.

Lan, Y., Thomson, S. J., Huang, Y., Hoffmann, W. C., & Zhang, H. (2010). Current status and future directions of precision aerial application for site-specific crop management in the USA. *Computers and Electronics in Agriculture, 74*(1), 34–38.

Li, L., Fan, Y., Huang, X., & Tian, L. (2016). Real-time UAV weed scout for selective weed control by adaptive robust control and machine learning algorithm. In *2016 ASABE annual international meeting* (p. 1). American Society of Agricultural and Biological Engineers.

Li, Y., Al-Sarayreh, M., Irie, K., Hackell, D., Bourdot, G., Reis, M. M., & Ghamkhar, K. (2021). Identification of weeds based on hyperspectral imaging and machine learning. *Frontiers in Plant Science, 11*, 611622. https://doi.org/10.3389/fpls.2020.611622

Liao, J., Zang, Y., Luo, X., Zhou, Z., Lan, Y., Zang, Y., Gu, X., Xu, W., & Hewitt, A. J. (2019). Optimization of variables for maximizing efficacy and efficiency in aerial spray application to cotton using unmanned aerial systems. *International Journal of Agricultural and Biological Engineering, 12*(2), 10–17. https://doi.org/10.25165/j.ijabe.20191202.4288

Lord, D., Desjardins, R. L., Dube, P. A., & Brach, E. J. (1985). Variations of crop canopy spectral reflectance measurements under changing sky conditions. *Photogrammetric Engineering and Remote Sensing, 51*(6), 689–695.

Lou, Z., Xin, F., Han, X., Lan, Y., Duan, T., & Fu, W. (2018). Effect of unmanned aerial vehicle flight height on droplet distribution, drift and control of cotton aphids and spider mites. *Agronomy, 8*(9), 187. https://doi.org/10.3390/agronomy8090187

Louargant, M., Jones, G., Faroux, R., Paoli, J. N., Maillot, T., Gée, C., & Villette, S. (2018). Unsupervised classification algorithm for early weed detection in row-crops by combining spatial and spectral information. *Remote Sensing, 10*(5), 761. https://doi.org/10.3390/rs10050761

Makam, S., Komatineni, B. K., Meena, S. S., & Meena, U. (2024). Unmanned aerial vehicles (UAVs): An adoptable technology for precise and smart farming. *Discover Internet of Things, 4*, 12. https://doi.org/10.1007/s43926-024-00066-5

Martin, D. E., Woldt, W. E., & Latheef, M. A. (2019). Effect of application height and ground speed on spray pattern and droplet spectra from remotely piloted aerial application systems. *Drones, 3*(4), 83.

Matese, A., & Di Gennaro, S. F. (2018). Practical applications of a multisensor UAV platform based on multispectral, thermal and RGB high resolution images in precision viticulture. *Agriculture, 8*(7), 116.

Meesaragandla, S., Jagtap, M. P., Khatri, N., Madan, H., & Vadduri, A. A. (2024). Herbicide spraying and weed identification using drone technology in modern farms: A comprehensive review. *Results in Engineering, 21*, 101870. https://doi.org/10.1016/j.rineng.2024.101870

MicaSense. (2018). *MicaSense knowledge base*. Retrieved from https://support.micasense.com

Monteiro, A., & Santos, S. (2022). Sustainable approach to weed management: The role of precision weed management. *Agronomy, 12*(1), 118. https://doi.org/10.3390/agronomy12010118

Motohka, T., Nasahara, K. N., Oguma, H., & Tsuchida, S. (2010). Applicability of green-red vegetation index for remote sensing of vegetation phenology. *Remote Sensing, 2*(10), 2369–2387.

Nikolić, N., Rizzo, D., Marraccini, E., Gotor, A. A., Mattivi, P., Saulet, P., Persichetti, A., & Masin, R. (2021). Site and time-specific early weed control is able to reduce herbicide use in maize-a case study. *Italian Journal of Agronomy, 16*(4), 1780.

Nobre, F. L. D. L., Santos, R. F., Herrera, J. L., Araújo, A. L. D., Johann, J. A., Gurgacz, F., Siqueira, J. A. C., & Prior, M. (2023). Use of drones in herbicide spot spraying: A systematic review. *Advances in Weed Science, 41*, e020230014.

Olsen, A., Konovalov, D. A., Philippa, B., Ridd, P., Wood, J. C., Johns, J., Banks, W., Girgenti, B., Kenny, O., Whinney, J., & Calvert, B. (2019). DeepWeeds: A multiclass weed species image dataset for deep learning. *Scientific Reports, 9*(1), 2058. https://doi.org/10.1038/s41598-018-38343-3

Pahikkala, T., Kari, K., Mattila, H., Lepistö, A., Teuhola, J., Nevalainen, O. S., & Tyystjärvi, E. (2015). Classification of plant species from images of overlapping leaves. *Computers and Electronics in Agriculture, 118*, 186–192.

Papp, L., Van Leeuwen, B., Szilassi, P., Tobak, Z., Szatmári, J., Árvai, M., Mészáros, J., & Pásztor, L. (2021). Monitoring invasive plant species using hyperspectral remote sensing data. *Landscape, 10*(1), 29.

Paul, R. A. I., Palanisamy, M. A., Peramaiyan, P., Kumar, V., Bagavathiannan, M., Gurjar, B., Vijayakumar, S., Djanaguiraman, M., Pazhanivelan, S., & Ramasamy, K. (2024). Spray volume optimization with UAV-based herbicide application for effective droplet deposition and weed control in direct-seeded rice. *Frontiers in Agronomy, 6*, 1491842. https://doi.org/10.3389/fagro.2024.1491842

Peña, J. M., Torres-Sánchez, J., de Castro, A. I., Kelly, M., & López-Granados, F. (2013). Weed mapping in early-season maize fields using object-based analysis of unmanned aerial vehicle (UAV) images. *PLoS One, 8*(10), e77151.

Peña, J. M., Torres-Sánchez, J., Serrano-Pérez, A., De Castro, A. I., & López-Granados, F. (2015). Quantifying efficacy and limits of unmanned aerial vehicle (UAV) technology for weed seedling detection as affected by sensor resolution. *Sensors, 15*(3), 5609–5626.

Pérez-Ortiz, M., Peña, J. M., Gutiérrez, P. A., Torres-Sánchez, J., Hervás-Martínez, C., & López-Granados, F. (2015). A semi-supervised system for weed mapping in sunflower crops using unmanned aerial vehicles and a crop row detection method. *Applied Soft Computing, 37*, 533–544.

Pflanz, M., Nordmeyer, H., & Schirrmann, M. (2018). Weed mapping with UAS imagery and a bag of visual words based image classifier. *Remote Sensing, 10*(10), 1530.

Pooja, K., Saravanane, P., Nadaradjan, S., Sundaravaradhan, S., Muthukumarasamy, S., & Vijayakumar, S. (2021). Effect of cultivars and different weed management practices on yield, nutrient uptake and soil micro-flora in dry direct-seeded rice under coastal deltaic ecosystem. *Oryza-An International Journal of Rice, 58*(1), 21–25. https://doi.org/10.35709/ory.2021.58.1.4

Qin, W., Xue, X., Zhou, L., Zhang, S., Sun, Z., Kong, W., & Wang, B. (2014). Effects of spraying parameters of unmanned aerial vehicle on droplets deposition distribution of maize canopies. *Transactions of the Chinese Society of Agricultural Engineering, 30*(5), 50–56.

Qin, W. C., Qiu, B. J., Xue, X. Y., Chen, C., Xu, Z. F., & Zhou, Q. Q. (2016). Droplet deposition and control effect of insecticides sprayed with an unmanned aerial vehicle against plant hoppers. *Crop Protection, 85*, 79–88. https://doi.org/10.1016/j.cropro.2016.03.018

Rai, N., Mahecha, M. V., Christensen, A., Quanbeck, J., Zhang, Y., Howatt, K., Ostlie, M., & Sun, X. (2023). Multi-format open-source weed image dataset for real-time weed identification in precision agriculture. *Data in Brief, 51*, 109691.

Rajmis, S., Karpinski, I., Pohl, J. P., Herrmann, M., & Kehlenbeck, H. (2022). Economic potential of site-specific pesticide application scenarios with direct injection and automatic application assistant in northern Germany. *Precision Agriculture, 23*(6), 2063–2088.

Ramesh, K., Kumar, S. V., Upadhyay, P. K., & Chauhan, B. S. (2021). Revisiting the concept of the critical period of weed control. *The Journal of Agricultural Science, 159*(9-10), 636–642.

Rasmussen, J., Nielsen, J., Streibig, J. C., Jensen, J. E., Pedersen, K. S., & Olsen, S. I. (2019). Pre-harvest weed mapping of Cirsium arvense in wheat and barley with off-the-shelf UAVs. *Precision Agriculture, 20*, 983–999.

Richardson, B., Rolando, C. A., Kimberley, M. O., & Strand, T. M. (2019). Spray application efficiency from a multi-rotor unmanned aerial vehicle configured for aerial pesticide application. *Transactions of the ASABE, 62*(6), 1447–1453.

Sa, I., Popović, M., Khanna, R., Chen, Z., Lottes, P., Liebisch, F., Nieto, J., Stachniss, C., Walter, A., & Siegwart, R. (2018). WeedMap: A large-scale semantic weed mapping framework using aerial multispectral imaging and deep neural network for precision farming. *Remote Sensing, 10*(9), 1423. https://doi.org/10.3390/rs10091423

Sapkota, R., Stenger, J., Ostlie, M., & Flores, P. (2023). Towards reducing chemical usage for weed control in agriculture using UAS imagery analysis and computer vision techniques. *Scientific Reports, 13*(1), 6548.

Scherrer, B., Sheppard, J., Jha, P., & Shaw, J. A. (2019). Hyperspectral imaging and neural networks to classify herbicide-resistant weeds. *Journal of Applied Remote Sensing, 13*(4), 044516.

Serrano, L., Filella, I., & Penuelas, J. (2000). Remote sensing of biomass and yield of winter wheat under different nitrogen supplies. *Crop Science, 40*(3), 723–731.

Shekhawat, K., Rathore, S. S., & Chauhan, B. S. (2020). Weed management in dry direct-seeded rice: A review on challenges and opportunities for sustainable rice production. *Agronomy, 10*(9), 1264. https://doi.org/10.3390/agronomy10091264

Singh, V., Latheef, M., Martin, D., & Bagavathiannan, M. V. (2019). Evaluation of remotely piloted aerial application systems (RPAAS) for herbicide application. In *Proceedings of the 59th annual meeting of the Weed Science Society of America* (p. 172). Weed Science Society of America.

Singh, V., Rana, A., Bishop, M., Filippi, A. M., Cope, D., Rajan, N., & Bagavathiannan, M. (2020). Unmanned aircraft systems for precision weed detection and management: Prospects and challenges. *Advances in Agronomy, 159*, 93–134.

Slaughter, D. C., Giles, D. K., Fennimore, S. A., & Smith, R. F. (2008). Multispectral machine vision identification of lettuce and weed seedlings for automated weed control. *Weed Technology, 22*(2), 378–384.

Smith, L. N., Zhang, W., Hansen, M. F., Hales, I. J., & Smith, M. L. (2018). Innovative 3D and 2D machine vision methods for analysis of plants and crops in the field. *Computers in Industry, 97*, 122–131.

Somers, B., Tits, L., Roberts, D., & Wetherley, E. (2016). Endmember library approaches to resolve spectral mixing problems in remotely sensed data: Potential, challenges, and applications. *Data Handling in Science and Technology, 30*, 551–577.

Teimouri, N., Jørgensen, R. N., & Green, O. (2022). Novel assessment of region-based CNNs for detecting monocot/dicot weeds in dense field environments. *Agronomy, 12*(5), 1167. https://doi.org/10.3390/agronomy12051167

Timmermann, C., Gerhards, R., & Kühbauch, W. (2003). The economic impact of site-specific weed control. *Precision Agriculture, 4*(3), 249–260.

Torres-Sánchez, J., López-Granados, F., De Castro, A. I., & Peña-Barragán, J. M. (2013). Configuration and specifications of an unmanned aerial vehicle (UAV) for early site specific weed management. *PLoS One, 8*(3), e58210.

Velumani, K., Lopez-Lozano, R., Madec, S., Guo, W., Gillet, J., Comar, A., & Baret, F. (2021). Estimates of maize plant density from UAV RGB images using faster-RCNN detection model: Impact of the spatial resolution. *Plant Phenomics, 2021*, 9824843.

Vijayakumar, S., Aravindan, S., Saravanane, P., & Sivashankari, M. (2021). Unmanned aerial vehicles policies evaluation and suggestion to boost its agriculture application in India. *Kerala Karshakan e-journal, 8*(7), 6–8.

Vijayakumar, S., Kumar, R. M., Sundaram, R. M., & Balasubramanian, P. (2022). Remote sensing based transformative crop insurance for rice. *Current Science, 123*(3), 254–255.

Vijayakumar, S., Madireddy, H., Bhusarapu, S. C., Kumar, R. M., & Sundaram, R. M. (2022). Drone application in rice cultivation: Experiences from ICAR-IIRR trails. *Indian Farming, 72*(12), 3–6.

Wiles, L. J. (2009). Beyond patch spraying: Site-specific weed management with several herbicides. *Precision Agriculture, 10*, 277–290.

Woebbecke, D. M., Meyer, G. E., Von Bargen, K., & Mortensen, D. A. (1995). Color indices for weed identification under various soil, residue, and lighting conditions. *Transactions of ASAE, 38*(1), 259–269.

Xing, F., An, R., Guo, X., & Shen, X. (2024). Mapping invasive noxious weed species in the alpine grassland ecosystems using very high spatial resolution UAV hyperspectral imagery and a novel deep learning model. *GIScience & Remote Sensing, 61*(1), 2327146. https://doi.org/10.1080/15481603.2024.2327146

Zhang, Y., Teng, P., Shimizu, Y., Hosoi, F., & Omasa, K. (2016). Estimating 3D leaf and stem shape of nursery paprika plants by a novel multi-camera photography system. *Sensors, 16*(6), 874.

Chapter 5
Drivers, Concepts, Processes, and Recent Advances in Integrated Weed Management

Wesley J. Everman and Diego Contreras

Abstract There is widespread renewed interest in integrated weed management practices across the globe. Integrated weed management is an established strategy that utilizes a range of chemical, nonchemical, and preventative weed management tools to manage weed populations in a comprehensive approach while minimizing adverse impacts on the environment and human health. The goal of integrated weed management is generally not to remove chemical weed control but rather to identify and implement complementary, effective biological, cultural, mechanical, preventative, and chemical practices, which can be used in a multidisciplinary approach. While the concept is not new, there are advances in technology and practices, which can be adopted and enhance the effectiveness of integrated weed management in small- and large-scale operations. Integrated weed management is a staple for weed management in many parts of the world; however, in areas with broadacre crop production, it remains underutilized and has not been widely adopted. Drivers of this new interest in integrated weed management range from increased chemical resistance to societal pressure for sustainable production systems or environmental protection concerns through government regulation. Integrated management approaches are necessary to diversify management, augment weed control and production practices, and maintain long-term sustainability in farming operations. While a collective adoption of integrated weed management strategies should be the desired objective, it is crucial to emphasize the economic viability and practicality of including these to individual growers. Historical, current, and future practices, which fall into each of the complementary integrated weed management practices, are identified and explained in this chapter.

W. J. Everman (✉)
Department of Crop and Soil Sciences, North Carolina State University, Raleigh, NC, USA

Department of Agronomy, Iowa State University, Ames, IA, USA
e-mail: weverman@iastate.edu

D. Contreras
Department of Crop and Soil Sciences, North Carolina State University, Raleigh, NC, USA

B. S. Chauhan, G. Mahajan (eds.), *Recent Advances in Weed Science*,
https://doi.org/10.1007/978-3-031-91014-2_5

Keywords IWM · Sustainable weed control · Nonchemical strategies · Weed resistance · Environmental impact of herbicides · Agricultural sustainability

1 Introduction

Integrated weed management is a strategy that utilizes a range of weed management tools to manage weed populations in a comprehensive approach while minimizing adverse impacts on the environment and human health. These tools can include chemical, nonchemical, and preventative measures integrated into a management plan (Fig. 5.1). Although the term integrated weed management was first described in the early 1980s, many weed control programs in the past 40 years have focused on chemical control with very little reliance on alternative and supplementary management options, allowing weeds to adapt to those weed management practices (Walker & Buchanan, 1982). The goal of integrated weed management is generally not to remove chemical weed control but rather to identify and implement complementary, effective biological, cultural, mechanical, preventative, and chemical practices, which can be used in a multidisciplinary approach. Owen et al. (2014) postulated that collective integrated weed management programs might more

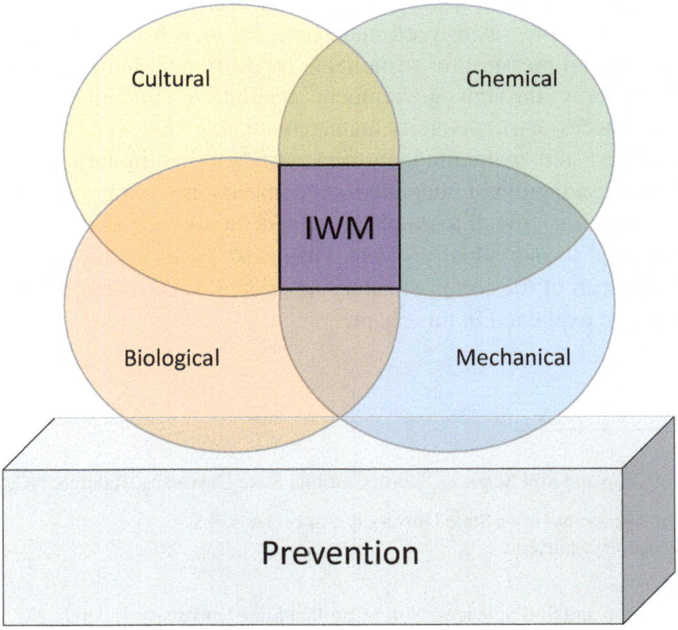

Fig. 5.1 Integrated weed management tools, including chemical, nonchemical, and preventative measures, combined within a comprehensive management plan

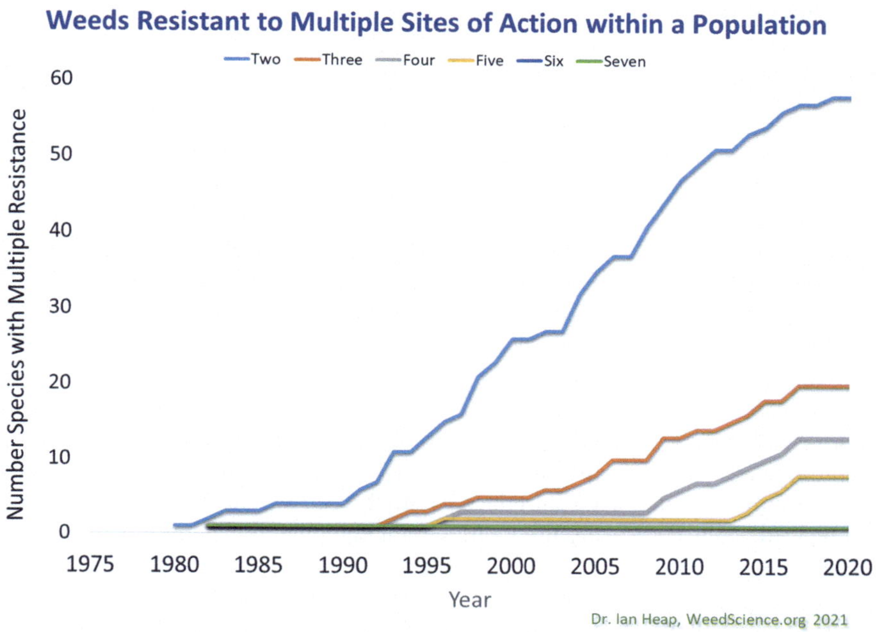

Fig. 5.2 Weeds resistant to multiple sites of action within a weed population (Heap 2024)

properly be redefined as integrated herbicide management in much of North America, due to the increasing evolution of herbicide-resistant weeds. As the prevalence of herbicide-resistant weed species continues to increase, and with the concerning trend of multiple-resistant weed species with resistance to four or more herbicides, it is increasingly important to identify and implement effective integrated weed management practices (Fig. 5.2). By adopting integrated weed control practices prior to widespread resistance in a region, farmers can avoid dramatic shifts in production practices needed to combat resistant weed species (Harker, 2013; Soltani et al., 2023).

The fact that many weed control practices commonly used by farmers are centered on reactionary control measures increases the challenges faced once herbicide resistance appears on the landscape (Dentzman & Jussaume, 2017; Hurley & Frisvold, 2016; Jussaume et al., 2022). Often the practices needed to control resistant weed species are cultural or mechanical, which could be implemented prior to resistance to preserve herbicides as a viable component in the weed management program (Owen et al., 2014). Integrated weed management is a staple for weed management in many parts of the world where industrial agriculture is not the dominant form of crop production; however, in areas with broadacre crop production, it remains underutilized and is not widely adopted. There is renewed interest in promoting integrated weed management to "keep weeds off balance" (Knezevic et al., 2017).

2 Components

2.1 Prevention

Preventative measures are often the first considerations when identifying an integrated weed management program. Steps should be taken to prevent the introduction, establishment, and spread of weed species into an area not currently infested with that species. While individual measures can be highly effective on their own, they often yield even better results when combined and implemented together. For example, the use of weed-free crop seed minimizes the risk of contamination by weed seeds, preventing the introduction and dispersal of weed seeds. Cleaning agricultural equipment and machinery before moving it between weed-infested to non-infested fields is crucial to prevent and minimize weed seed spread throughout a farm. When contracting custom harvesters or other agricultural operations, it is essential to ensure the equipment is thoroughly cleaned and free from weed seeds, as these may unintentionally introduce external weed species to fields. The same can be said about purchasing used agricultural equipment from areas with known infestations of weed species, as they should be thoroughly cleaned before introduction to the farm. Farms that use irrigation must be cautious and use water from a weed-free source to avoid carrying weed seeds and potentially spreading them across fields. Special caution should be taken with hay, livestock feed, manure, and mulch that is introduced to the farm from external suppliers as it may be contaminated with weed seeds. As the average farm size increases, so does the need for careful consideration of rented land and the previous weed management practices utilized. Renting weedy acres increases the chance for spread via equipment, and it becomes important to prevent introduced weeds from going to seed, including field borders and ditches. A strong monitoring program will help identify fields and areas of greatest concern and will aid growers in preventing introduced weeds from going to seed in a field. By offering regional problems, scenarios, and solutions, the preventive program may become more practical and accessible to growers.

Considering prevention as the foundation, four additional key components are utilized when developing a strong integrated weed management program. These are cultural, mechanical, biological, and chemical management practices. Using a variety of weed management tactics reduces the reliance on any single method, thereby lowering the risk of resistance and promoting overall effectiveness.

2.2 Cultural

A solid integrated weed management program starts with a competitive crop, which is achieved by creating conditions that are more favorable for crop growth to compete better with weeds. The ideal cultural weed management practices will promote optimal crop growth through various agronomic practices such as crop cultivar, crop rotation, crop row spacing, planting date, optimal fertility and pH, and cover crops. Crop competition as a weed control tool essentially means utilizing the crop to compete with weeds for the growth inputs light, water, and nutrients, with light being the most important. Thus, crop competition results in free weed control. For the most significant effect, practices that create conditions favorable for crop growth to allow the crop to get ahead of weeds and shade out weeds are utilized. Such practices include planting in narrow rows to reach canopy closure faster, proper fertilization and liming, proper seedbed preparation, proper seeding rate, proper planting date, use of high-quality seed, and insect, nematode, and disease control.

Crop row spacing can greatly affect crop competition with weeds, especially in soybean, wheat, corn, and cotton (Ethridge et al., 2022; Wilson et al., 2007). However, decisions on row spacing are often determined by equipment limitations or other production factors rather than optimal weed competition. While crop-weed competition is a primary factor in the adoption of narrow row spacing, row spacing has a limited influence on initial weed emergence (i.e., about the same number of weeds emerge initially in narrow rows as in wide rows). Still, growth of those weeds may be reduced in narrow rows (Hoffner et al., 2012; Mahoney et al., 2020). One of the most significant benefits of altering row spacing is that once the soil is shaded, further emergence and growth of weeds are significantly reduced. After initial weed control, such as a herbicide application or cultivation, narrow row spacing reduces weed "resurgence" and can reduce or eliminate the need for follow-up herbicide applications.

Crop cultivar selection may also affect competition with weeds. Although we often focus on weed competition effects on the crop, crops also compete with weeds in the field. Research has shown that some crops are more competitive with weeds; for example, corn is more competitive than peanuts when grown in competition with the same weed species (Zimdahl, 2004). Some research suggests some differences can be capitalized upon within a crop as well. Cultivars may vary in leaf area production, whereas cultivars with more leaf area tend to be more competitive. Greater competitive ability can be gained through varying partitioning of resources to obtain early development, increased speed of growth, greater plant height, and canopy architecture, which favors the crop (Andrew et al., 2015). However, most breeders focus on yield, and since yield is the result of reproductive growth while vegetative characteristics are what influence competitiveness, few breeders concentrate on this area.

A key component of cultural weed management is the presence of a healthy crop with optimal growth. Agronomic practices promoting crop growth will aid in competition with weeds such as crop rotation, soil nutrients, pH, and soil quality. A given crop tends to be plagued more by weed species that possess similar growth habits and thrive under the same cultural conditions provided for the crop. Implementing effective crop rotation helps break the cycle, impacts chemical control practices, and is a highly important aspect of an overall weed management program because some weeds are easier to control in one crop than in another (Merfield, 2019). An additional benefit of rotating crops is that herbicides and other management practices can be rotated to better control the total system. Rotation of herbicides and additional management practices helps to avoid weed population shifts and will help prevent herbicide resistance (Harker, 2013).

Managing soil nutrients, pH, and quality can significantly impact crop growth and the germination, growth, seed production, and seed quality of weed species. These factors can also impact other integrated weed management practices and competition dynamics between the crop and weeds (Blackshaw & Brandt, 2008; Humphreys, 1995; Humphreys et al., 1999; Merfield, 2019). Soil health should be managed for optimal crop growth and yield. However, the timing of soil amendment application can impact seed germination and crop/weed competition (Di Tomaso, 1995; Petersen, 2003).

Cover crops are a widely adopted practice of planting any crop to protect and improve the soil and provide weed control between cash crops. Cover crops are used as a smother crop to compete with the weeds infesting an area before establishing a cash crop. Small grains such as wheat and cereal rye are the most commonly used cover crops for weed suppression, but legumes, including crown vetch, hairy vetch, crimson clover, and white clover, are regularly used in mixtures, especially where soil cover and health are priorities. In most cases, the cover crop is terminated prior to planting the crop of interest, although some producers choose to plant into green cover crops. The cover crop helps prevent weed emergence and/or reduces the growth of weeds when it grows, such as winter annual weeds in a cereal cover crop (Hoffman & Regnier, 2006). After the cover crop is terminated, weed management is achieved because the residue excludes light, provides a physical barrier to weed emergence, and affects soil temperature and moisture (Teasdale et al., 2007). Allelopathic chemicals from the cover crop residue may also inhibit the germination/growth of weeds (de Albuquerque et al., 2011; Jabran et al., 2015). While this is not a major reason for planting cover crops in most instances, it is still an effective weed management tool, which warrants further research and collaboration to enhance the suppressive ability of cover crops.

2.3 Mechanical

Mechanical control includes techniques that physically remove or damage a weed, such as hoeing, tillage, mowing, mulching, and flame weeding. Mechanical control practices are utilized with some frequency on most commercial crop production acres, excluding long-term perennial crops or no-till production. Physical weed management is often the cheapest form of management and is a foundational practice for many growers around the world.

The simplest form of mechanical weed management is hand pulling and hoeing. It can be quite effective; however, it must be repeated several times throughout the growing season. Cost of manual labor is the biggest factor in adoption, being used extensively in developing countries where labor is abundant and cheap, but it can be very expensive and is practical only for high-value crops in some countries. For example, hand pulling is generally not used in agronomic crops in the United States, although hand weeding problem fields occasionally accompanies chemical control failures, and troublesome weeds like glyphosate-resistant Palmer amaranth may be exceptions.

Tillage is a widely used mechanical weed management tactic because it destroys weeds by burial, mutilation, or uprooting. Tillage can be divided into primary, such as seedbed preparation, and secondary tillage, which generally refers to in-crop cultivation. Primary tillage is used to bury or incorporate crop residues to produce a clean seedbed and, in the process, bury weed seed below the emergence zone. For some species with short dormancy and persistence, primary tillage can be a valuable tool. However, recent concerns about soil and water quality have negatively affected the use of tillage, reducing the frequency of tillage in many operations (Brady & Weil, 2008).

In-crop cultivation can be an important component of a weed management system; however, depending upon cultivation alone is very risky, primarily because adverse weather can keep one out of the field at critical periods, allowing weeds to become established and more difficult to control. Additionally, cultivation alone is often not adequate for heavy weed infestations but can be an effective supplement to herbicides and other management practices. In-crop cultivation works best on hot, dry soils when weeds are small and easily uprooted. Care should be taken to cultivate shallowly to avoid crop root damage and to avoid bringing up untreated soil from below the herbicide-treated zone if used in conjunction with herbicides. Cultivation may present challenges by promoting additional weed germination and soil moisture loss; it may enhance disease problems in crops such as peanuts, and it may cause physical damage to crops. Additionally, cultivation may not remove or bury weeds in the crop row, and narrow-row planting may eliminate the option to cultivate. Significant improvements have been made in cultivators in recent years. Some rather sophisticated cultivation equipment is now available, especially for specialty crops.

Mowing can be utilized in perennial cropping systems such as pastures, orchards, utility rights of way, and roadsides where the crop can outgrow the weeds and canopy over the area mowed. Mowing is more effective for tall-growing weeds and dicots, and repeated mowing may prevent seed production. Mowing is more effective on broadleaf weeds than on grasses because grasses have a protected growing point in the sheath near the soil surface. Even if the growing point is mowed off in a grass species, the plant can sometimes produce a tiller from the crown region and persist. Some weeds may adapt to mowing in highly managed crops like turf and hay crops by adopting a prostrate growth habit. An example is yellow foxtail, which is mostly erect in row crops; however, in alfalfa, which is mowed several times per season, yellow foxtail will adapt to mowing by growing in a more prostrate manner. For perennial weeds, mowing is most effective when stored food reserves are low, usually just after full leaf development. To be effective on perennial weeds, mowing must be repeated, with the objective of depleting food reserves in underground storage organs.

Mulching is the use of a physical barrier to smother weeds and prevent germination. Materials commonly used to mulch crops include straw, grass clippings, wood chips, paper, and black plastic to exclude light from weeds. Mulching is expensive and difficult to apply to large acreage. Therefore, the most practical uses are in high-value horticultural crops, ornamentals, and home gardens. While existing weeds may be smothered, weed seed can be introduced through mulches, and if the mulch is not biodegradable, it must eventually be disposed of, which can be expensive.

Burning and heat have been used as a cheap and effective management option in noncrop areas to remove hardwoods from pine forests and railroad rights of way. It must be repeated in order to be effective and can be dangerous to neighboring farms and properties. Flame cultivation or flame weeding can be very useful for controlling small broadleaf weeds and can be successfully utilized when chemical control or in-crop cultivation is not an option (Knezevic et al., 2017).

2.4 Biological

Biological control refers to the use of a living organism to reduce weed populations to economically acceptable levels. The greatest challenge to the adoption of biological weed management is that biocontrol agents need to be host-specific and not harm crops or other beneficial vegetation. There have been examples of success in natural areas; for example, the use of the *Cactoblastis cactorum* moth to control prickly pear in Australia is an often-cited classic example of biological control. Prickly pear cactus was introduced into Australia in 1839 in a flowerpot. By 1925, when the control project started, prickly pear infested about 60 million acres. In 1925, eggs of *Cactoblastis cactorum* were introduced into Australia, and by 1932, most of the cacti were dead. Fungi can also be used to manage weeds through a classical approach, initial inoculation of the weed with self-sustaining pathogenic fungi, or a mycoherbicide approach, making an annual application of endemic or foreign pathogenic fungi in a manner similar to the use of herbicides. The classical approach to the use of fungi for weed control is better suited to situations where annual applications of a pathogen are not economically feasible, such as the control of perennial weeds in aquatic, pasture, rangeland, and waste areas. In the classical approach, persistence and efficient natural dispersal of the pathogen are vital to the successful suppression or control of weeds. Two examples of control by the classical approach are control of skeletonweed in Australia by the fungal rust *Puccinia chondrillina* and suppression of curly dock in Europe by another fungal rust *Uromyces rumicis*.

There have been several attempts to commercialize mycoherbicides for use in various field and fruit crops with limited success. Control by mycoherbicides is usually less than can be achieved by chemical herbicides, although there are exceptions. Mycoherbicides are living products applied as aqueous sprays to the weed foliage in the same manner as postemergence herbicides. Because the "active ingredient" is a living organism, extra care is required in shipping, storage, and handling. Therefore, shelf life can be a problem with mycoherbicides. Additionally, sprayers used for mycoherbicide application must be cleaned out carefully because residues of active or inert ingredients of conventional pesticides may be toxic to the fungi. This is especially true for a sprayer that is also used for fungicide applications. Also, fungicides applied to crops to protect against pathogenic diseases can kill the mycoherbicide. In some cases, the mycoherbicide is effective only when applied under certain environmental conditions, such as having dew on the leaves for a specified number of hours.

Herbivory is a tactic utilizing fish and grazing animals such as geese, goats, cows, and pigs to manage undesirable plants. Grass carp are used to control aquatic weeds and can be extremely effective on vascular plants. Grazing animals have been used effectively for certain weed control situations. It is essential that grazing animals be selective, that is, that they find undesirable plants more palatable than the other vegetation present. Geese, particularly the White Chinese breed, have been used to control immature grasses and nutsedge in orchards, vineyards, nursery crops, and broadleaf row crops such as beans, asparagus, cotton, potatoes, and strawberries.

The concept of biological control has been widely publicized, so the general public views it as a viable and readily available alternative to pesticides. Although there have been some outstanding successes, particularly in rangelands and aquatic sites, biological control is generally not feasible for intensive agricultural areas. Inherent in biological control is acceptance of a low level of the weed population so the biological control agent can survive. This may not be acceptable in intensively managed systems. Biological control is better suited to nonintensive production systems such as rangeland or to natural areas.

Disadvantages of biological weed control for intensively managed areas include a narrow spectrum of control of a single species, delayed effectiveness, which may miss the critical period for weed control, unsuitable environment due to intensive management through tillage, crop rotation, or heavy pesticide use, incompatibility with other pest management practices, and special handling. Agricultural fields or commercial turf areas are typically infested with several weed species. Hence, even if the biological control agent is used for one species, some other control method (usually chemical) will still be needed to control other species.

2.5 Chemical

Chemical weed control is the predominant tool utilized in much of North and South American broad-acre crop production, as well as in other industrialized nations in Europe and Asia. Herbicide use has been over-relied on for decades in these regions, leading to herbicide-resistant weed biotypes and an increasing incidence of multiple-resistant weed species. The overuse of herbicides has been driven by the cost and simplicity of treating large areas in crop production. Combined with high levels of efficacy and general crop safety, adoption has been widespread, especially with the introduction of genetically engineered crops like corn, soybean, cotton, and canola, among others. There are many modes of action for weed management in various crops, and many integrated weed management programs were centered more on rotating herbicides or implementing various chemical weed management practices in the aforementioned integrated herbicide management approach. The heavy reliance on relatively few herbicides and the lack of rotation of genetically engineered crops tolerant to a single herbicide, such as glyphosate in glyphosate-tolerant corn, cotton, or soybean, has led to weed population shifts and an increase in herbicide-resistant weeds in countries where these crops are common (Owen et al., 2014).

3 Drivers of Integrated Weed Management

Strategies employing integrated weed management practices are not novel ideas. However, broad adoption and implementation remain challenging in most production systems due to both social and economic factors, complexity, and knowledge intensity (Bastiaans et al., 2008; Liebman et al., 2016). Interestingly, results of experiments and modeling studies suggest that integrated weed management strategies can be successful at the farm or microeconomic scale, yet it is not clear how such systems would work on larger scales if they involve diversification of cropping systems and integration of livestock production, with associated shifts in supply and demand relationships and possible changes in farm revenue (Liebman et al., 2016).

There are several key drivers in the adoption and use of integrated weed management, including the use of chemical weed management, farm profitability and efficiency, farmer perceptions and lifestyle, environmental impacts, herbicide resistance, and consumer preferences. In fact, it has been argued that growers are stuck within a farming ideology that focuses on the reduction of time and labor with a goal of increasing profitability through an ever-increasing efficiency of scale (Dentzman & Jussaume, 2017; Weisberger et al., 2024). Unsurprisingly, the wide range of weed management practices and a lack of homogeneity are due to a mix of factors reflecting socioeconomic factors that are both internal and external to the farm (Jussaume et al., 2022). The social dynamics within farming communities also significantly influence the success of integrated weed management. Community engagement plays a pivotal role in knowledge sharing and collective action. For example, farmers who collaborate can pool resources, share experiences, and disseminate best practices related to weed management. This collective approach can lead to innovative solutions that individual farmers might not have developed on their own.

Furthermore, local extension services and agricultural organizations can facilitate training sessions on integrated weed management, providing farmers with the knowledge and skills needed to implement these strategies effectively. Successful integrated weed management implementation often hinges on the willingness of farmers to adopt new practices, which can be influenced by the perceptions and experiences of their peers. If influential community members adopt integrated weed management, it can create a ripple effect, encouraging others to follow suit. Although farmers may adopt integrated weed management to effectively manage weeds, productivity and profitability are important factors in a farming operation. Farmers and agribusinesses face significant economic pressures to maximize crop yields and minimize costs.

The use of herbicides continues to increase worldwide, primarily due to reduced management time, resulting in a positive impact on farmers' lifestyles. Herbicides also provide alternative positive impacts through improved soil and water conservation, implementation of reduced tillage practices, reduced farm labor and increased labor efficiency, and ultimately improve farm profitability (Gianessi, 2013; Gianessi & Reigner, 2007; Liebman et al., 2016; Pannell et al., 2011; Zimdahl, 2013).

Unfortunately, the use and overuse of herbicides have led to the development of herbicide-resistant weed populations, which complicates weed management and necessitates the adoption of diverse strategies to prevent or delay resistance. Currently, weed resistance to herbicides has been documented for populations of at least 273 weed species worldwide and for 21 of the 31 known groups of herbicide chemistries (Table 5.1) (Heap, 2024). Most recently, concerns about resistance to multiple modes of action, or stacked resistance, have taken center stage with resistance in some of our most troublesome weeds worldwide, including *Amaranthus*, *Echinochloa*, *Lolium*, and *Poa* species due to target site resistance, genetic resistance, and metabolic resistance becoming increasingly prevalent (Gaines et al.,

Table 5.1 Worldwide frequency of resistant dicots and monocots by herbicide site of action

HRAC group	Site of action	Dicots	Monocots	Total
2	Inhibition of acetolactate synthase	107	68	175
5	PSII inhibitors: Serine 264 binders	53	34	87
9	Inhibition of enolpyruvyl shikimate phosphate synthase	28	31	59
1	Inhibition of acetyl CoA carboxylase	0	51	51
4	Auxin mimics	35	9	44
22	PS I electron diversion	23	10	33
14	Inhibition of protoporphyrinogen oxidase	13	4	17
3	Inhibition of microtubule assembly	2	10	12
15	Very long-chain fatty acid synthesis inhibitors	2	8	10
34	Inhibition of lycopene cyclase	1	5	6
10	Inhibition of glutamine synthetase	1	5	6
6	PSII inhibitors: histidine 215 binders	3	2	5
12	Phytoene desaturase inhibitors	4	1	5
27	Inhibition of hydroxyphenyl pyruvate dioxygenase	4	1	5
29	Inhibition of cellulose synthesis	0	4	4
0	Unknown	0	4	4
13	Inhibition of deoxy-D-xylulose phosphate synthase	0	3	3
0	Antimicrotubule mitotic disrupter	0	3	3
23	Inhibition of microtubule organization	0	1	1
0	Nucleic acid inhibitors	1	0	1
0	Cell elongation inhibitors	0	1	1
	Totals	277	255	532

2020; Heap, 2024; Tranel et al., 2011; Yu & Powles, 2014). Combine the increase in multiple herbicide resistance with the lack of a few new classes of chemistry in the past 40 years, and you get rapidly dwindling chemical weed control options (Gaines et al., 2020; Heap, 2024). There are cases of target-site and nontarget-site resistance mechanisms to most herbicide classes (Gaines et al., 2020).

Herbicide resistance in weeds is also a difficult area-wide socioeconomic problem because resistance genes are not sessile; they can spread from one farm to another through the movement of pollen and seeds (Ervin & Jussaume, 2014). Thus, decisions and actions by individual farmers can contribute to resistance problems on surrounding farms due to the adoption of key resistance management practices being low on both owned and rented land (Frisvold et al., 2020). The heterogeneity of grower attitudes and behaviors results in a wide range of integrated weed management practices around the world, highlighting the fact that there is no "correct way" to manage weeds (Jussaume et al., 2022; Weisberger et al., 2024). Growers traditionally adopted integrated weed management practices for sustainable reasons, integrating values of environmental stewardship (Floress et al., 2017; Pereira et al., 2016; Thompson et al., 2015). The environmental impact of traditional weed management practices, such as heavy reliance on herbicides, has raised concerns about potential impacts on soil degradation, water contamination, and loss of biodiversity. Integrated weed management emphasizes sustainable practices that reduce environmental harm and promote ecosystem health (Liebman et al., 2016). Integrated weed management is further complicated by the need to limit environmental damage caused by control measures themselves, especially where tillage is utilized (Buckley & Han, 2014).

Though a wealth of details about each approach to increase integrated weed management adoption has been identified throughout the years, in most regions of the world, broad-scale use of these and other approaches has never occurred (Chantre & Cardona, 2014; Liebman et al., 2016; Owen et al., 2014). Liebman et al. (2016) detailed factors that are responsible for the failure to move effectively from the science of integrated weed management to practice: (1) insufficient and/or inappropriate government policy instruments to alter farmer and land manager behavior; (2) a lack of market mechanisms to motivate farmers to change their crop protection practices; (3) a paucity of social infrastructure with which to support relevant learning and decision-making by farmers and land managers; and (4) the different lifestyles and values of landowners and managers that lead them to favor particular management systems and to differ in attitudes to adopt change (Jordan et al., 2006; Mohler et al., 2001; Pannell et al., 2011).

While the availability of herbicides in some regions is lost due to resistance, in other regions, such as the European Union, the removal of various herbicides from the marketplace through regulation has occurred due to societal concerns over environmental and human health impacts (European Commission, 2009; Liebman et al., 2016). Integrated weed management principles and practices fit well with these increasing regulatory requirements by promoting practices that reduce pesticide use and incorporate alternative management strategies. While there are many ecological reasons for integrated weed management use, it is clear that regulatory policies are critical to the adoption of integrated weed management practices (Powles & Gaines, 2016). Government regulations regarding pesticide use, environmental protection, and agricultural practices significantly influence how farmers manage their operations. For instance, strict regulations on herbicide application can compel farmers to explore alternative weed management strategies, such as cultural practices, mechanical methods, and biological controls.

In many countries, environmental protection agencies set limits on the types and quantities of herbicides that can be used, often based on their potential impact on human health and ecosystems. These regulations may encourage farmers to adopt integrated weed management as a more sustainable alternative. Additionally, policies promoting organic farming and sustainable agricultural practices further support the implementation of integrated weed management by providing incentives for reduced chemical inputs.

However, regulatory frameworks can also pose challenges. In some regions, the lack of clear guidelines or support for integrated weed management practices can lead to uncertainty among farmers. They may hesitate to invest in new methods if they are unsure of their long-term viability or compliance with existing regulations. Therefore, effective communication and well-structured policies are essential to facilitate the transition to integrated weed management.

A novel driver of integrated weed management has emerged recently, with a growing consumer demand for sustainably produced food, which includes reduced pesticide residues. Integrated weed management supports this demand by integrating practices that minimize chemical use and enhance environmental stewardship. Increasing consumer awareness and demand for sustainably produced food also drives the adoption of integrated weed management. Today's consumers are more concerned about the environmental impact of agricultural practices, including pesticide use and soil health. Retailers and food companies increasingly seek to promote sustainability in their supply chains, which often translates to a preference for suppliers who employ environmentally friendly practices (Gullickson, 2015; Liebman et al., 2016). Market access can be a significant incentive for farmers to implement integrated weed management. As a result, many farmers are motivated to partner with processors and retailers to adopt integrated weed management as a way to align their practices with consumer preferences for organic and sustainably sourced products. For example, companies in England have worked with farmers to better control weeds through crop rotation, careful seedbed preparation, weed mapping for targeted herbicide application, hand-weeding patches of new weed species, and choice of herbicide products for peas (Williamson & Buffin, 2005). This consumer-driven demand for sustainability encourages farmers to explore integrated weed management as a viable strategy to meet market expectations while reducing their environmental footprint. The company also conducts research and provides technical assistance to growers. Its weed management protocols are part of a broader effort to reduce pesticide use; over the previous 20 years, pesticide application (kg a.i. per ha) has been halved (Williamson & Buffin, 2005).

Clearly, the drivers of integrated weed management are diverse and have adjusted over time to accommodate shifting cultural, environmental, and economic climates.

The addition of an increased societal interest in sustainability and food products grown through alternative management practices will continue to influence the role of integrated weed management in food and fiber production. A balanced and complementary combination of weed management practices will be needed to facilitate adoption and provide sustainable solutions. In addition, for IWM to succeed, researchers must be able to provide scalable models for larger farms, implement financial incentives, and use behavioral psychology to overcome farmer resistance to changes in their weed management practices. The identification of digital tools in conjunction with unified policies for adoption could be developed through the creation of international consortia to work on IWM.

4 Recent Advances in IWM

There have been significant advances in integrated weed management tools through innovative concepts and the use and application of new technology to improve accuracy and efficacy. These advances are breathing new life into old concepts and practices to make them more viable in the twenty-first century. Many of the concepts are based on the traditional cultural, mechanical, and chemical control mechanisms; however, we are using advances in technology to use them in a more precise and efficient manner, clearing hurdles to adoption. For example, there are commercial applications of precision tillage, precision spray, precision lasers, and the use of unmanned aerial vehicles (UAVs) for weed detection, mapping, and treatment.

In the context of weed control, precision tillage focuses on minimizing soil disturbance while effectively managing weed populations (Kurstjens, 2007). This method uses targeted tillage techniques that disrupt weed growth without unnecessarily affecting the surrounding soil and crops. Farmers can identify specific areas of a field that require weed management, enabling them to focus tillage efforts where weeds are the most problematic rather than the entire field. The availability of optical sensing technologies offers opportunities to implement real-time targeted tillage for weed control (Walsh et al., 2020). This method helps maintain the core soil conservation principles of the crop production system. Reduced soil disturbance in nonweed-infested areas helps preserve any residual soil-applied herbicide activity and reduces the risk of promoting new weed growth. The "Weed Chipper," a 6-meter-wide prototype, achieved 90% weed control with low soil disturbance (1.8%) over areas with low weed densities, validating the effectiveness of site-specific tillage as a weed management option in conservation cropping systems (Walsh et al., 2020).

Precision spraying for weed control involves targeted application of herbicides using advanced technologies and techniques to minimize chemical use while maximizing effectiveness. This approach may enhance the efficiency of weed management by focusing on specific problematic areas of the field, reducing herbicide output, and lowering costs for farmers. Precision spraying is achieved through multiple components, which include the use of artificial intelligence, machine vision, high-resolution cameras, and modern nozzle technologies (Bhakta et al., 2019; Vijayakumar et al., 2023). Sprayers that contain variable rate technology allow adjustments of herbicide application rates based on specific field conditions, such as weed density and weed growth stage and/or size. Boom sprayers with sectional control allow operators to manage individual nozzles or sections, allowing for precise applications only where needed. John Deere's See & Spray™ sprayers include sensors and imaging systems that allow for real-time detection and herbicide applications of weeds within a cropping system.

Precision laser weed removal machines are an innovative technology that utilizes lasers to target and destroy weeds with high precision. This method offers an alternative to traditional herbicide applications and other mechanical weeding techniques. These machines use focused laser beams to destroy the cellular structure of weeds, effectively eliminating them without harming the surrounding crops. These machines are equipped with cameras and sensors that allow them to selectively target individual weeds. The effectiveness of laser weed control depends on several factors, including wavelength, laser power, exposure time, and spot size (Mathiassen et al., 2006; Vijayakumar et al., 2023). Another alternative to herbicides for weed control, the Weed Zapper is designed to control weeds using electrical currents. The idea of using electricity to control weeds has been around for a while (Diprose et al., 1980). The Weed Zapper uses high-voltage electric currents to kill weeds by disrupting their cellular structures. Schreier et al. (2022) conducted studies looking at the effectiveness of weed control through electricity, which ranged between 50 and 97%. They also found that weed seed viability could be reduced between 54 and 80% if the weeds contained seed at the time of electrocution. However, the Weed Zapper does have some limitations, as targeted weeds ideally would be taller than the crop or injury could occur (Schreier et al., 2022).

Commonly known as drones, UAVs have become increasingly important in agriculture over the past 20 years. They have transformed the way farmers manage crops, monitor fields, and even apply pesticides. Early uses of UAVs in agriculture and specifically in weed management focused on aerial photography and mapping, allowing farmers to visualize their field conditions and weed distribution from the sky, offering an efficient alternative to traditional scouting methods. UAVs are more versatile than piloted agricultural planes for capturing high-resolution images at low altitudes and slow speeds (Huang et al., 2018). Studies have also shown that through the use of UAV imagery and computational models, we are able to differentiate weeds from crops successfully (Sanders et al., 2021). As UAVs became more advanced, they began to play a more direct role in weed management. Equipped with sprayers, UAVs can deliver herbicides with high precision and consistent spray coverage. The ability to apply variable rates and provide site-specific weed management has further enhanced efficiency and reduced pesticide waste. Another advantage of UAVs is that there could be practically no mechanical injury to the crop while spraying. However, these UAVs are limited with the amount of spray volume they can carry, reducing the spray output possible. This can bring some problems as certain herbicide chemistries have certain minimum requirements for total spray volume. Spray drones also require coarser droplets to avoid physical drift caused by the drone's propulsion system, and these droplets could increase crop injury (Chen et al., 2019). Another limitation is the short period of battery lifetime, which reduces the amount of area that can be covered by a single UAV.

In addition to technological advances, there are examples of new techniques to manage weeds through harvest weed seed control (HWSC). Seed retention at maturity in certain annual weed species allows for a significant amount of total weed seed production to be harvested when these species are present during grain crop harvest (Walsh, Broster, & Powles, 2018). Harvest weed seed control (HWSC) can be defined as a compendium of weed management strategies aimed at minimizing the spread and seed bank of weeds during a crop's harvest process, using specific techniques to capture, remove, and/or destroy these weed seeds before they can disperse back into fields. There has been widespread adoption of harvest weed seed control practices in western Australia, where rigid ryegrass has developed resistance to many of the available chemical options growers used to control weeds. The loss of effective herbicides drove the innovation and development of a suite of harvest weed seed control practices.

The most common HWSC practices include narrow windrow burning, chaff lining, chaff carts, direct baling of chaff, and impact mills. Weed seed kill rates of 99% for both annual ryegrass and wild radish have been observed through the narrow-windrow burning of crop residues from wheat, canola, and lupin (Walsh & Newman, 2007). Chaff lining is the second most popular HWSC strategy among Australian growers; however, there is limited information available on its effectiveness (Walsh et al., 2017). Chaff cart collection systems have demonstrated their ability to effectively collect and remove substantial amounts of annual ryegrass, wild radish, and wild oats seeds (Shirtliffe & Entz, 2005; Walsh & Powles, 2007); however, the challenges associated with managing the large volumes of collected chaff have been the main reason for the low adoption of this method (Walsh et al., 2017). Direct baling effectively captures weed seeds and packages harvest residues for livestock feed, with previous studies indicating up to 95% removal of annual ryegrass seeds (Walsh & Powles, 2007); however, the absence of suitable markets for the baled material has constrained its adoption among Australian producers (Walsh et al., 2017). Studies on seed impact mills have indicated that weed seed destruction rates range from 93% to 100% in research conducted in Australia and the United States (Schwartz-Lazaro et al., 2017; Walsh, Broster, Schwartz-Lazaro, et al., 2018). These practices have become common in small grain and pulse crop production areas of western Australia and areas of the Great Plains of Canada and the United States have started to adopt these practices as well.

An area with increased interest in the past few years is the development of bioherbicides and gene editing to control weed species. Plant viruses have been previously studied as means of biological weed control (Epstein et al., 1997; Izhevskii et al., 1981; Randles, 1986); however, there is only one virus-based bioherbicide registered with the United States' Environmental Protection Agency (Charudattan, 2024). Gene editing involves the direct genetic control of agricultural pests through various techniques aimed at introducing and propagating novel, fitness-reducing genetic modifications within pest populations (Neve, 2018). Proof of concept for gene editing has been established for controlling disease-vectoring insects, but research is now exploring the potential applications of gene drives in agricultural weed management (Neve, 2018). These technologies, among others, are several novel products that are being investigated in hopes of bringing an alternative management tool to the market.

5 Conclusion

There is widespread renewed interest in integrated weed management practices across the globe. While the drivers of this interest range from increased chemical resistance to societal pressure for sustainable production systems or environmental protection concerns through government regulation, it is clear that integrated management approaches are necessary to diversify management, augment control and production practices, and maintain long-term sustainability in farming operations. While a collective adoption of integrated weed management strategies should be the desired objective, it is crucial to emphasize the economic viability and practicality of including these to individual growers. The novel approaches to integrated weed management through machine learning and novel control platforms will increase the economic feasibility and subsequently increase adoption on a wide scale. It is important to note that different options presently available range in costs, benefits, and feasibility, and there should be a concerted effort to communicate these to individual growers. Smaller growers can still benefit from affordable options yet practical weed control options, and larger growers have the opportunity to incorporate some more complex strategies. However, regardless of the different management options available, the unilateral recommendation should be to include multiple strategies in a weed management program.

The adoption of integrated weed management tools can be further expanded through focus on policies that incentivize a holistic approach for weed management and support access to these. These include economic incentives, farmer outreach, and education programs, as well as collaboration with community leaders in the agricultural sector. Overcoming barriers to the integration of digital tools, such as infrastructure limitations and knowledge gaps, will be critical to expanding their use in practical settings by growers. Even though some of the drivers of adoption are worrisome, the future of integrated weed management is bright.

References

Andrew, I. K. S., Storkey, J., & Sparkes, D. L. (2015). A review of the potential for competitive cereal cultivars as a tool in integrated weed management. *Weed Research, 55*(3), 239–248. https://doi.org/10.1111/wre.12137

Bastiaans, L., Paolini, R., & Baumann, D. T. (2008). Focus on ecological weed management: What is hindering adoption? *Weed Research, 48*, 481–491.

Bhakta, I., Phadikar, S., & Majumder, K. (2019). State-of-the-art technologies in precision agriculture: A systematic review. *Journal of the Science of Food and Agriculture, 99*(11), 4878–4888.

Blackshaw, R. E., & Brandt, R. N. (2008). Nitrogen fertilizer rate effects on weed competitiveness is species dependent. *Weed Science, 56*, 743–747.

Brady, N. C., & Weil, R. R. (2008). *The nature and properties of soil* (14th ed.). Pearson Education.

Buckley, Y. M., & Han, Y. (2014). Managing the side effects of invasion control. *Science, 344*, 975–976.

Chantre, E., & Cardona, A. (2014). Trajectories of French field crop farmers moving toward sustainable farming practices: Change, learning, and links with the advisory services. *Agroecology and Sustainable Food Systems, 38*, 573–602.

Charudattan, R. (2024). Use of plant viruses as bioherbicides: The first virus-based bioherbicide and future opportunities. *Pest Management Science, 80*(1), 103–114.

Chen, Y., Qi, H., Li, G., & Lan, Y. (2019). Weed control effect of unmanned aerial vehicle (UAV) application in wheat field. *International Journal of Precision Agricultural Aviation, 2*(2), 25–31.

de Albuquerque, M. B., dos Santos, R. C., Lima, L. M., Melo Filho, P. A., Nogueira, R. M. C., & da Camara, C. A. G. (2011). Allelopathy, an alternative tool to improve cropping systems. A review. *Agronomy for Sustainable Development, 31*(2), 379–395. https://doi.org/10.1051/agro/2010031

Dentzman, K., & Jussaume, R. (2017). The ideology of U.S. agriculture: How are integrated management approaches envisioned? *Society & Natural Resources, 30*(11), 1311–1327. https://doi.org/10.1080/08941920.2017.1295498

Diprose, M.F., Benson, F.A. and HACKAM, R. (1980). Electrothermal control of weed beet and bolting sugar beet. *Weed Research, 20*, 311–322. https://doi.org/10.1111/j.1365-3180.1980.tb01625.x

Di Tomaso, J. M. (1995). Approaches for improving crop competitiveness through the manipulation of fertilization strategies. *Weed Science, 43*(3), 491–497.

Epstein, A. H., Hill, J. H., & Nutter, F. W. (1997). Augmentation of rose rosette disease for biocontrol of multiflora rose (Rosa multiflora). *Weed Science, 45*(1), 172–178.

Ervin, D., & Jussaume, R. (2014). Integrating social science into managing herbicide-resistant weeds and associated environmental impacts. *Weed Science, 62*, 403–414.

Ethridge, S. R., Locke, A. M., Everman, W. J., Jordan, D. L., & Leon, R. G. (2022). Response of maize, cotton, and soybean to increased crop density in heterogeneous planting arrangements. *Agronomy, 12*(5), 1238. https://doi.org/10.3390/agronomy12051238

European Commission. (2009). *EU action on pesticides*. Directorate General for Health and Food Safety.

Floress, K., de Jalón, G., Church, S., Babin, N., Ulrich-Schad, J., & Prokopy, L. (2017). Toward a theory of farmer conservation attitudes: Dual interests and willingness to take action to protect water quality. *Journal of Environmental Psychology, 53*, 73–80.

Frisvold, G. E., Albright, J., Ervin, D. E., Owen, M. D. K., Norsworthy, J. K., Dentzman, K. E., Hurley, T. M., Jussaume, R. A., Gunsolus, J. L., & Everman, W. (2020). Do farmers manage weeds on owned and rented land differently? Evidence from US corn and soybean farms. *Pest Management Science, 76*, 2030–2039.

Gaines, T. A. S. O., Duke, S., Morran, C. A. G., Rigon, P. J., Tranel, A. K., & Dayan, F. E. (2020). Mechanisms of evolved herbicide resistance. *Journal of Biological Chemistry, 295*(30), 10307–10330. https://doi.org/10.1074/jbc.REV120.013572

Gianessi, L. P. (2013). The increasing importance of herbicides in worldwide crop production. *Pest Management Science, 69*, 1099–1105.

Gianessi, L. P., & Reigner, N. P. (2007). The value of herbicides in U.S. crop production. *Weed Technology, 21*, 559–566.

Gullickson, G. (2015). Corporate sustainability guidelines will change the way you farm. *Successful Farming, 2015*, 47–49.

Harker, K. N. (2013). Slowing weed evolution with integrated weed management. *Canadian Journal of Plant Science, 93*, 759–764. https://doi.org/10.4141/CJPS2013-049

Heap, I. The International Herbicide-Resistant Weed Database. Online. Thursday, October 10, 2024. Available www.weedscience.org

Hoffman, M. L., & Regnier, E. E. (2006). Contributions to weed supression from cover crops. In H. P. Singh, D. R. Batish, & R. K. Kohli (Eds.), *Handbook of sustainable weed management* (pp. 51–67). The Haworth Press.

Hoffner, A. E., Jordan, D. L., York, A. C., Dunphy, E. J., & Everman, W. J. (2012). Influence of soybean (Glycine max) population and herbicide program on palmer amaranth (*Amaranthus palmeri*) control, soybean yield, and economic return. *ISRN Agronomy, 2012*, 947395.

Huang, Y., Reddy, K. N., Fletcher, R. S., & Pennington, D. (2018). UAV low-altitude remote sensing for precision weed management. *Weed Technology, 32*(1), 2–6.

Humphreys, J. (1995). *Investigations into aspects of the dynamics of Rumex obtusifolius L. (broad-leaved dock) populations in Grassland (PhD)*. University College Dublin.

Humphreys, J., Jansen, T., Culleton, N., Macnaeidhe, F. S., & Storey, T. (1999). Soil potassium supply and Rumex obtusifolius and Rumex crispus abundance in silage and grazed grassland swards. *Weed Research, 39*, 1–13. https://doi.org/10.1046/j.1365-3180.1999.00123.x

Hurley, TM., Frisvold, G. (2016) Economic Barriers to Herbicide-Resistance Management. *Weed Science,* 64(SP1), 585–594. https://doi.org/10.1614/WS-D-15-00046.1

Izhevskii, S., Livshits, A., Murusidze, G., & Gogoladze, G. (1981). Prospects of using alke strain in the integrated control of Solanum carolinense in tea plantations. *Subtropicheskie Kul'tury, 4*, 60–65.

Jabran, K., Mahajan, G., Sardana, V., & Chauhan, B. S. (2015). Allelopathy for weed control in agricultural systems. *Crop Protection, 72*, 57–65.

Jordan, N., Niemi, H., Simmons, S., Becker, R., Gunsolus, J., & White, S. (2006). Learning groups for implementation of integrated weed management: Principles and practical guidelines. In H. P. Singh, D. R. Batish, & R. K. Kohli (Eds.), *Handbook of sustainable weed management* (pp. 825–854). Food Products Press.

Jussaume, R. A., Dentzman, K., Frisvold, G., Ervin, D., & Owen, M. (2022). Factors that influence on-farm decision-making: Evidence from weed management. *Society & Natural Resources, 35*(5), 527–546. https://doi.org/10.1080/08941920.2021.2001123

Knezevic, S. Z., Jhala, A., & Datta, A. (2017). Integrated weed management. In B. Thomas, B. G. Murray, & D. J. Murphy (Eds.), *Encyclopedia of applied plant sciences* (2nd ed., pp. 459–462). Academic Press. https://doi.org/10.1016/B978-0-12-394807-6.00231-8

Kurstjens, D. A. (2007). Precise tillage systems for enhanced non-chemical weed management. *Soil and Tillage Research, 97*(2), 293–305.

Liebman, M., Baraibar, B., Buckley, Y., Childs, D., Christensen, S., Cousens, R., Eizenberg, H., Heijting, S., Loddo, D., Merotto, A., Renton, M., & Reimens, M. (2016). Ecologically sustainable weed management: How do we get from proof-of-concept to adoption? *Ecological Applications, 26*(5), 1352–1369.

Mahoney, D. J., Jordan, D., Hare, A. T., Roma-Burgos, N., Jennings, K. M., Leon, R. G., Vann, M. C., Everman, W. J., & Cahoon, C. W. (2020). The influence of soybean population and postemergence herbicide application timing on in- and subsequent-season palmer amaranth (*Amaranthus palmeri*) control and economic returns. *Weed Technology, 35*(1), 106–112.

Mathiassen, S. K., Bak, T., Christensen, S., & Kudsk, P. (2006). The effect of laser treatment as a weed control method. *Biosystems Engineering, 95*(4), 497–505.

Merfield, C. N. (2019). Chapter 5: Integrated weed management in organic farming. In S. Chandran, M. R. Unni, & S. Thomas (Eds.), *Woodhead publishing series in food science, technology and nutrition, organic farming* (pp. 117–180). Woodhead Publishing. https://doi.org/10.1016/B978-0-12-813272-2.00005-7

Mohler, C. L., Liebman, M., & Staver, C. P. (2001). Weed management: The broader context. In M. Liebman, C. L. Mohler, & C. P. Staver (Eds.), *Ecological management of agricultural weeds* (pp. 494–518). Cambridge University Press.

Neve, P. (2018). Gene drive systems: Do they have a place in agricultural weed management? *Pest Management Science, 74*(12), 2671–2679.

Owen, M. D. K., Beckie, H. J., Leeson, J. Y., Norsworthy, J. K., & Steckel, L. E. (2014). Integrated pest management and weed management in the United States and Canada. *Pest Management Science, 71*, 357–376. https://doi.org/10.1002/ps.3928

Pannell, D. J., Marshall, G. R., Barr, N., Curtis, A., Vanclay, F., & Wilkinson, R. (2011). Understanding and promoting adoption of conservation practices by rural landholders. In D. J. Pannell & F. Vanclay (Eds.), *Changing land management: Adoption of new practices by rural landholders* (pp. 11–38). CSIRO.

Pereira, M., Fairweather, J., Woodford, K., & Nuthall, P. (2016). Assessing the diversity of values and goals amongst Brazilian commercial-scale progressive beef farmers using Q-methodology. *Agricultural Systems, 144*, 1–8.

Petersen, J. (2003). Weed: Spring barley competition for applied nitrogen in pig slurry. *Weed Research, 43*(1), 33–39.

Powles, S. B., & Gaines, T. A. (2016). Exploring the potential for regulatory change to encourage diversity in herbicide use. *Weed Science, 64*, 649–654.

Yu, Q. Powles, S., (2014). Metabolism-Based herbicide resistance and cross-resistance in crop weeds: A threat to herbicide sustainability and global Crop production. *Plant Physiology, 166*(3), 1106-1108. https://doi.org/10.1104/pp.114.242750

Randles, J. W. (1986). Susceptibility of Echium plantagineum L. to tobacco mosaic, alfalfa mosaic, tobacco ringspot, and tobacco necrosis viruses. *Australasian Plant Pathology, 15*, 74–77.

Sanders, J. T., Jones, E. A., Minter, A., Austin, R., Roberson, G. T., Richardson, R. J., & Everman, W. J. (2021). Remote sensing for Italian ryegrass [*Lolium perenne* L. ssp. multiflorum (lam.) Husnot] detection in winter wheat (*Triticum aestivum* L.). *Frontiers in Agronomy, 3*, 687112.

Schreier, H., Bish, M., & Bradley, K. W. (2022). The impact of electrocution treatments on weed control and weed seed viability in soybean. *Weed Technology, 36*(4), 481–489.

Schwartz-Lazaro, L. M., Norsworthy, J. K., Walsh, M. J., & Bagavathiannan, M. V. (2017). Efficacy of the integrated Harrington seed destructor on weeds of soybean and rice production systems in the southern United States. *Crop Science, 57*(5), 2812–2818.

Shirtliffe, S. J., & Entz, M. H. (2005). Chaff collection reduces seed dispersal of wild oat (Avena fatua) by a combine harvester. *Weed Science, 53*(4), 465–470.

Soltani, N., Shropshire, C., & Sikkema, P. H. (2023). Integrated weed management strategies for the depletion of multiple herbicide-resistant waterhemp (*Amaranthus tuberculatus*) seed in the soil seedbank. *Weed Technology, 37*, 108–112. https://doi.org/10.1017/wet.2023.15

Teasdale, J. R., Brandsæter, L. O., Calegari, A., & Skora Neto, F. (2007). Cover crops and weed management. In M. K. Upadhyaya & R. E. Blackshaw (Eds.), *Non-chemical weed management: Principles, concepts & technology* (pp. 49–64). CABI.

Thompson, A., Reimer, A., & Prokopy, L. (2015). Farmers' views of the environment: The influence of competing attitude frames on landscape conservation efforts. *Agriculture and Human Values, 32*, 385–399.

Tranel, P. J., Riggins, C. W., Bell, M. S., & Hagar, A. G. (2011). Herbicide resistances in Amaranthus tuberculatus: A call for new options. *Journal of Agricultural and Food Chemistry, 59*, 5808–5812.

Vijayakumar, V., Ampatzidis, Y., Schueller, J. K., & Burks, T. (2023). Smart spraying technologies for precision weed management: A review. *Smart Agricultural Technology, 6*, 100337.

Walker, R. H., & Buchanan, G. A. (1982). Crop manipulation in integrated weed management systems. *Weed Science, 30*, 17–24.

Walsh, M., & Newman, P. (2007). Burning narrow windrows for weed seed destruction. *Field Crops Research, 104*(1-3), 24–30.

Walsh, M., Ouzman, J., Newman, P., Powles, S., & Llewellyn, R. (2017). High levels of adoption indicate that harvest weed seed control is now an established weed control practice in Australian cropping. *Weed Technology, 31*(3), 341–347.

Walsh, M. J., Broster, J. C., & Powles, S. B. (2018). iHSD mill efficacy on the seeds of Australian cropping system weeds. *Weed Technology, 32*(2), 103–108.

Walsh, M. J., Broster, J. C., Schwartz-Lazaro, L. M., Norsworthy, J. K., Davis, A. S., Tidemann, B. D., Beckie, H. J., Lyon, D. J., Soni, N., Neve, P., & Bagavathiannan, M. V. (2018). Opportunities and challenges for harvest weed seed control in global cropping systems. *Pest Management Science, 74*(10), 2235–2245.

Walsh, M. J., & Powles, S. B. (2007). Management strategies for herbicide-resistant weed populations in Australian dryland crop production systems. *Weed Technology, 21*(2), 332–338.

Walsh, M. J., Squires, C. C., Coleman, G. R., Widderick, M. J., McKiernan, A. B., Chauhan, B. S., Peressini, C., & Guzzomi, A. L. (2020). Tillage based, site-specific weed control for conservation cropping systems. *Weed Technology, 34*(5), 704–710.

Weisberger, D., Ray, M. A., Basinger, N. T., & Thompson, J. J. (2024). Chemical, ecological, other? Identifying weed management typologies within industrialized cropping systems in Georgia (U.S.). *Agriculture and Human Values, 41*, 935–953. https://doi.org/10.1007/s10460-023-10530-7

Williamson, S., & Buffin, D. (2005). Towards safe pest management in industrialized agricultural systems. In J. Pretty (Ed.), *The pesticide detox: Towards a more sustainable agriculture* (pp. 212–225). Earthscan.

Wilson, D. G., York, A. C., & Jordan, D. L. (2007). Effect of row spacing on weed management in glufosinate-resistant cotton. *Weed Technology, 21*, 489–495.

Zimdahl, R. L. (2004). *Weed-crop competition: A review*. Blackwell Publishing.

Zimdahl, R. L. (2013). *Fundamentals of weed science* (4th ed.). Elsevier.

Chapter 6
Adaptive Weed Management in Regenerative Agriculture: Opportunities and Challenges

Seyed Vahid Eslami

Abstract Regenerative agriculture (RA) offers a transformative approach to farming, addressing critical global challenges such as environmental degradation, soil health, and the overuse of chemical inputs. Unlike conventional agricultural practices, RA emphasizes the restoration of soil ecosystems, biodiversity, and the minimization of synthetic chemicals. This chapter explores the role of RA in effective weed management, contrasting it with traditional methods that rely heavily on herbicides and intensive tillage. By integrating techniques such as cover cropping, crop rotation, agroforestry, and rotational grazing, RA fosters soil health and ecological balance while reducing weed competition through holistic and nature-based solutions. Despite the promising potential of RA for sustainable weed control, challenges remain, particularly in the transition to herbicide-free systems. This chapter outlines both the opportunities and challenges in managing weeds within RA systems, emphasizing the need for innovative strategies that acknowledge the complex interplay among soil health, biodiversity, and weed dynamics. Through a broader perspective on weed management, RA offers a novel framework for overcoming conventional agricultural limitations and promoting long-term agricultural resilience.

Keywords Sustainable weed management · Soil health restoration · Biodiversity · Nonchemical weed control · Cover cropping · Crop rotation · Agroforestry · Agricultural resilience

S. V. Eslami (✉)
Department of Plant Production & Genetics, Faculty of Agriculture, University of Birjand, Birjand, Iran
e-mail: sveslami@birjand.ac.ir

B. S. Chauhan, G. Mahajan (eds.), *Recent Advances in Weed Science*,
https://doi.org/10.1007/978-3-031-91014-2_6

1 Introduction

The global food system is facing various crises and troubles, such as hunger, poverty, obesity, the environmental impact of industrial farming, excessive reliance on chemical inputs like fertilizers and pesticides, food quality issues, environmental degradation, loss of biodiversity, exploitative labor practices, and animal welfare issues. In response, regenerative agriculture (RA) has emerged as a compelling alternative, rooted in principles that prioritize ecological health, social fairness, and economic resilience. RA gained prominence in the late 1970s, notably through the efforts of the Rodale Institute in the early 1980s. Robert Rodale defined RA as a system that improves land and soil biological production while maintaining economic and ecological stability with minimal environmental impact and pesticide-free food production. The agronomist Richard Harwood highlighted that regenerative agriculture (RA) differs significantly from traditional organic farming by focusing on the concepts of interconnectedness, biological balance, and minimizing disruptive practices (Giller et al., 2021).

Recognizing the vital role of soil in maintaining the Earth's carbon balance, regenerative agriculture draws inspiration from the traditional farming practices of indigenous communities globally, which prioritize biodiversity and soil health. These methods are gaining traction among Western scientists and major food corporations, who are increasingly convinced that RA can rejuvenate soil health, enhance watersheds, and boost the soil's capacity to sequester carbon, thereby helping to combat global warming.

Presently, conventional farming methods such as tillage and monocropping have led to insufficient carbon in our soils and excessive carbon in the atmosphere. Regenerative agriculture aims to rectify this imbalance (Anonymous, 2024).

RA shares many similarities with other sustainable agricultural practices like agroecology, organic farming, and conservation agriculture, all emphasizing soil health and restoration. Agroecology, similar to RA, uses an integral system approach to enhance farm ecology and reduce agricultural impact. Agroecology adopts a more politically charged and activist approach, placing a significant emphasis on social values. In contrast, regenerative agriculture (RA) occasionally addresses social factors but generally lacks a strong political focus. When compared to conservation agriculture, RA is often viewed as an evolution that embraces a more comprehensive perspective, including socioeconomic sustainability. While RA shares some practices with organic farming, like crop rotation, there are key distinctions. Organic farming aims to minimize agriculture's environmental impact through stringent regulations, such as prohibiting synthetic fertilizers and pesticides, without necessarily striving for positive ecological outcomes. On the other hand, RA prioritizes achieving results like enhanced soil health, with less concern for the specific methods employed to attain these objectives.

Defining RA conceptually can be complex, yet its practical application is more straightforward, guided by a clear set of principles and techniques. These principles include maximizing ground cover, ensuring continuous living roots in the soil, minimizing soil disturbance, promoting plant diversity, and integrating livestock into holistic management systems. Together, these practices translate into various farming and ranching techniques designed to improve soil health and support regenerative management systems. These techniques include cover cropping, reduced or no tillage, compost or manure application, crop diversity and rotation, rotational grazing, and living mulch or crop residues, among others. In addition, practices such as intercropping, alley cropping, agroforestry, silvopasture, biochar, and perennial cropping play a significant role in enhancing soil health and can be viewed as integral components of a regenerative management system.

Weeds present a significant challenge in agriculture, and effectively managing them is essential for successful crop production (Gilchrist, 2021). Traditional or conventional agriculture uses tillage equipment to achieve weed control, and it also increases fineness of tilth and soil smoothness, aeration, artificial porosity, friability, and optimum moisture content. Further, well-cultivated soil facilitates the subsequent sowing and covering of the seeds (Baker et al., 2007). This intensive tillage-based agricultural system, however, causes several problems such as deterioration of physical, chemical, and biological properties of soil that lead to soil erosion, compaction, and reduced microbial fauna (Johansen et al., 2012). In response to the need for improved soil health, no-till (NT) farming has been adopted, which has led to a significant increase in the use of herbicides for weed management.

However, a major concern with no-till (NT) farming is the heavy reliance on herbicides, resulting in environmental pollution, the emergence of herbicide-resistant weeds, and potential risks to human health. Nevertheless, it is possible to reduce herbicide usage in no-till crop production by implementing an integrated approach. This strategy involves enhancing crop competitiveness, utilizing seed bank reduction technologies, practicing crop rotation, and applying biological control methods (Kumar & Aloke, 2020; Swanton & Weise, 1991; Summers et al., 2021).

As it was mentioned, regenerative agriculture shares many similarities with other sustainable farming practices, such as conservation agriculture and organic farming. The strategies employed in practices such as no-till farming, reduced soil disturbance, cover cropping, intercropping, perennial cropping, crop rotations, alley cropping, agroforestry, and rotational grazing significantly influence weed communities and their dynamics. As a result, when transitioning to regenerative agriculture, similar weed-related challenges can be anticipated due to the implementation of these sustainable practices.

Regenerative agriculture encourages a shift in perspective regarding weeds and their roles within agricultural ecosystems. Unlike conventional farming systems that often view weeds as adversaries to be eliminated, regenerative approaches acknowledge that certain weeds can provide valuable benefits. These include offering habitats for pollinators, enhancing soil health, and even serving as nutritious forage options. By exploring these alternative perspectives on weeds, we can begin to understand the nuanced complexities of regenerative weed management and appreciate the potential for innovative, ecologically minded strategies that challenge conventional farming norms.

There is some ongoing, yet subdued, discussion regarding the use of herbicides and synthetic fertilizers in RA. Conventional farmers adopting a regenerative, no-till strategy often depend heavily on herbicides for weed management. In contrast, organic regenerative farmers utilize a broader range of alternative methods that are less reliant on chemicals and generally require more manual labor (Wozniacka 2019). RA generally focuses on practices that improve soil health, biodiversity, and ecological balance, often preferring natural and biological methods of weed control over the use of chemical interventions (LaCanne & Lundgren, 2018; Schreefel et al., 2020). While herbicide use is not completely prohibited within all frameworks of regenerative agriculture, it is typically minimized or avoided whenever feasible to promote a more sustainable and an eco-friendly farming approach (Rhodes, 2017). The primary goal is to lessen reliance on synthetic chemicals, thereby fostering healthier ecosystems and enhancing long-term agricultural resilience (Gosnell et al., 2020).

Rather than depending on herbicides, RA advocates for alternatives such as cover cropping, crop rotation, agroforestry, mulching, and the integration of livestock to control weeds (Schreefel et al., 2020; Lal, 2020). In some cases, herbicides may be used selectively and in limited quantities, especially during transition periods or when other weed control methods have not been successful (LaCanne & Lundgren, 2018). However, the primary objective remains to move toward practices that reduce chemical inputs and prioritize holistic, nature-based solutions (Gosnell et al., 2020).

As we strive for agricultural sustainability, regenerative practices have emerged as a focal point. Within this framework, effective weed management plays a vital role in maintaining soil health, promoting biodiversity, and ensuring economic resilience. Unlike conventional systems that often depend on extensive herbicide use or intensive tillage, regenerative agriculture advocates for a more integrated and holistic approach to weed control. This chapter will delve into the opportunities that regenerative agriculture presents for effective weed management. Unlike conventional approaches that categorize weed management into cultural, mechanical, chemical, and biological control methods, this chapter aims to provide a broader perspective on how regenerative agricultural practices can inherently influence weed dynamics. Due to the limited number of case studies specifically focused on weed management within regenerative agriculture (RA) systems, this discussion will highlight both the potential benefits and challenges associated with these practices. By integrating insights into the unique principles of RA, such as enhanced soil health, increased biodiversity, and reduced reliance on chemical inputs, the chapter will outline how these approaches can contribute to sustainable weed management. The aim is to highlight the innovative potential of RA to reshape weed management strategies and address the ecological complexities of weed–crop interactions.

2 Weed Community Dynamics in Regenerative Agriculture, Opportunities and Challenges for Weed Management

As previously noted, integrating strategies like intercropping, agroforestry, grazing, cover cropping, and reduced or no soil disturbance into agricultural systems can result in changes to weed communities. Some weed species may be diminished or eradicated, while others may establish themselves within the system. This section addresses the shifts in weed communities that occur under regenerative agriculture (RA) practices, focusing on the impacts of reduced soil disturbance, agroforestry, crop rotations, and the integration of livestock and cover crops. These changes in weed populations present a range of opportunities and challenges for effective weed management. We will delve into the dynamics of weed communities in RA systems, highlighting key alterations in species composition and behavior, while also examining the potential benefits and obstacles for sustainable weed control.

2.1 Agroforestry

Agroforestry, as defined by the United States Department of Agriculture (USDA), is the integration of trees and shrubs into farming systems to provide environmental, economic, and social benefits. It enhances soil organic carbon, improves nutrient availability and fertility, and promotes soil microbial dynamics, contributing to overall soil health (Dollinger & Jose, 2018). Common practices include alley cropping (trees planted with crops in between), forest farming (cultivating shade-tolerant specialty crops under forest cover), riparian buffers (vegetation strips along waterways), silvopasture (trees combined with pasture and livestock), and windbreaks (tree barriers reducing wind impact) (Gold & Garrett, 2009; Hillbrand et al., 2017; Fownes & Raynor, 1993; Mudge & Gabriel, 2014).

Agroforestry systems improve nutrient cycling, reduce soil erosion, and enhance carbon sequestration (Koyejo et al., 2021; Casanova-Lugo et al., 2023). Shade trees promote biodiversity, soil fertility, and drought resistance while supporting weed and pest control (Tscharntke et al., 2011). Traditional weed control methods, such as manual labor or herbicide use, are labor-intensive, costly, and ecologically unsustainable, reducing biodiversity (Caamal-Maldonado et al., 2001; Dudley & Alexander, 2017). Agroforestry offers sustainable alternatives for weed control by incorporating woody and legume species, which suppress weeds through shading, competition for resources, and allelopathic effects (Atangana et al., 2014; Boinot et al., 2019). For instance, coffee plantations with shade trees like *Erythrina poeppigiana* exhibit fewer weeds compared to full-sun plantations (Gutiérrez Castillo et al., 2004).

Agroforestry significantly enhances biodiversity, including floral, faunal, and soil microbial diversity, compared to monocropping systems. Soil organisms, such as arbuscular mycorrhizal fungi, are more abundant in agroforestry systems, offering benefits like improved soil structure, nutrient cycling, and reduced weed populations (Udawatta et al., 2019; Shukla et al., 2012; Bainard et al., 2010).

Agroforestry can alter soil conditions, impacting weed dynamics. Increased soil fertility may strengthen crop growth, enhancing competition with weeds. However, weeds may also outcompete crops due to their responsiveness to nutrient increases. Practices like alley cropping suppress weeds by shading and reducing light availability, while tree pruning (e.g., *Leucaena leucocephala*, *Erythrina* spp.) used as manure inhibits weed growth and adds organic matter (Sharma & Verma, 2019). Nonetheless, agroforestry's weed suppression depends on long-term system establishment, as initial years often fail to curb weeds, particularly in semi-arid areas.

Lacerda et al. (2013) demonstrated that combining short-cycle leguminous species with afforestation effectively controlled weeds without herbicides in a degraded pasture in Brazil. These systems leveraged mixed plantations of oil palm, manioc, and forest species to manage weeds sustainably.

Agroforestry reduces both parasitic and nonparasitic weeds while increasing crop yields (Scott & Freckleton, 2022). In Kenya, alley cropping with *Faidherbia albida* and *Leucaena leucocephala* reduced weed biomass in maize and green gram by 42–98% compared to monocrops (Jama et al., 1991; Jama & Getahun, 1992). Similarly, in Costa Rica, *Erythrina poeppigiana* and *Gliricidia sepium* hedgerows decreased maize weed biomass by 52% and 28%, respectively (Rippin et al., 1994). Hedgerows are particularly effective against problematic weeds like speargrass (*Imperata cylindrica*). In Nigeria, hedgerows of *L. leucocephala* and *G. sepium* reduced speargrass population by 51–67% and rhizome biomass by up to 96% (Anoka et al., 1991), while similar results were observed in Indonesia (Rao et al., 1998).

Weed control in alley cropping is achieved through shading, mulch from pruning, and allelopathy. Shading by *G. sepium* and *L. leucocephala* caused 31% and 25% speargrass rhizome mortality, respectively, and altered weed species composition over time (Anoka et al., 1991; Ong & Rao, 2001). In Nigeria, shifts from speargrass to other weeds like *Chromolaena odorata* and *Rottboellia cochinchinensis* were observed (Anoka et al., 1991).

Agroforestry also reduces crop stress by limiting weed competition. For example, termite damage on maize was lower after agroforestry systems compared to natural fallows, attributed to reduced weed competition (Sileshi & Mafongoya, 2003).

Trees in agroforestry systems release natural chemicals through roots, leaves, or litter, exhibiting allelopathic effects that suppress weeds and reduce competition for resources like water, nutrients, and sunlight. This suppression minimizes the need for herbicides, promoting sustainable farming and improving crop productivity by reducing weed pressure (Sileshi et al., 2008). Selecting trees with minimal impact on crops is essential for coexistence.

Species within the genus *Eucalyptus* are known for their allelopathic properties, attributed to essential oils containing monoterpenes, phenols, and other compounds (Batish et al., 2008; Zhang & Fu, 2010). Studies show that *Eucalyptus grandis*, *Eucalyptus urophylla*, and hybrids like *E. urophylla* × *grandis* have minimal inhibitory effects on maize germination and growth, making them suitable for intercropping with crops like maize, paddy, and wheat in irrigated systems (El-Khawas & Shehata, 2005; Espinosa-Garcia et al., 2008; Singh et al., 2021). These hybrids are valued for their rapid growth, ability to suppress weeds, resilience, and adaptability to diverse conditions.

Mango (*Mangifera indica L.*) is an important tree in gardens, providing food, medicine, fuel, timber, and apiculture benefits. Its allelopathic effects, studied for over two decades, show that mango leaf extracts, leachates, and residues inhibit germination and growth of various plants due to phytotoxic substances such as coumaric, vanillic, and gallic acids, among others (Kato-Noguchi & Kurniadie, 2020). These compounds, released through leachates and residue decomposition, offer potential applications in weed management, reducing reliance on herbicides.

In a field study, Oluwatobi and Olorunmaiye (2014) found lower weed species distribution and higher biodiversity in plots intercropped with okra compared to control plots. Similarly, a meta-analysis by Pumariño et al. (2015) revealed significantly lower weed abundance in agroforestry systems, attributed to shading, improved soil quality, and moderated microclimates (Nestel & Altieri, 1992; Sileshi et al., 2008). Enhanced soil nutrients and organic material from trees promote crop competitiveness, reducing parasitic and nonparasitic weeds (Barrios et al., 1998; Carson, 1989; Midega et al., 2013). Covering soil with plant litter was also found to suppress witchweed (*Striga* spp.) by lowering germination rates.

Dias et al. (2019) assessed weed pressure in an agrosilvopastoral system with corn, palisadegrass (*Brachiaria brizantha* cv. Marandu), perennial horse gram (*Macrotyloma axillare* cv. Java), and eucalyptus. Crop arrangements with eucalyptus, corn, and palisadegrass showed the lowest weed dry mass, while corn monoculture had the highest weed incidence. All arrangements contained dicot weeds, but monocot weeds were only present in treatments with horse gram and in corn monoculture. The low weed incidence in palisadegrass plots is likely due to its competitive ability to suppress other species (Fernandez et al., 2002). Machado et al. (2011) noted higher weed incidence in sorghum monoculture compared to consortia with *B. brizantha* cv. Xaraés, attributing this to the rapid growth of *B. brizantha*. In Nigeria, Oluwatobi (2020) intercropped two accessions of tomato and eggplant within immature oil palm alleys. The control plot exhibited the highest weed species richness at 6 weeks after intercropping, while the juvenile oil palm/tomato intercrop had the lowest Simpson's Index of Diversity (0.734). Intercropping resulted in a greater abundance of broadleaf weeds than grass species, with common species including *Mariscus alternifolius* and *Euphorbia heterophylla*. Recent research in Galicia, Spain, found that combining broadleaf trees like *Juglans regia L.* (walnut) with wheat (*Triticum aestivum*) enhances cereal production by reducing light availability in the understory, limiting weed establishment (Ferreiro-Domínguez et al., 2021). Conversely, Kaushal and Verma (2003) reported that *Grewia optiva* negatively impacted wheat growth beneath its canopy but improved with distance from the tree trunk. Choosing suitable tree species for agroforestry systems is crucial for maximizing positive interactions while minimizing negative effects (Gonçalves et al., 2021; Liu et al., 2018).

Casanova-Lugo et al. (2023) found that alley cropping systems involving *Leucaena leucocephala*, *Guazuma ulmifolia*, and *Moringa oleifera* with corn altered weed community composition, conserved biodiversity, and reduced dominant weed species without affecting corn yield. The *L. leucocephala*–corn system showed significant sustainability benefits, serving as a refuge for native plant species.

Nitrogen-fixing trees in agroforestry systems play a dual role by improving soil fertility and suppressing weeds through shading and competition. Species like *Gliricidia sepium* and *L. leucocephala* reduce weed density due to their dense canopies and rapid growth (Steppler & Nair, 1987). However, effects vary by species; for instance, *G. sepium* may also promote certain weeds through nitrogen facilitation (Callaway, 1995). While mulch from species like *Gliricidia* and *Senna siamea* is effective in reducing weed biomass (Kamara et al., 2000), others, like *L. leucocephala* with rapidly decomposing leaves, are less effective for weed suppression (Orwa et al., 2022). Mechanical weed control, such as manual removal or hoeing, remains necessary to maximize agroforestry's positive effects on crop yields (Baier et al., 2023).

Nitrogen-fixing trees reduce weed density and diversity through shading and allelopathic effects, as their dense canopies limit light availability and their leaf litter suppresses weed germination. However, careful selection and management are necessary to minimize competition with crops. While these trees can effectively suppress weeds, they require protection from weed competition during establishment, which can be achieved by soil improvement and applying sheet mulch at planting (Elevitch & Wilkinson, 2024).

Transitioning from monoculture to agroforestry affects weed dynamics by introducing diverse weed communities and altering microclimates through increased shade and humidity. This complexity requires holistic management approaches combining cultural, mechanical, and biological controls such as mowing, grazing, and cover crops (Liebman & Staver, 2001; Shirtliffe & Benaragama, 2014). In agroforestry, weed suppression is influenced by shading, organic matter, and allelopathy. However, the mechanisms behind these effects, particularly allelopathy, are not well understood and require further research to optimize tree–crop combinations and manage initial weed growth. Agroforestry systems can effectively reduce weed abundance through competition for resources like light and nutrients. Despite these benefits, long-term effects and species interactions need more study to refine weed management strategies.

2.2 Lowered Soil Disturbance

The reduction or elimination of soil tillage as a prescribed strategy in RA often results in increased weed density and species diversity, as farmers lose their mechanical tools against weeds, and creates a favorable environment for certain weed species that thrive in undisturbed soil. Mishra & Singh (2012) noted that, over time, zero-till (ZT) systems tended to support higher weed growth compared to conventional tillage (CT) in rice–wheat rotations. Although weed growth in the first year was not significantly greater under continuous ZT than under CT, by the third year, total weed dry mass was noticeably higher in the ZT system. In a 6-year field study, similarly, Steponavičienė et al. (Steponavičienė et al., 2021) reported an increase in the abundance of weed species in reduced and zero-till systems compared to those with deep ploughing. Research by Simon et al. (2019) and Chetan et al. (2022) also found that decreasing the intensity and depth of soil disturbance led to higher weed densities, particularly of perennial species. In a three-year study, Minhas et al. (2023) demonstrated that weed infestation levels and weed biomass in wheat crops were the highest under zero tillage (ZT), followed by minimum tillage (MT), and were the lowest under conventional tillage (CT). Woźniak and Soroka (2015) found that reduced tillage significantly enhanced the biomass and diversity of weed species in pea crops. In another experiment, these researchers observed that reduced tillage significantly enhanced weed diversity in spring wheat (Woźniak & Soroka, 2018). When tillage is eliminated, the growing environment tends to favor different types of weeds. In no-till (NT) systems, weed seeds remain on the soil surface, favoring the growth of perennial weeds and annual grasses that can germinate on the surface. In contrast, conventional tillage (CT) systems, which involve higher soil disturbance, tend to favor annual broadleaf species (Hume et al., 1991; Moyer et al., 1994; Bàrberi et al., 2001; Menalled et al., 2001; Tuesca et al., 2001; Taa et al., 2004). The lack of soil disturbance in NT systems may lead to an increase in small-seeded annual weeds that require light for germination (Moyer et al., 1994; Swanton et al., 1999). Studies have shown higher densities of small-seeded species such as *Amaranthus tuberculatus*, *Chamaesyce humistrata*, and *Lamium amplexicaule* in NT systems compared to CT systems (Govindasamy et al., 2020). Winter annual weeds are also more prevalent in NT systems (Barberi & Lo, 2001). Zhang and Wu (2021) found that, in NT systems with crop residue retention, graminoid weeds accounted for more than 85% of total weed biomass, likely due to increased soil moisture and reduced disturbance, compared to less than one-third in other treatments. The shallow-rooted nature of graminoid weeds limits their access to deeper soil water, whereas tap-rooted forbs can access deeper water, allowing them to endure drought conditions. No-till cropping systems can increase the persistence of perennial weeds due to the lack of disturbance to their underground structures (Armengot et al., 2016; Zhang & Wu, 2021). Kumar et al. (2022) found that annual weeds were more prevalent in conventional tillage (CT), while perennial weeds dominated in no-till (NT). The study also noted a shift toward greater monocot dominance and a slight decrease in dicots in NT systems.

While no-till (NT) systems can lead to an increase in perennial weeds, they do not consistently provide favorable conditions for them (Nichols et al., 2015). Surprisingly, some studies have shown that minimal soil disturbance may be less favorable for certain perennials like *Asclepias syriaca*, *Cyperus esculentus*, and *Helianthus tuberosus*, which proliferate when their underground parts are disturbed (Sharma et al., 2023; Swanton et al., 1993). In a 12-year study, Sharma et al. (2023) found that NT permanent beds in wheat fields reduced the occurrence of *C. esculentus* compared to conventional tillage (CT), possibly due to its unique ecological characteristics. Unlike *Cyperus rotundus*, *C. esculentus* grows quickly, germinates in crowded conditions, and has small tubers that can sprout immediately after soil disturbance. Reduced tillage practices can lead to the selection of weed species that complement the life cycles of agronomic crops (Chancellor, 1985; Dorado et al., 1999). In some cases, transitioning to no-till within crop rotations did not increase perennial weed growth, highlighting the importance of crop rotation in weed management, especially in reduced tillage systems (Derksen et al., 1993; Blackshaw et al., 2001; Tuesca et al., 2001).

Reduced tillage systems may also pose risks related to the increased presence of invasive weed species (Mayor & Maillard, 1995; Tørresen et al., 2003; Légère et al., 2005; Feledyn-Szewczyk et al., 2020). Studies have shown that stopping tillage can lead to a rapid increase in invasive weeds (Travlos et al., 2018). Feledyn-Szewczyk et al. (2020) found that invasive species like *Conyza canadensis* were more prevalent in reduced tillage and no-till systems compared to traditional ploughing. Similarly, Woźniak (2018) observed significant increases in *Echinochloa crus-galli* and *Viola arvensis* when harrowing replaced ploughing. Cussans (2022) also reported the emergence of invasive species like *Echinochloa crus-galli* following the adoption of conservation agriculture in the UK. Moreover, Grey et al. (2017) demonstrated that tillage alone could effectively manage the invasive potential of *Pennisetum purpureum*.

In regenerative agriculture (RA) systems, leaving seeds on the soil surface can lead to increased seed decay and predation. Weed seeds that remain at or near the surface often experience higher mortality rates, likely due to accelerated physiological aging or predation. Additionally, higher biological activity in minimally disturbed soils may promote seed decay and degradation. Even without seed predators, seeds from species like wild-proso millet (*Panicum miliaceum L.*), redroot pigweed (*Amaranthus retroflexus L.*), and hairy nightshade (*Solanum sarrachoides* Sendtner) lose viability faster when near the surface compared to those buried deeper (Peachey & Mallory-Smith, 2007, 2011).

In RA systems, prolonged reduction in soil disturbance can lead to compaction, which impacts crop and weed growth. Minhas et al. (2023) found that conventional tillage resulted in looser and more porous soils, while no-till (NT) had the opposite effect, potentially leading to bare soil and greater weed infestations. Yu et al. (2020) reported that tillage reduced soil bulk density (BD) by 5.19% and improved soil porosity by 5.69%. Similarly, Peixoto et al. (2020) and Rai et al. (2018) observed that minimal tillage lowered BD, soil carbon, and penetration resistance, whereas increased tillage intensity enhanced water infiltration and porosity.

Soil compaction can also influence weed competition, with some species more tolerant than crops. Shekoofa et al. (2020) noted that *Eleusine indica*, a common weed, thrived in low-oxygen environments caused by compaction. Mia et al. (2023) observed that some weeds were less affected by compaction than crops, and Everaarts (1992) found that planter-induced compaction did not hinder weed growth. Reintam and Kuht (2012) reported that while perennial species like *Spergula arvensis* and *Tripleurospermum perforatum* were less impacted by compaction, annual species like *Fumaria officinalis* experienced significant declines, with *Chenopodium album* increasing its biomass in compacted conditions.

Compaction also affects root development. Azam et al. (2014) found that garden pea roots remained near the soil surface in compacted soils, while *Acacia salicina*, a perennial, had more evenly distributed roots. Place et al. (2008) discovered that weeds such as *Amaranthus palmeri* and *Senna obtusifolia* penetrated compacted layers more effectively than soybean crops, accessing nitrogen more efficiently. Even when root growth was restricted, weed shoot growth remained stable, unlike soybeans, whose shoot growth declined under compaction. McCurdy et al. (2022) similarly found that *Juncus tenuis*, a perennial weed, persisted in compacted soils.

To manage compaction, some researchers propose occasional tillage or shallow cultivation to control perennials and cover crops (McPheeters et al., 2022; Mia et al., 2023; Server, 2021). This practice, known as strategic tillage (ST), involves periodic disturbance, typically every 4–10 years. While ST has been suggested, it deviates from the major strategies in RA (low soil disturbance) and can disrupt long-term system sustainability. Instead, use of cover crops with deep roots or rotating cash crops that reduce compaction is recommended as a more sustainable approach.

In no-till system (the pillar of RA), weed density and diversity increase as the main tillage-based weed management tool is lost. Many studies report higher weed growth in ZT systems compared with conventional systems, with a particular increase in perennial and annual grasses after 3 years of cropping. A common problem with no-till systems is compaction; this can compound weed problems and make weed management more difficult as many weed species are tolerant to low oxygen levels and can tolerate root restriction. Managing weeds in RA systems is a complex challenge; therefore, new approaches or a combination of old approaches is needed depending on the weed pressure and soil compaction issues. In RA systems, cover crops, strategic tillage, and crop rotations are among some of the main approaches used to manage weeds.

2.3 Cover Crops and Living Mulches

In regenerative agriculture (RA), cover crops play a key role in managing weeds by maximizing ground cover. This practice is known to enhance competition among crops, cover crops, and weeds, thereby suppressing weed seed germination and seedling emergence through the creation of a physical barrier that alters light availability and quality (Teasdale, 2018). Studies like those by Weisberger et al. (2023) in the southeastern USA have demonstrated the effectiveness of cover crops in reducing weed density. The accumulation of surface residues also helps conserve soil moisture and reduce light penetration, contributing to lower soil surface temperatures, as noted by Cox et al. (1990). For instance, Govindasamy et al. (2020) found that weeds such as *Amaranthus tuberculatus* and *Sorghum halepense* experienced delayed emergence in no-till systems due to reduced soil temperatures.

Refsell and Hartzler (2009) observed that it took 35 days for half of *A. tuberculatus* seedlings to emerge in no-till compared to 10 days in conventional tillage.

On the flip side, the environmental changes induced by cover crops can promote the growth of weed species that tolerate cooler temperatures, moisture, and shade. Weeds like *Amaranthus palmeri*, *Caperonia palustris*, and *Poa annua* are more likely to thrive under such conditions (Yasin et al., 2017). Wind-dispersed species are also more prevalent in no-till systems, as crop residues can trap their seeds (Derksen et al., 1994; Tuesca et al., 2001). Furthermore, certain cover crops can serve as hosts for parasitic weeds, increasing their presence. For instance, *Orobanche minor* and *Cuscuta campestris* have been observed growing on vetch and crimson clover cover crops, respectively, in the UK (Cussans, 2022).

The effects of cover crops on weed suppression can vary significantly depending on factors such as the cover crop species, timing of planting, density, biomass, and how the cover crop is managed. Climate and rotation with cash crops also play a role in shaping weed communities. In a Spanish vineyard, *Festuca arundinacea* and *Hordeum vulgare* were found to be particularly effective in suppressing *Cynodon dactylon*, while *F. arundinacea* was associated with fewer weed species overall (Recasens et al., 2023). This perennial cover crop created a simplified weed community, largely due to the lack of soil disturbance and tolerance to mechanical shredding. However, some weeds, like *Malva sylvestris* and *Picris echioides*, were able to withstand these control methods.

Despite the widespread belief in the weed-suppressive power of cover crops, there are cases where their use has led to increased weed densities. The efficacy of cover crops in weed management depends on multiple factors, including crop type, climate conditions, and the interaction between cover crops and weed species. Additionally, research shows that using a mixture of cover crop species tends to be more effective in controlling weeds than single-species cover crops (Fernando & Shrestha, 2023). Thus, while cover crops are a cornerstone of weed management in RA, their success requires careful planning and management tailored to specific ecological and farming conditions.

As previously noted, RA emphasizes maintaining soil cover year-round, primarily to prevent weeds from germinating and thriving without competition. One effective method for achieving this is by introducing a living mulch to the fields (Larsen, 2023). Unlike annual cover crops that are typically terminated before the main crop is planted, perennial living mulches are crops planted either before or alongside the main crop and are kept in place throughout the growing season, (Andrews et al., 2018). This mulch remains as a ground cover throughout the growing season and into subsequent crop rotations. This technique also shows promise for managing weeds in the understory of tree rows within regenerative systems (Jacquot & Parveaud, 2022) (Fig. 6.1). Living mulch offers multiple benefits, including weed suppression, regulating soil temperature, reducing evaporation, improving soil structure, preventing erosion, and aiding in nutrient retention and addition (Larsen, 2023). The species most often recommended for use as living mulches are those from the legume family—clover (*Trifolium repens*, *Trifolium pratense*, *Trifolium fragiferum*, and *Trifolium subterraneum L.*), alfalfa (*Medicago sativa L.*), winter

Fig. 6.1 Living mulches as understory in tree rows. Left: Yarrow as cover crop under peach trees (France); Middle: *Alchemilla* as living mulch under apple trees (Poland); Right: *Phuopsis stylosa* cultivated on apple tree rows (France). (Adapted from Jacquot and Parveaud (2022))

vetch (*Vicia villosa* Roth.), bird's-foot trefoil (*Lotus corniculatus L.*), pink serradella (*Ornithopus sativus* Brot)—and from the grasses family—perennial ryegrass (*Lolium perenne L.*), smooth-stalked meadowgrass (*Poa pratensis L.*), red fescue (*Festuca rubra L.*), and also wheat, rye, and barley. Others include winter rape (*Brassica napus L.*), charlock (*Sinapis arvensis L.*), tansy phacelia (*Phacelia tanacetifolia* Benth.), pot marigold (*Calendula officinalis L.*)—low-growing varieties and spreading tagetes (*Tagetes patula L.*) (Cougnon et al., 2022).

Cultivating cereals on living legume mulch is one of the several strategies used to lessen reliance on nitrogen fertilizers and herbicides in arable crop rotations (Cougnon et al., 2022). Specifically, growing winter cereals on perennial legume living mulches, such as white clover (*Trifolium repens L.*) or alfalfa (*Medicago sativa L.*), offers an approach to reduce the use of mineral nitrogen fertilizers and herbicides in winter cereal production (Amosse et al., 2013; Turlin, 2016). The goal is for the living mulch to suppress weed growth during the period between cereal harvest and sowing while also meeting part of the nitrogen needs of the cereals (Fig. 6.2).

Following the cereal harvest, the living mulch acts as a cover crop during the summer until the next cereal crop is established. Without light competition from the cereal crop, the mulch can develop quickly after harvest, assuming water availability. During this time, mowing or chopping the cover (leaving the cut material to decompose in the field) can help manage or slow the growth of perennial weeds like Canada thistle (*Cirsium arvense*) (Favreliere et al., 2020). Another option is cutting and harvesting the mulch for livestock feed or biogas production, although this removes nutrients from the field unless the manure or digestate is returned. Additionally, cutting or grazing the mulch before or shortly after cereal sowing is recommended to minimize competition for light between the mulch and the emerging crop (Jones & Clements, 1993). Sportelli et al. (2022) found that autonomous

Fig. 6.2 Maize production with Kentucky bluegrass as ground cover ((**a**) 31 May, (**b**) 13 June, and (**c**) 10 July 2018). Kentucky bluegrass was established in the fall of 2017. A grass-free zone was created in the spring of 2018. Maize hybrid, DeKalb DB57-75RIB (refuge in bag), was planted at 32,000 seeds per acre on April 30, 2018. Maize grown with the Kentucky bluegrass cultivar "Shannon" yielded at least 90% of the grain yield obtained from the non-PGC (Perennial Ground Cover) control plots without using any additional agronomic practices to control interspecific competition. (Adapted from Moore et al. (2019))

mowers improved organic vegetable production while offering energy savings compared to flaming and conventional tilling. The use of autonomous mowers maintained acceptable fresh marketable yields and reduced weed dry biomass, all with lower energy consumption.

The presence of living mulch can affect crop–weed competition, weed population dynamics, and weed life cycles. In addition to directly competing with weeds for resources like space, light, moisture, and nutrients, living mulch may also reduce weed seed banks by promoting invertebrate predation and microbial decay of weed seeds in the soil. This permanent groundcover lowers the red light ratio reaching the soil, which can inhibit weed seed germination and seedling growth, helping to reduce early-season weed interference and yield losses in maize crops (Gallandt et al., 2005; McCravy & Lundgren, 2011; Menalled et al., 2007) Using permanent ground cover systems in maize–soybean production could support resilient, ecologically based weed management strategies, cutting down herbicide use and minimizing its negative impact on soil, water, and habitats (Norsworthy et al., 2012). Additionally, reducing herbicide reliance may help prevent or delay the evolution of herbicide-resistant weed species, a major challenge for growers, with annual losses in maize and soybean production in North America estimated to be US$43 billion (Moore et al., 2019). Li et al. (2023) discovered that using perennial white clover (*Trifolium pratense L.*) as living mulch significantly enhanced the diversity and evenness of the soil bacterial community, traits that are commonly linked to improved soil health.

Despite all the benefits of establishing a living mulch system, Cougnon et al. (2022) noted that competition between main crops and the living mulch can cause significant yield losses if not properly managed. Even when competition between the crop and groundcover is minimized by separating their growing environments, some competitive effects still persist (Flynn et al., 2013). These effects are most pronounced during the crop's establishment phase when the groundcover is growing at its fastest rate. During this time, the groundcover impacts the crop's growth and development in a way similar to the effects seen when crops are established alongside weed species (Dzvene et al., 2023). Carof et al. (2007) investigated the impact of undersowing wheat with living mulches on wheat grain yield over three growing seasons. The treatments included wheat grown in conventionally tilled soil, no-till soil, and no-till soil with different living mulches, including red fescue, sheep's fescue, alfalfa, bird's-foot-trefoil, black medic, and white clover. The study found that using living mulches during wheat cropping reduced wheat yield by 19–81% compared to wheat grown without mulches. Cougnon et al. (2022) pointed out that the perennial legume varieties currently used as living mulch are bred for forage production. They suggested that a dedicated breeding program could develop mulch-specific varieties with less impact on cereal yields, enabling cereal production with reduced nitrogen and herbicide inputs. Overall, mulch–crop competition can lead to three potential outcomes for cash crop yield. In resource-limited environments, yield may decline as living mulch density increases (Pouryousef et al., 2015). Conversely, where weed pressure is high, yield may increase with greater living mulch density (Kaneko et al., 2011). In some cases, there may be no difference at either low or high living mulch densities, suggesting that the competitive effects of living mulch on the cash crop are similar to those of the weeds (Mohammadi, 2010).

Living mulches offer numerous environmental, agronomic, and economic benefits. However, proper management is necessary to prevent competition between service crops and cash crops (Sportelli et al., 2022). Wiggans et al. (2012) studied groundcover suppression strategies for no-till and zone-till maize grown with Kentucky bluegrass (*Poa pratensis L.*) and red fescue (*Festuca rubra L.*). The strategies involved preplant burn-down with paraquat followed by one or two post-emergence glyphosate applications. In the first year, a wet spring delayed maize planting, allowing the groundcover to recover, but maize in zone-tilled bluegrass suppressed with paraquat performed nearly as well as conventionally grown maize. In the following 2 years, some groundcover treatments matched the performance of conventional maize. They concluded that bluegrass, suppressed with paraquat in spring, was the least competitive, allowing maize yields comparable to conventional methods. The key finding was that despite spring benefits, groundcover behaves like a weed during maize establishment and needs suppression for maize to establish properly. The benefits of living mulches are highly dependent on how they are managed, including factors such as the species used, sowing time, seeding rates, tillage practices, weed composition, water and nutrient availability, and climatic conditions. The extent of these advantages is directly linked to the amount of biomass they produce (Sportelli et al., 2022).

In RA, cover crops as a physical barrier can suppress weeds through the alteration of light availability for weed seed germination and seedling emergence. However, there are some weed species that respond positively to cool and moist environmental conditions and, therefore, might be induced by cover crops. Knowledge gaps exist that could be addressed by experimentation with more species-specific cover crop management that relates the form, design, and timing of cover crops to crop and weed species and climate and cropping systems so that we can more clearly understand the factors that drive weed suppression and how to optimize weed-control cover crops.

2.4 Crop Diversification

Monoculture cropping depletes both soil organic matter and microbial diversity (Balota et al., 2004; Maillard et al., 2016), leading to increased weed infestations, worsened crop health, and ultimately reduced grain yield and quality (Hernández-Restrepo et al., 2016; Tillmann et al., 2017; Calzarano et al., 2018; Woźniak, 2020). While monocultures may simplify weed communities, leading to dominance by a few species (Moyer et al., 1994; Blackshaw et al., 2001), this often increases the risk of herbicide-resistant weeds. In contrast, diverse crop rotations, a key principle of regenerative agriculture, help disrupt weed cycles and decrease weed pressure (Minhas et al., 2023).

Woźniak (2020) found that weed density and biomass were significantly lower in rotation systems compared to continuous monoculture of durum wheat. Similarly, Minhas et al. (2023) observed the lowest weed infestations in sorghum–wheat rotations and the highest in fallow-wheat systems. The allelopathic effects of crops like sorghum and sunflower further reduced weed establishment. Shahzad et al. (2016) and Naeem et al. (2021a) reported similar benefits in wheat and barley rotations, respectively.

Long-term crop rotations can preserve weed biodiversity and prevent the dominance of noxious species, contributing to the protection of native segetal weeds threatened by intensive farming (Mhlanga et al., 2015). Nikolić et al. (2018) observed that crop rotations with winter wheat significantly influenced weed community composition without increasing invasive species. Weisberger et al. (2019) conducted a meta-analysis, revealing that more diverse rotations reduce weed density by 49%, especially under zero-tillage conditions.

Steponavičienė et al. (2021) found that crop rotation promoted weed species from various communities, creating new temporal niches and varying weed life cycles (Derrouch et al., 2021). Naeem et al. (2021a) highlighted that different crops in rotation can promote or suppress specific weed species, depending on factors such as allelopathy, soil disturbance, and nutrient uptake. For example, cotton's deep roots allow it to access nutrients from deep soil layers, affecting weed competition (Riaz Marral et al., 2020). Mung bean–barley rotations, which improve soil nutrients through nitrogen fixation, also enhanced weed diversity (Shahzad et al.,

2016). In barley-based systems, Naeem et al. (2021b) found that diverse crop rotations, such as sorghum–barley and maize–barley, effectively suppressed both broadleaf and grassy weeds due to their allelopathic properties.

2.5 Integration of Livestock

In regenerative agriculture, livestock plays a crucial role in maintaining and enhancing ecosystem health. By integrating livestock into farming systems, regenerative practices harness natural processes to improve soil fertility, enhance biodiversity, and support the sustainable management of land. Livestock contribute to nutrient cycling by returning organic matter to the soil through manure, which enriches soil structure and boosts microbial activity. Furthermore, the movement of animals across the land helps mimic natural grazing patterns, promoting diverse plant growth and preventing the overdominance of single species. Beyond their ecological benefits, livestock integration supports economic resilience for farmers by diversifying income sources and reducing dependency on external inputs, aligning with the principles of regenerative agriculture that prioritize long-term sustainability over short-term gains (Soussana & Lemaire, 2014; Teague & Barnes, 2017). These practices have the potential to increase yield, reduce costs, create new business opportunities, and enhance public perception of production agriculture. Animals can be managed in a way that stimulates plant photosynthesis and, thus, soil carbon sequestration and transforms plant residue from hard-to-decompose, carbon-rich material into low-carbon material that soil can use more readily. By restricting grazing duration and rotating animals through a series of paddocks, overgrazing is prevented, manure is recycled, and ecosystem and carbon benefits available through holistic livestock management accrue (Gilchrist, 2021). The grazing patterns of sheep and cattle can significantly boost grass production and carbon sequestration in rangelands. Regenerative rotational grazing of sheep has been shown to increase topsoil organic carbon by up to 3.6% and enhance springtime grass production by 30% (Díaz De Otálora et al., 2021).

Despite the potential benefits, incorporating grazing into hay and dryland grain production for weed and insect control has received considerably less attention (Hatfield et al., 2006). Many weeds are valuable forage plants, meeting most livestock nutritional requirements (Dora et al., 2008). Livestock can graze weeds directly or consume mown weeds, chaff, and screenings. Weeds can be incorporated into the diets of cattle and sheep to help meet their nutritional needs. Mature beef cows require about 7–8% crude protein, while sheep need around 9% for maintenance. Both species require approximately 55% of total digestible nutrients (TDNs) for maintenance (Vickers et al., 2017). There are weed species like Canada thistle and kochia (*Kochia scoparia*) that fall well within these nutritional ranges. Canada thistle is a nutritious plant, with nutritional quality comparable to grass-legume mixed pastures and even alfalfa. Kochia is a problematic, herbicide-resistant weed in western Canada and the United States, causing yield losses and harvesting issues due to its rapid spread and resistance to multiple herbicides (Geddes, 2022). However, kochia can be managed by integrating it into crop–livestock systems as a high-protein forage, potentially replacing alfalfa in beef and dairy diets. A study at the Lethbridge Research and Development Centre found that the timing of kochia harvest significantly affects its feed value and composition, with early-harvested kochia having degradability similar to alfalfa. Glyphosate-resistant kochia produced higher yields under herbicide treatments, suggesting that it could be an effective forage option, providing both weed control and nutritional benefits (Geddes, 2022). Additionally, weeds are resilient and often remain available as a forage source during drought conditions when other forages are scarce (Vickers et al., 2017).

Research on leafy spurge (*Euphorbia esula*) illustrates that grazing can enhance herbicide efficacy and insect biocontrol (Lym et al., 1997). Sheep grazing has shown potential to reduce weed recruitment following herbicide treatments. These findings suggest that grazing can enhance the effectiveness of integrated pest management (IPM) systems while reducing herbicide reliance (Frost & Launchbaugh, 2003). Livestock also provides indirect benefits in a weed management system by reclaiming otherwise useless resources like screenings, chaff, and mown patches, making these practices more viable for producers. Integrating livestock grazing into crop residue management offers several benefits for weed control and overall farm productivity (Frick & Johnson, 2002). Grazing crop residues can significantly reduce the number of weed seeds added to the seedbank, thereby helping to manage weed populations. Additionally, grazing can be an effective way to dispose of weed seed-contaminated fodder while simultaneously gaining nutritional value from it. Weed seeds within crop residues can provide a substantial portion of the nutritional value for livestock, although the feed value of the residue may vary. Post-harvest grazing also helps to reduce stubble burdens, potentially mitigating crop establishment problems in subsequent planting seasons. By incorporating grazing into crop residue management, farmers can enhance weed control, improve livestock nutrition, and streamline field preparation for future crops.

Grazing the animals together in a pasture can lead to more even consumption of pasture forage, which reduces weed pressure, and moving animals on a regular basis prevents overgrazing that reduces pasture health. Intensive grazing is the practice of keeping more animals on a smaller space but moving them to new pastures on a daily or more frequent interval. This allows for more forage to be consumed more evenly and greatly improves the soil health and reduces weed pressure (Ohletz, 2022). Livestock can enhance farming efficiency by consuming weeds or other forage plants and recycling the nutrients from these plants through manure, which supports other parts of the agroecosystem. Livestock waste, including both urine and manure, contributes to this nutrient recycling (Gassner & Dobie, 2022). Khan et al. (2013) analyzed 16 weed species in Pakistan and found that many, such as *Dichanthium annulatum*, *Avena fatua*, and *Amaranthus viridis*, are rich in essential minerals. Broadleaf weeds generally had higher mineral content than grasses, suggesting their potential as nutrient sources in livestock diets. Grasses like *A. fatua* provide good fiber, while broadleaf weeds like *A. viridis* offer substantial protein and minerals, benefiting both livestock and crop growth.

Different livestock species vary in their ability to eat and control weed species. Therefore, it is essential to align the animal's nutritional needs with land management goals. Sheep prefer forbs and broadleaf plants and are effective in controlling range weeds (Hatfield et al., 2006). Cattle are naturally not prolific weed eaters due to their large mouths and tongues, which make it difficult to strip leaves and consume small flower heads. Additionally, their digestive and metabolic systems are less effective at detoxifying harmful plant compounds found in many weedy forbs. However, some practitioners have managed to train cattle to eat broadleaf weeds despite these limitations (De Bruijn & Bork, 2006). Cattle can be useful for reducing biomass or removing volunteer grain. In contrast, sheep naturally prefer forbs over grasses and grasses over shrubs, making them suitable for consuming weedy forbs. Yet, sheep also graze on grasses, which some cattle producers may not prefer, limiting their use for weed control in cattle pastures. In legume crops, sheep will graze out grassy weeds. Goats, on the other hand, generally prefer shrubs and woody plants like aspen or rose over forbs or grasses and also eat thistles, making them less competitive with cattle for grasses but potentially undesirable in areas with shrubs and small trees, such as wooded riparian zones. Goats can also effectively remove weeds or volunteer crop plants from stubble. Geese can control grassy weeds in garden plots, and weeder geese (at five to six geese per hectare) can be used once crops are too large for birds to eat. Hogs, at 24 animals per hectare, can control perennial weeds between cropping seasons (Frick & Johnson, 2002; Hatfield et al., 2006).

Although livestock breed might affect their efficiency in weed control, few studies have explored the weed-eating abilities of different sheep or goat breeds (Popay & Field, 1996). Research in North Dakota examined leafy spurge consumption among various sheep breeds, finding only minor and inconsistent differences, indicating that most common breeds can graze leafy spurge effectively. White-face sheep, which form tighter flocks, are better for herded situations, while black-face sheep are more suited to areas with permanent fencing, limiting their large-scale weed control use. Goat breeds also differ in their herding tendencies, which can be influenced by training and production settings. Other breed selection factors include hardiness and meat or fiber characteristics, although these are secondary when the primary goal is weed control (Glasser et al., 2009; Launchbaugh et al., 1997; Olson & Launchbaugh, 2006; Pritz et al., 1997). Sheep of any breed, age, or background can be used for weed grazing if they are at an immature stage or if volunteer grain is the predominant weed. Research at Montana State University indicated that cull ewes, yearling rams, and wethers effectively removed weeds and volunteer crop plants (Hatfield et al., 2006).

The age of animals can influence their willingness to consume weeds. Young animals, particularly those learning from adult females, are more likely to graze weeds due to curiosity (Launchbaugh et al., 1997; Ralphs & Provenza, 1999; Grings et al., 2001; Mellado et al., 2005). For example, yearling sheep exposed to leafy spurge as lambs were more likely to graze it in early summer, suggesting that using experienced sheep can be beneficial for weed control (Olson & Launchbaugh, 2006).

The effectiveness of livestock in weed control also depends on the plant species and agricultural context. Cattle are suited for managing grasses, while goats excel at controlling invasive woody plants like juniper (*Juniperus scopulorum*) and mesquite (*Prosopis glandulosa*) due to their browsing ability (Michalsky et al., 2022). Sheep are effective against herbaceous weeds, particularly in areas where chemical control is challenging. Selecting the right species is crucial, as diet preferences vary among breeds and individuals based on nutritional needs and past experiences. Animals can be trained to target specific weeds, but their choices are ultimately driven by nutritional benefits, which change with age, sex, body condition, and physiological state.

The season significantly impacts the effectiveness of grazing for weed control. Grazing during early growth stages, typically in spring, is more effective in suppressing weed establishment and reducing seed production (Adams, 2019). Conversely, grazing in the dormant season helps manage perennial weeds by targeting regrowth (Chojnacky, 2014). Timing grazing to coincide with vulnerable weed stages maximizes its impact, contributing to long-term control. For example, intensive spring grazing can effectively manage palatable weeds like cheatgrass (*Bromus tectorum L.*) (Arterburn, 2020). Grazing programs should align with periods when the desired plant community is resilient to grazing.

The season and intensity of defoliation significantly affect plant regrowth post-grazing. Early in the growing season, plants are more resilient to grazing, but as they mature, grazing becomes more harmful (Bailey et al., 2019). Grazing should target weeds when they are most palatable and vulnerable, often during the bolting phase. However, this timing must consider the palatability and susceptibility of both weeds and desirable species to avoid giving weeds a competitive edge (Hendrickson & Olson, 2006). Adjusting the grazing season can help reduce weed seed production and minimize damage to desired vegetation.

Managed grazing involves systematically rotating livestock to different pastures, preventing overgrazing by allowing plants adequate recovery time. This practice limits livestock ability to selectively graze, leading to increased grazing of weeds, particularly when they are younger and more palatable. Regular grazing events typically disadvantage weeds, promoting the growth of more productive plants like grasses and legumes and eventually leading to weeds being "grazed out" of the pasture (Synnestvedt, 2022).

The primary factor influencing stocking rates is the density of weed infestation and the palatability of the plants. Sparse infestations of nutritious, palatable plants like spotted knapweed may be effectively managed with light stocking rates of sheep, leveraging their preference for these plants. Conversely, denser infestations or less palatable weeds may require heavier stocking rates to ensure more uniform forage utilization. In extremely dense infestations, animals are often concentrated (mob-stocked) to completely remove all forage, achieved through herding or temporary fencing until the desired results are attained (Walker et al., 2006).

A 2006 study in Canada evaluated three cattle grazing systems for controlling Canada thistle: (1) continuous or season-long grazing (SL), (2) short-duration (SD) rotational grazing, and (3) high intensity–low frequency (HILF) rotational grazing, which is similar to adaptive multi-paddock (AMP) grazing (De Bruijn & Bork, 2006). The findings revealed that SL grazing either maintained or worsened Canada thistle infestations and reduced forage yield. In contrast, HILF grazing significantly reduced Canada thistle shoot density, biomass, and flowering, outperforming the SD system. Two intense defoliations per year over 2–3 years nearly eliminated Canada thistle stems, with the remaining shoots being mainly vegetative and of higher forage quality under HILF grazing. The study concluded that high intensity–low frequency grazing could be an effective method for biological weed control. Adaptive multi-paddock (AMP) grazing involves high livestock densities for short periods followed by extended forage rest to stimulate accelerated grass growth. This system is not rigidly scheduled but adapts based on land and livestock responses, mimicking the natural patterns of wild ruminant herds driven by predation and food availability. Encouraging and training cattle to consume weed species, such as through AMP grazing, could provide environmental benefits by reducing herbicide and fuel use, while also improving cattle health and welfare due to the nutritional and medicinal properties of weeds.

Oil palm agriculture heavily relies on herbicides, which can harm biodiversity and ecosystems, necessitating alternative weed management strategies (Umar et al., 2023). Integrating cattle grazing into oil palm plantations offers a sustainable alternative, reducing herbicide use and greenhouse gas emissions while boosting rural economies. This approach involves rotational grazing, where cattle are moved every 6–8 days to control weeds biologically, enhancing biodiversity and managing nutrients (Azhar et al., 2021). This eco-weeding system provides farmers with additional income by reducing weeds and insect pests, serving as a viable alternative to chemical treatments in industrial plantations.

The duration of grazing depends on the specific weed and the surrounding vegetation. The objective is to graze in a way that damages the weed while supporting the growth of desirable plants. This often involves grazing small areas intensively for a short period and then moving livestock to a new area once the defoliation goals are achieved. It is also common to return later in the season to graze regrowth when desirable species are dormant (Launchbaugh & Walker, 2006).

Effective weed control through grazing typically requires several consecutive years, initially focusing on suppressing weeds over 2–5 years. Once weed populations are reduced, grazing intensity and frequency can be decreased for maintenance (Voth, 2024). Targeted grazing is a long-term strategy, not a quick fix. It is crucial to consider the health and well-being of livestock, as certain weeds such as common tansy, poison hemlock, and St. John's wort can be toxic. Managing forage crops and pasture systems, such as through rotational grazing or dietary supplements, can mitigate these risks.

While integrating livestock into a regenerative farming system offers several benefits, such as reducing weed pressure, poor management can lead to significant problems and undermine the advantages of this strategy. Integrating livestock into crop systems can effectively manage weeds and improve soil health, but its success depends on proper management. Well-managed livestock can reduce weed pressure, lower the need for herbicides, and enhance soil fertility by adding nutrients and organic matter. However, poor management risks crop damage, soil compaction, and increased runoff, while allowing problematic weeds to thrive. Livestock grazing, like any land use, can lead to unintended harm rather than restoration. Poor grazing management practices have been linked to the introduction and spread of invasive plants, degrading native plant communities and reducing their competitive ability (Frost & Launchbaugh, 2003). Larson et al. (2021) compared weed community responses in a sheep-grazed organic cropping system with two other systems: high-input conventional no-till and tilled organic, over a 5-year crop rotation. The study found that both the conventional no-till and tilled organic systems suppressed weed biomass and reduced species richness and diversity. In contrast, the grazed organic system led to higher weed biomass, species richness, and diversity. Initially, weed communities in the two organic systems were distinct from those in the conventional no-till system, but over time, the grazed organic system became unique from both the tilled and conventional systems. Weed biomass and diversity were the highest in sweet clover and the lowest in winter wheat, with safflower and lentil showing similar weed characteristics. The study suggests that while integrating livestock into cropping systems can enhance plant biodiversity, the increase in weed biomass makes grazing unsuitable as the sole weed management strategy, highlighting the need for it to be a part of a broader integrated weed management approach.

Effective weed management with livestock involves intensive rotational grazing to ensure even plant consumption, although it can be labor-intensive. Livestock are selective grazers, often preferring palatable plants over weeds, which necessitates careful planning to prevent overgrazing of desirable species. High-intensity rotational grazing forces animals to consume all available plants, preventing regrazing during regrowth. This approach helps maintain a balance between grazing and plant regrowth, reducing weed competition. Additional strategies like mowing or aversion training may be needed for less palatable weeds. Strategic placement of resources can encourage grazing in weedy areas, while protecting sensitive zones like riparian areas through fencing.

Despite all benefits, livestock can act as seed dispersal agents by transporting seeds attached to their coats or by ingesting and passing viable seeds through their digestive systems, as seen with weeds like houndstongue (*Cynoglossum officinale L.*) and cheatgrass (Frost & Launchbaugh, 2003). Grazing during seed set phases risks spreading seeds, but this can be minimized by avoiding infested areas during flowering and seeding. Temporarily confining animals can ensure complete seed passage before moving them to uninfested areas. While digestion reduces viable seeds, those that survive are deposited in nutrient-rich manure, enhancing germination chances (Frick & Johnson, 2002). For example, up to 22% of spotted knapweed (*Centaurea stoebe*) seeds remain viable after passing through sheep and mule deer digestive tracts (Frost & Launchbaugh, 2003). Furthermore, soil compaction is a concern when integrating livestock into cropping systems. Studies have shown that cattle grazing wheat fields can compact the soil to some degree, while sheep have less impact. Murphy et al. (1995) compared cattle and sheep grazing on smooth-stalked meadowgrass-dominant white clover sward and found that, at similar stock densities (32 animal units per acre), soil compaction was 81% greater with cattle than with sheep (Hatfield et al., 2006).

While there is a good understanding of the short-term benefits of managed grazing, however, there are still some knowledge gaps regarding the long-term effects of seed dispersal by livestock and uneven grazing of weeds and the potential for soil compaction. More research is needed to provide information to farmers on how to reduce the number of viable weed seeds spread by grazing animals, by grazing at the right time of year to avoid seed set and through confinement periods. Research could also look at how to improve rotational grazing systems by integrating complementary weed control methods (e.g., mowing and behavioral training) and to identify livestock stocking densities that minimize soil compaction, especially in situations where livestock are integrated into cropping systems.

2.6 *Predation and Seed Decay in Regenerative Agriculture*

Beneficial organisms have the potential to play a significant role in weed management. Regenerative agriculture plays a crucial role in attracting and supporting a diverse array of beneficial organisms, which collectively contribute to effective weed management (Lal, 2020; Rhodes, 2017). Encouraging these creatures involves maintaining their habitat, such as reducing tillage, preserving shelterbelts and wooded refuges, and leaving unbroken native land. Soil biological agents can affect competitive relationships between crops and weeds. Some agents, such as rhizobial inoculants for legumes or the fungal organism *Penicillium bilaiae* to solubilize inorganic phosphorus, improve crop growth. Arbuscular mycorrhizal fungi (AMF) are symbiotic organisms that colonize the roots of many plant species, forming a mutually beneficial relationship. AMF can benefit plants by facilitating nutrient uptake, improving growth and yield, and benefiting some species like cereals and legumes over those that do not associate with them, such as wild mustard, lambsquarters, wild buckwheat, tame mustard, canola, and quinoa (Frick & Johnson, 2002). AMF can influence weed communities in agroecosystems through various mechanisms, such as altering the relative abundance of mycotrophic AMF-host weed species compared to nonmycotrophic nonhost species (Jordan et al., 2000; Stampe & Daehler, 2003). During competition, both weeds and crops adjust their interactions with soil microorganisms, often relying more heavily on associations with soil symbiotic microbiota to enhance their growth (Massenssini et al., 2014; Trinchera & Warren Raffa, 2023).

The interaction with AMF has also been shown to reduce weed seed germination. A pot experiment involving *Striga hermonthica* and sorghum showed a 62% reduction in witchweed emergence and a 30% increase in sorghum yield after inoculation with *Glomus mosseae* (Gworgwor & Weber, 2003). Comparable results were observed in another study on the effects of AMF on witchweed in sorghum and maize cropping systems, which reported a 30% reduction in witchweed shoots and more than a 50% decrease in its presence in both maize and sorghum fields

(Lendzemo et al., 2005). Similarly, a microcosm study demonstrated that the seed germination of *Orobanche* and *Phelipanche* species was decreased in the presence of root exudates from pea plants colonized by *G. mosseae* and *Glomus intraradices* (Fernandez Aparicio et al., 2010). Another microcosm experiment with sunflower found that total weed biomass was reduced by 47% following AMF inoculation of the main crop (Rinaudo et al., 2010).

Studies have shown that high populations of arbuscular mycorrhizal fungi (AMF) tend to thrive in soils where specific conditions are met. These conditions include (i) avoiding bare-soil fallow periods, (ii) using minimal tillage and limiting inputs of synthetic fertilizers and certain high-phosphorus animal manures, and (iii) reducing the frequency of rotating to crops that are poor or nonhosts of AMF (Jordan et al., 2000). Research has demonstrated that maize planted following a winter cover crop experienced quicker and more extensive AMF colonization compared to maize planted after a winter fallow (Kabir & Koide, 2002; Deguchi et al., 2007). Additionally, extended fallow periods and tillage practices can reduce the capacity of AMF to colonize crops. Conventional high-input cropping systems frequently fail to meet conditions required for AMF flourish, which can significantly diminish both the diversity and abundance of AMF (Jordan et al., 2000; Schipanski et al., 2014; Trinchera & Warren Raffa, 2023). If AMF negatively impact weeds while having neutral or beneficial effects on crops under certain cropping system conditions, they could serve as a form of conservation biocontrol against weeds. To measure the effects of AMF on crops and agricultural weeds, Vatovec et al. (2005) examined how seedlings of 14 agronomic weed species responded to AMF inoculants, with samples taken from three different cropping systems: organic, transitional-organic, and high-input/conventional. The response of weed biomass to AMF varied

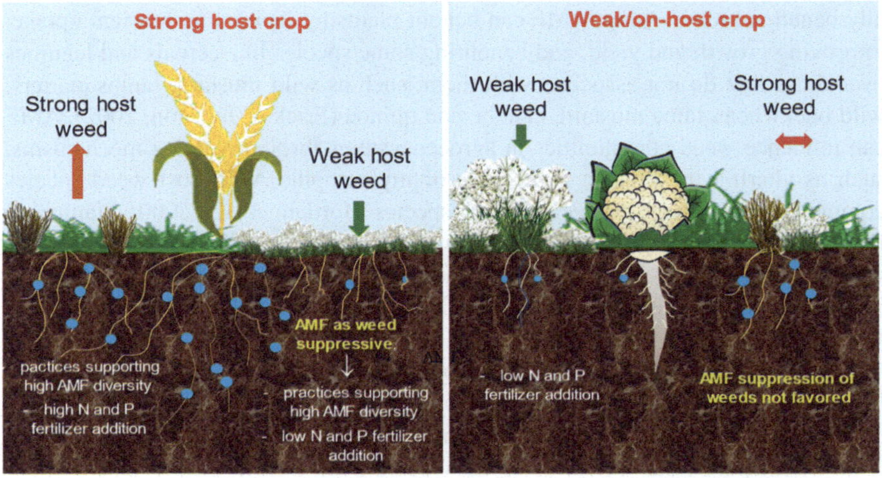

Fig. 6.3 Management strategies for using arbuscular mycorrhizal fungi (AMF) in weed control based on the host status of the plants and the composition of the crop–weed community. (Adapted from Vatovec et al. (2005))

significantly among species, influenced by specific traits, particularly whether they were strong, weak, or nonhost species. Additionally, the weed biomass response to AMF inoculation differed notably across cropping systems, indicating that more conservative management practices, such as organic farming, could significantly impact AMF–weed interactions (Fig. 6.3).

Two types of weed suppressive effects were identified across different cropping systems: (i) a direct effect, where AMF inoculation reduced the biomass of weak host weeds by favoring strong host weeds, and (ii) an indirect effect, where AMF enhanced the competitive ability of strong host crops (Trinchera & Warren Raffa, 2023).

Li et al. (2016) found that crops generally exhibit a greater growth response to AMF than agricultural weeds, especially with phosphorus fertilizer. Both C3 and C4 crops responded more positively than their weed counterparts, and mixed AMF communities were more beneficial than single species. In maize production, Li et al. (2019) observed that certain AMF taxa negatively affected specific weeds without harming maize, suggesting tailored inoculant use to exploit AMF's selective influence on weeds. This supports the strategic management of AMF for the biocontrol of agricultural weeds.

By emphasizing practices such as minimal soil disturbance, cover cropping, crop rotation, agroforestry, and organic soil amendments, regenerative agriculture creates a habitat that is conducive to the proliferation of these beneficial creatures (Schreefel et al., 2020; Wezel et al., 2015). Reduced tillage maintains soil structure and organic matter, fostering an environment where soil microbes and AMF can thrive, enhancing nutrient cycling and outcompeting weeds for resources (Van Der Heijden & Horton, 2009). Previous research has demonstrated that agricultural soils with high biological activity often possess weed-suppressive qualities. Nondisturbed soils generally exhibit higher biological activity than disturbed soils due to differences in soil nutrient levels, micro- and macro-elements, organic matter, pH, and conductivity (Parr et al., 1992; Kremer & Li, 2003; Bissett et al., 2011). As a result, it is anticipated that seed decay rates will differ more significantly between disturbed and nondisturbed soils, with nondisturbed soils exhibiting stronger weed-suppressive properties (Rowell & Florence, 1993).

Nikolić et al. (2020) studied the degradation of seeds from six common weed species in a no-till (NT) field and an adjacent undisturbed buffer zone in Italy. They found that seed decay and microbial cellulolytic activity were higher in the NT field soil than in the undisturbed buffer zone. Higher microbial activity in the NT field, which targets the lignin and cellulose in seed coats, likely contributed to increased seed decay. In contrast, the lower microbial activity in the buffer zone resulted in less seed degradation.

Agroforestry, a key strategy in regenerative agriculture, integrates trees and shrubs into agricultural landscapes, enhancing biodiversity and providing habitats for beneficial insects and wildlife that naturally control weed populations (Jose, 2009; Torralba et al., 2016). Forest species can form a mycorrhizal network that facilitates the exchange of nutrients and defense signals between plants (Simard et al., 2012). The shade and leaf litter from trees improve soil conditions, supporting beneficial soil organisms (Cardinael et al., 2017). Agroforestry systems create complex ecosystems that disrupt weed growth, reduce weed seed viability, and promote beneficial plant growth (Eichhorn et al., 2006). This approach promotes a balanced ecosystem where beneficial organisms suppress weeds, reducing the need for chemical herbicides and supporting long-term sustainability (Schipanski et al., 2014).

Other soil organisms can be detrimental to weeds. Seed-borne bacteria may reduce dormant weed seed populations, while rhizobacteria can reduce the vigor of grass weeds in cereal crops. Applying microbial agents to control weeds might prove useful in the future, and reducing tillage can foster the growth of these bacteria, which thrive on crop residues in cooler, moister environments (Frick & Johnson, 2002). In regenerative agriculture, crop residues left on the soil surface can increase fungal pathogen infections and/or seed predation, leading to a lower overall weed seed density compared to conventional farming systems. Typically, the habitat quality for potential predators is superior in no-till (NT) fields compared to conventionally tilled (CT) fields. Because seedbank depletion in NT systems relies heavily on seed predation and seed mortality, minimizing the addition of new seeds to the seedbank is crucial for reducing overall weed biomass (Moonen & Barberi, 2004).

Helpful insects can be encouraged by providing habitat, such as shelterbelts and ground cover. Maintaining the weed seeds on the soil surface due to low-soil disturbance and ground covered with residues from cover crops and cash crops in regenerative systems would provide a suitable environment for the activity of seed predators. Tillage impacts seed predation by (1) burying weed seeds beyond the reach of predators, (2) killing seed predators during the process, or (3) destroying their critical habitats, thereby reducing their survival (Brainard et al., 2013). Harrison et al. (2003) found that around 90% of giant ragweed (*Ambrosia trifida L.*) seeds on the soil surface in a no-till maize field were removed by predators. Additionally, *Harpalus rufipes* can consume up to 90% of the post-dispersal seeds of certain weed species (Zhang, 1993). In rainfed cereal fields in northeastern Spain, populations of harvester ants (*Messor barbarus L.*) are notably abundant, particularly in no-till and minimally tilled fields (Baraibar et al., 2009, 2011). Tillage disrupts weed seed predation, as ants and most other seed predators generally do not excavate to access buried seeds (Baraibar et al., 2017). Ghosh et al. (2023) compared weed seedbank dynamics between conventional tillage (CT) and no-till (NT) systems, finding that seedbank density was lower in NT systems. The reduction in seed numbers in NT systems can be attributed to factors such as effective chemical control, increased seed predation, and natural seed decay. With minimal soil disturbance, weed seeds tend to accumulate on the surface, making them more accessible to predators, which can enhance seed removal rates. Additionally, the absence of tillage may promote higher populations of seed predators. In NT fields, the number, diversity, and activity of seed-consuming organisms tend to be higher compared to conventionally tilled fields (Blubaugh & Kaplan, 2015), possibly due to improved habitat conditions (Baraibar et al., 2009) and lower mortality rates of weed seed predators (Shearin et al., 2007).

Seed predation is usually the highest when weed seeds are left on the soil surface and sufficient residue cover is present, as seen in NT systems. This environment can favor certain granivorous species, potentially shifting the composition of the weed seedbank (Nichols et al., 2015). Generalist predators, such as ground beetles and crickets, can reduce weed seed emergence by approximately 5–15% (White et al., 2007). Larger animals, including rodents and birds, can also consume substantial quantities of weed seeds. When seeds are buried and inaccessible to predators, they may be attacked by soil pathogens. For example, buried *Avena fatua* seeds showed increased mortality with higher soil moisture levels, reaching 55–88% after 2 years when soil moisture increased from 6 to 24% (Mickelson & Grey, 2006).

Increased weed seed predation has been observed with cover crops, as ground cover provides a habitat for seed predators (Teasdale, 2018). In fact, crop residues provide habitat and invite the diversity of beneficial insects, birds, and a wide range of invertebrates (e.g., earthworms, small rodents, birds, carabid beetles, and field crickets) (Andersen, 2003). High numbers of insects, snails, slugs, and voles in heavy residues of rye and hairy vetch contribute to weed seedling reduction. Some carabid insects show a preference for species of foxtail, aiding in weed control (Frick & Johnson, 2002). According to Kumar et al. (2022), leaving crop residues on the soil surface increases the population sizes of earthworms, ants, and overall floral and faunal diversity as the residues break down. Kelton et al. (2011) reported that using high-residue cover crops can reduce the number of weed seeds in the top 7.6 cm of the soil seedbank. Armengot et al. (2016) found that intensive cropping with frequent tillage in conventional-till (CT) systems, without the presence of crop residues, can decrease the activity and abundance of seed predators. Therefore, the presence of crop residues in regenerative agriculture (RA) can lead to an increase in predator populations over time, resulting in lower weed seed densities compared to CT systems.

Additionally, the diversity of crops planted in RA systems supports increased weed seed predation. Crop rotations that include phenologically different species have been shown to enhance predation pressure, and the timing of the crop cycle can influence the seasonality of seed predation (Heggenstaller et al., 2009; Westerman et al., 2011). Furthermore, specific crops appear to promote the predation of different weed seeds (Honek et al., 2003; Menalled et al., 2006; O'Rourke et al., 2006). For example, O'Rourke et al. (2006) observed that, in maize and soybean fields, invertebrates exhibited higher predation rates on *Setaria faberi* seeds compared to triticale–alfalfa and pure alfalfa fields. Seed predation rates were higher in reduced-input plots than in conventionally managed soybean fields. Invertebrates consumed less than 30% of seeds in July, early August, and early October, but in late August and September, they consumed up to 80–90% of seeds in maize and soybean fields. The proportion of *Setaria faberi* seeds consumed by invertebrates, compared to all seed predators, increased in late summer and was significantly higher in maize and soybean fields than in triticale–alfalfa and alfalfa treatments. These findings suggest that incorporating crops with diverse growth periods in rotation can enhance the consumption of a broader range of seeds by predators (Nichols et al., 2015).

Maintaining high soil biological activity through proper organic soil management in regenerative agriculture can potentially reduce the lifespan of weed seedbanks, as soil microorganisms are known to promote weed seed decay. Furthermore, incorporating a legume or other cover crops might lead to two possible outcomes: It could either stimulate weed seed germination by increasing soil nitrate levels or enhance the decomposition of weed seeds or seedlings, as soil microorganisms break down green manure residues (Menalled & Schonbeck, 2009).

Although AMF and seed predators help to control weeds by increasing nutrient uptake and suppressing weed populations in RA, a lot of research is needed to identify the specific mechanisms behind the AMF action in diverse weed communities, how to maximize them in agricultural systems, and the use of microbial agents to control weed populations. In particular, future research should characterize AMF species interactions with crops and weeds, investigate novel microbial applications, and implement field trials to evaluate how long-term regenerative agricultural practices affect weed dynamics and soil health.

3 Conclusion

In regenerative agriculture, effective weed management emerges as a multifaceted strategy involving the integration of cover crops, living mulch, rotation, low-soil disturbance, livestock, and agroforestry. These practices collectively contribute to a dynamic approach that not only suppresses weed growth but also enhances overall soil health and ecosystem resilience.

Cover crops play a pivotal role in weed management by suppressing weed emergence and growth through competition for light, water, and nutrients. They also enhance soil organic matter, improve soil structure, and foster beneficial soil microbial communities. By reducing soil erosion and nutrient leaching, cover crops contribute to a more stable soil environment that supports the growth of crops over weeds. This practice, however, requires careful selection and management to ensure that cover crops do not themselves become problematic weeds. Living mulch, similarly, offers substantial benefits by creating a continuous plant cover that outcompetes weeds for resources and prevents weed seed germination. Its ability to maintain soil moisture and suppress weed growth can be particularly effective when integrated into diverse cropping systems. The effective implementation of living mulch relies on choosing the right species and managing their growth to balance weed suppression with crop productivity.

Crop rotation and low-soil disturbance are integral components of regenerative agriculture that further enhance weed management. Crop rotation disrupts weed life cycles by varying the types of crops grown, which can limit the establishment and spread of weed species adapted to specific crops. This practice also reduces reliance on chemical inputs and promotes biodiversity. Low-soil disturbance, achieved through minimal tillage or no-till practices, preserves soil structure and organic matter, supports beneficial microorganisms, and reduces the germination of weed seeds by keeping them on the soil surface where they are more accessible to predators and microbial activity.

Livestock integration into farming systems adds another layer of weed management by utilizing grazing and browsing to control weed populations. Different livestock species, including cattle, sheep, goats, and poultry, offer varied weed control mechanisms through selective feeding and trampling. Properly managed grazing not only reduces weed biomass but also contributes to soil health by enhancing nutrient cycling and promoting plant diversity.

Agroforestry practices complement these strategies by introducing trees and shrubs into agricultural landscapes. These systems enhance biodiversity, provide habitat for beneficial organisms, and create microclimates that support soil health. Agroforestry systems can disrupt weed growth patterns, reduce weed seed viability, and improve competitive conditions for crops. The presence of trees and shrubs also contributes to improved soil structure, moisture retention, and organic matter content, further supporting the overall ecosystem's ability to manage weeds.

In conclusion, the integration of these diverse practices in regenerative agriculture creates a holistic approach to weed management that leverages the natural processes of ecosystems. By combining cover crops, living mulch, crop rotation, low-soil disturbance, livestock, and agroforestry, farmers can foster resilient and sustainable agricultural systems that not only manage weeds effectively but also enhance soil health and promote long-term productivity. This interconnected approach underscores the importance of managing agricultural landscapes in a way that harmonizes with natural processes and supports the broader goals of regenerative agriculture.

While regenerative agriculture offers many advantages, transitioning to herbicide-free systems poses several challenges and uncertainties. A key concern is how this shift will impact crop yields across different soil types and climates, as well as long-term productivity. The economic viability of these systems remains unclear due to a lack of comprehensive data, making it difficult for growers to assess long-term profitability. This is exacerbated by limited research on the costs and returns associated with herbicide-free farming, which complicates large-scale implementation. Furthermore, the effectiveness of weed management strategies in these systems varies significantly with local conditions, highlighting the need for integrated approaches that combine cultural, mechanical, and biological controls.

References

Adams, R. P. (2019). Juniperus of Canada and the United States: Taxonomy, key, and distribution. *Lundellia., 2019*(21), 1–9.

Amosse, C., Jeuffroy, M.-H., Celette, F., & David, C. (2013). Relay-intercropped forage legumes help to control weeds in organic grain production. *Eur J Agron, 49*, 158–167. https://doi.org/10.1016/j.eja.2013.04.002

Andersen, A. (2003). Long-term experiments with reduced tillage in spring cereals. II. Effects on pests and beneficial insects. *Crop Protection, 22*, 147–152.

Andrews, J. S., Sanders, Z. P., Cabrera, M. L., Saha, U. K., & Hill, N. S. (2018). Nitrogen dynamics in living mulch and annual cover crop corn production systems. *Agronomy Journal, 110*, 1309–1317. https://doi.org/10.2134/agronj2017.10.0609

Anoka, U. A., Akobundu, I. O., & Okonkwo, S. N. C. (1991). Effect of Gliricidia sepium (Jacq.) Steud. and Leucaena leucocephala (Lam.) de Wit on the development of *Imperata cylindrica* (L.) Raeuschel. *Agroforestry Systems, 16*, 1–12.

Anonymous. The dirt on climate change: Regenerative agriculture and health care; 2024. Available from: https://us.noharm.org/news/dirt-climate-change-regenerative-agriculture-and-health-care.

Armengot, L., Blanco-Moreno, J. M., Barberi, P., Bocci, G., Carlesi, S., Aendekerk, R., Berner, A., et al. (2016). Tillage as a driver of change in weed communities: A functional perspective. *Agriculture, Ecosystems and Environment, 222*, 276–285.

Arterburn J. The role of grazing in integrated weed management; 2020. Accessed from: https://www.agproud.com/articles/53638-the-role-of-grazing-in-integrated-weed-management.

Atangana, A., Khasa, D., Chang, S., & Degrande, A. (2014). Ecological interactions and productivity in agroforestry systems. In A. Atangana, D. Khasa, S. Chang, & A. Degrande (Eds.), *Tropical agroforestry* (pp. 151–172). Springer.

Azam, G., Grant, C. D., Murray, R. S., Nuberg, I. K., & Misra, R. K. (2014). Comparison of the penetration of primary and lateral roots of pea and different tree seedlings growing in hard soils. *Soil Research, 52*, 87–96.

Azhar, B., Tohiran, K. A., Nobilly, F., Zulkifli, R., Syakir, M. I., Ishak, Z., & Razi, N. (2021). Time to revisit oil palm-livestock integration in the wake of United Nations sustainable development goals (SDGs). *Frontiers in Sustainable Food Systems, 5*, 640285.

Baier, C., Gross, A., Thevs, N., & Glaser, B. (2023). Effects of agroforestry on grain yield of maize (*Zea mays* L.)- A global meta-analysis. *Frontiers in Sustainable Food Systems, 7*, 1167686. https://doi.org/10.3389/fsufs.2023.1167686

Bailey, D., Mosley, J. C., Estell, R. E., Cibils, A. F., Horney, M., Hendrickson, J. R., Walker, J. W., et al. (2019). Targeted livestock grazing: Prescription for healthy rangelands. *Rangeland Ecology & Management, 72*, 865–877.

Bainard, L. D., Klironomos, J. N., & Gordon, A. M. (2010). Arbuscular mycorrhizal fungi in tree-based intercropping systems: A review of their abundance and diversity. *Pedobiologia, 54*, 57–61.

Baker, C. J., Saxton, K. E., Ritchie, W. R., Chamen, W. C. T., Reicosky, D. C., Ribeiro, F., Justice, S. E., & Hobbs, P. R. (2007). In C. J. Baker & K. E. Saxton (Eds.), *No-tillage seeding in conservation agriculture* (2nd ed.). FAO and CAB International.

Balota, E. L., Kanashiro, M., Filho, A. C., Andrade, D. S., & Dick, R. P. (2004). Soil enzyme activities under long-term tillage and crop rotation systems in subtropical agroecosystems. *Brazilian Journal of Microbiology, 35*, 300–306.

Baraibar, B., Westerman, P. R., Carrión, E., & Recasens, J. (2009). Effects of tillage and irrigation in cereal fields on weed seed removal by seed predators. *Journal of Applied Ecology, 46*, 380–387.

Baraibar, B., Torra, J., & Westerman, P. R. (2011). Harvester ant (*Messor barbarus* (L.)) density as related to soil properties, topography and management in semi-arid cereals. *Applied Soil Ecology, 51*, 60–65.

Baraibar, B., Canadell, C., Torra, J., Royo-Esnal, A., & Recasens, J. (2017). Weed seed fate during summer fallow: The importance of seed predation and seed burial. *Weed Science, 65*, 515–524. https://doi.org/10.1614/WS-D-16-00031.1

Barberi, P., & Lo Cascio, B. (2001). Long-term tillage and crop rotation effects on weed seedbank size and composition. *Weed Research, 41*, 325–340.

Bàrberi, P., Bonari, E., Mazzoncini, M., García-Torres, L., Benites, J., Martínez, J. (2001). Weed density and composition in winter wheat as influenced by tillage systems. First World Congress on Conservation Agriculture, Madrid, Spain, 451–455.

Barrios, E., Kwesiga, F., Buresh, R. J., Sprent, J. I., & Coe, R. (1998). Relating preseason soil nitrogen to maize yield in tree legume-maize rotations. *Soil Science Society of America Journal., 62*, 1604–1609.

Batish, D. R., Singh, H. P., Kohli, R. K., & Kaur, S. (2008). Eucalyptus essential oil as a natural pesticide. *Forest Ecology and Management, 256*, 2166–2174.

Bissett, A., Richardson, A. E., Baker, G., & Thrall, P. H. (2011). Long-term land use effects on soil microbial community structure and function. *Applied Soil Ecology, 51*, 66–78.

Blackshaw, R., Larney, F., Lindwall, C., Watson, P., & Derksen, D. (2001). Tillage intensity and crop rotation affect weed community dynamics in a winter wheat cropping system. *Canadian Journal of Plant Science, 81*, 805–813.

Blubaugh, C. K., & Kaplan, I. (2015). Tillage compromises weed seed predator activity across developmental stages. *Biological Control, 81*, 76–82.

Boinot, S., Poulmarc'h, J., Mézière, D., Lauri, P. E., & Sarthou, J. P. (2019). Distribution of overwintering invertebrates in temperate agroforestry systems: Implications for biodiversity conservation and biological control of crop pests. *Agriculture, Ecosystems and Environment, 285*, 106630. https://doi.org/10.1016/j.agee.2019.106630

Brainard, D. C., Peachey, R. E., Haramoto, E. R., Luna, J. M., & Rangarajan, A. (2013). Weed ecology and nonchemical management under strip-tillage: Implications for northern U.S. vegetable cropping systems. *Weed Technology, 27*, 218–230.

Caamal-Maldonado, J. A., Jiménez-Osornio, J., Torres-Barragá, A., & Anaya, A. L. (2001). The use of allelopathic legume cover and mulch species for weed control in cropping systems. *Agronomy Journal, 93*, 27–36. https://doi.org/10.2134/agronj2001.93127x

Callaway, R. M. (1995). Positive interactions among plants. *The Botanical Review, 61*, 306–349.

Calzarano, F., Stagnari, F., D'Egidio, S., Pagnani, G., Galieni, A., Di Marco, S., Metruccio, E. G., et al. (2018). Durum wheat quality, yield and sanitary status under conservation agriculture. *Agriculture, 8*, 140.

Cardinael, R., Chevallier, T., Barthès, B. G., Saby, N. P. A., Parent, T., Dupraz, C., & Bernoux, M. (2017). Impact of alley cropping agroforestry on stocks, forms and spatial distribution of soil organic carbon- A case study in a Mediterranean context. *Geoderma, 289*, 29–38.

Carof, M., Tourdonnet, S. D., Saulas, P., Floch, D. L., & Roger-Estrade, J. (2007). Undersowing wheat with different living mulches in a no-till system. I. Yield analysis. *Agronomy for Sustainable Development., 27*, 347–356.

Carson, A. G. (1989). Effect of intercropping sorghum and groundnut on density of *Striga hermonthica* in The Gambia. *Tropical Pest Management, 35*, 130–132.

Casanova-Lugo, F., Lara-Perez, L. A., Dzib-Castillo, B., Caamal-Maldonado, B. J., Ramírez-Barajas, P. J., Cetzal-Ix, W. R., & Estrada-Medina, H. (2023). Alley cropping agroforestry systems change weed community composition and reduce dominant weed species associated with corn in southern Mexico. *Agriculture, Ecosystems and Environment, 350*, 108471.

Chancellor, R. (1985). Changes in the weed flora of an arable field cultivated for 20 years. *Journal of Applied Ecology, 22*, 491–501.

Chetan, F., Rusu, T., Chetan, C., Urdă, C., Rezi, R., Simon, A., & Bogdan, I. (2022). Influence of soil tillage systems on the yield and weeds infestation in the soybean crop. *Land., 11*, 1–13.

Chojnacky, D. C. (2014). Updated generalized biomass equations for North American tree species. *Forestry, 87*, 129–151.

Cougnon, M., Durand, J. L., Julier, B., Barre, P., & Litrico, I. (2022). Using perennial plant varieties for use as living mulch for winter cereals. A review. *Agronomy for Sustainable Development, 42*, 110. https://doi.org/10.1007/s13593-022-00844-x

Cox, W. J., Zobel, R. W., Van Es, H. M., & Otis, D. J. (1990). Tillage effects on some soil physical and corn physiological characteristics. *Agronomy Journal, 82*, 806.

Cussans, J. (2022). The threat of weeds for regenerative agriculture. *Groundswell Agriculture*. https://www.youtube.com/watch?v=Zb9kEUsVhYA. Accessed February 03, 2024

De Bruijn, S. L., & Bork, E. W. (2006). Biological control of Canada thistle in temperate pastures using high density rotational cattle grazing. *Biological Control, 36*, 305–315.

Deguchi, S., Shimazaki, Y., Uozumi, S., Tawaraya, K., Kawamoto, H., & Tanaka, O. (2007). White clover living mulch increases the yield of silage corn via arbuscular mycorrhizal fungus colonization. *Plant and Soil, 291*, 291–299.

Derksen, D. A., Lafond, G. P., Thomas, A. G., Loeppky, H. A., & Swanton, C. J. (1993). Impact of agronomic practices on weed communities: Tillage systems. *Weed Science, 41*, 409–417.

Derksen, D. A., Thomas, A. G., Lafond, G. F., Loeppky, H. A., & Swanton, C. L. (1994). Impact of agronomic practices on weed communities: Fallow within tillage systems. *Weed Science, 42*, 184–194.

Derrouch, D., Chauvel, B., Cordeau, S., & Dessaint, F. (2021). Functional shifts in weed community composition following adoption of conservation agriculture. *Weed Research, 62*(2), 103–112.

Dias, R. C., Santos, M. C., Evaristo, A. B., Santos, J. B., Frazão, L. A., & Silva, L. D. (2019). Weed management in agrosilvopastoral systems containing corn, palisadegrass, java, and eucalyptus. *Agroforestry Systems, 93*, 1339–1346. https://doi.org/10.1007/s10457-018-0235-5

Díaz De Otálora, X., Epelde, L., Arranz, J., Garbisu, C., Ruiz, R., & Mandaluniz, N. (2021). Regenerative rotational grazing management of dairy sheep increases springtime grass production and topsoil carbon storage. *Ecol Indic., 125*, 107484. https://doi.org/10.1016/j.ecolind.2021.107484

Dollinger, J., & Jose, S. (2018). Agroforestry for soil health. *Agroforestry Systems, 92*, 213–219. https://doi.org/10.1007/s10457-018-0223-9

Dora, G., Mendoza, S., Serrano, V., Bah, M., Pelz, R., Balderas, R., & Leon, F. (2008). Proximate composition, mineral content, and antioxidant properties of 14 Mexican weeds used as fodder. *Weed Biology and Management, 8*, 291–296.

Dorado, J., Del Monte, J., & Lopez-Fando, C. (1999). Weed seedbank response to crop rotation and tillage in semiarid agroecosystems. *Weed Science, 47*, 67–73.

Dudley, N., & Alexander, S. (2017). Agriculture and biodiversity: A review. *Biodiversity, 18*, 45–49. https://doi.org/10.1080/14888386.2017.1351892

Dzvene, A. R., Tesfuhuney, W. A., Walker, S., & Ceronio, G. (2023). Management of cover crop intercropping for live mulch on plant productivity and growth resources: A review. *Air, Soil and Water Research, 16*, 1–12.

Eichhorn, M. P., Paris, P., Herzog, F., Incoll, L. D., Liagre, F., Mantzanas, K., Mayus, M., et al. (2006). Silvoarable systems in Europe–past, present and future prospects. *Agroforestry Systems, 67*, 29–50.

Elevitch, G., Wilkinson, K. (2024). Nitrogen Fixing Trees - Multipurpose Pioneers. Available from: https://www.agroforestry.org/free-publications/nitrogen-fixing-trees (accessed 20th July 2024).

El-Khawas, S. A., & Shehata, M. M. (2005). The allelopathic potentialities of *Acacia nilotica* and *Eucalyptus rostrata* on monocot (*Zea mays* L.) and dicot (*Phaseolus vulgaris* L.) plants. *Biotechnology, 4*, 23–34.

Espinosa-Garcia, F. J., Martınez-Hernandez, E., & Quiroz-Flores, A. (2008). Allelopathic potential of *Eucalyptus* spp plantations on germination and early growth of annual crops. *Allelopathy Journal, 21*, 25–38.

Everaarts, A. P. (1992). Response of weeds to the method of fertilizer application on low-fertility acid soils in Suriname. *Weed Research, 32*, 391–397.

Favreliere, E., Ronceux, A., Pernel, J., & Meynard, J. M. (2020). Nonchemical control of a perennial weed, Cirsium arvense, in arable cropping systems. A review. *Agronomy for Sustainable Development, 40*, 31. https://doi.org/10.1007/s13593-020-00635-2

Feledyn-Szewczyk, B., Smagacz, J., Kwiatkowski, C. A., Harasim, E., & Woźniak, A. (2020). Weed flora and soil seed bank composition as affected by tillage system in three-year crop rotation. *Agriculture, 10*, 186.

Fernandez Aparicio, M., García-Garrido, J., Ocampo, J. A., & Rubiales, D. (2010). Colonisation of field pea roots by arbuscular mycorrhizal fungi reduces *Orobanche* and *Phelipanche* species seed germination. *Weed Research, 50*, 262–268.

Fernandez, O. N., Vignolio, O. R., & Requesens, E. C. (2002). Competition between corn (*Zea mays*) and bermudagrass (*Cynodon dactylon*) in relation to the crop plant arrangement. *Agronomie, 22*, 293–305.

Fernando, M., & Shrestha, A. (2023). The potential of cover crops for weed management: A sole tool or component of an integrated weed management system? *Plants., 12*, 752.

Ferreiro-Domínguez, N., Papadopoulus, P., Rigueiro-Rodíguez, A., & Mosquera-Losada, M. R. (2021). Wheat varieties established under walnut of different ages in Galicia (NW Spain). In *Proceedings of the 5th EURAF conference*, Nuoro.

Flynn, E. S., Moore, K. J., Singer, J., & Lamkey, K. R. (2013). Evaluation of grass and legume species as perennial ground covers in corn production. *Crop Science, 53*, 611–620.

Fownes, J. H., & Raynor, W. C. (1993). Seasonality and yield of breadfruit cultivars in the indigenous agroforestry system of Pohnpei, Federated States of Micronesia. *Trop Agric., 70*, 103–109.

Frick, B., & Johnson, E. (2002). Livestock and other beneficial organisms for weed management. Research report 2002, Canada-Saskatchewan. *Agri-Food Innovation Fund.*

Frost, R. A., & Launchbaugh, K. L. (2003). Prescription grazing for rangeland weed management. *Rangelands, 25*, 43–47.

Gallandt, E. R., Molloy, T., Lynch, R. P., & Drummond, F. A. (2005). Effect of cover-cropping systems on invertebrate seed predation. *Weed Science, 53*, 69–76.

Gassner A, Dobie P. eds. Agroforestry: A primer. Design and management principles for people and the environment. Bogor, Indonesia: Center for International Forestry Research (CIFOR) and Nairobi: World Agroforestry (ICRAF), 2022. doi:10.5716/cifor-icraf/BK.25114.

Geddes, C. (2022). Integrated crop-livestock systems may be effective tool for managing herbicide-resistant kochia. *Lethbridge Research and Development Centre*. Available from: https://canadianagronomist.ca/integrated-crop-livestock-systems-may-be

Ghosh, S., Das, T. K., Nath, C. P., Bhatia, A., Biswas, D. R., Bandyopadhay, K. K., Yeasin, M. D., et al. (2023). Weed seedbank, above-ground weed community and crop yields under conventional and conservation agriculture practices in maize–wheat–mungbean rotation. *Weed Research, 63*, 270–281.

Gilchrist J. The Promise of Regenerative Agriculture. Final report, 2021, E2, a program of NRDC.

Giller, K. E., Hijbeek, R., Andersson, J. A., & Sumberg, J. (2021). Regenerative agriculture: An agronomic perspective. *Outlook on Agriculture, 50*, 13–25.

Glasser, T. A., Ungar, E. D., Landau, S. Y., Perevolotsky, A., Muklada, H., & Walker, J. W. (2009). Breed and maternal effects on the intake of tannin-rich browse by juvenile domestic goats (*Capra hircus*). *Applied Animal Behaviour Science, 119*, 71–77. https://doi.org/10.1016/J.APPLANIM.2009.02.028

Gold, M. A., & Garrett, H. E. (2009). Agroforestry nomenclature, concepts, and practices. In *North American agroforestry: An integrated science and practice* (2nd ed., pp. 45–56). American Society of Agronomy.

Gonçalves, B., Morais, M. C., Pereira, S., Mosquera-Losada, M. R., & Santos, M. (2021). Tree–crop ecological and physiological interactions within climate change Contexts: A mini-review. *Frontiers in Ecology and Evolution, 9*, 661978. https://doi.org/10.3389/fevo.2021.661978

Gosnell, H., Charnley, S., & Stanley, P. (2020). Climate change mitigation as a co-benefit of regenerative ranching: Insights from Australia and the United States. *Interface Focus, 10*, 20190127.

Govindasamy, P., Sarangi, D., Provin, T., Hons, F. M., & Bagavathiannan, M. (2020). No-tillage altered weed species dynamics in a long-term (36-year) grain sorghum experiment in Southeast Texas. *Weed Science, 68*, 476–484.

Grey, T. L., Webster, T. M., Li, X., Anderson, W., & Cutts, G. S. (2017). Evaluation of control of napiergrass (Pennisetum purpureum) with tillage and herbicides. *Invasive Plant Science and Management, 8*, 393–400.

Grings, E. E., Short, R. E., Haferkamp, M. R., & Heitschmidt, R. K. (2001). Animal age and sex effects on diets of grazing cattle. *Journal of Range Management, 54*, 77–81. https://doi.org/10.2307/4003532

Gutiérrez Castillo, M., Harmand, J., & Dambrine, E. (2004). Disponibilidad de nitrógeno en el suelo bajo especies maderables y leguminosas usadas como sombra en sistemas de *Coffea arabica*. *Agroforesteria en Las Am., 11*, 69–76.

Gworgwor, N. A., & Weber, H. C. (2003). Arbuscular mycorrhizal fungi-parasite-host interaction for the control of *Striga hermonthica* (Del.) Benth. in sorghum [*Sorghum bicolor* (L.) Moench]. *Mycorrhiza, 13*, 277–281.

Harrison, S. K., Regnier, E. E., & Schmoll, J. T. (2003). Postdispersal predation of giant ragweed (*Ambrosia trifida*) seed in no-tillage corn. *Weed Science, 51*, 955–964.

Hatfield, P., Goosey, H., Lenssen, A., & Blodgett, S. (2006). Incorporating targeted grazing into farming systems. In K. Launchbaugh (Ed.), *Targeted grazing: A natural approach to vegetation management and landscape restoration* (pp. 129–141). ASI, A. Peischel and D.D. Henry, Jr..

Heggenstaller, A. H., Liebman, M., & Anex, R. P. (2009). Growth analysis of biomass production in sole-crop and double-crop corn systems. *Crop Science, 49*, 2215–2224.

Hendrickson, J., & Olson, B. (2006). Understanding plant response to grazing. In K. Launchbaugh (Ed.), *Targeted grazing: A natural approach to vegetation management and landscape restoration* (pp. 32–39). ASI, A. Peischel and D.D. Henry.

Hernandez-Restrepo, M., Groenewald, J. Z., Elliott, M. L., Canning, G., McMillan, V.E., & Crous, P. W. (2016). Take-all or nothing. *Studies in Mycology, 83*, 19–48.

Hillbrand, A., Borelli, S., Conigliaro, M., & Olivier, A. (2017). *Agroforestry for landscape restoration: Exploring the potential of agroforestry to enhance the sustainability and resilience of degraded landscapes*. Food and Agriculture Organization of the United Nations.

Honek, A., Martinkova, Z., & Jarosik, V. (2003). Ground beetles (*Carabidae*) as seed predators. *European Journal of Entomology, 100*, 531–544.

Hume, L., Tessier, S., & Dyck, F. (1991). Tillage and rotation influences on weed community composition in wheat (*Triticum aestivum* L.) in southwestern Saskatchewan. *Canadian Journal of Plant Science, 71*, 783–789.

Jacquot, M., & Parveaud, C. E. (2022). *Cover crops (living mulches) in the tree row of organic orchards: Why and how?* GRAB—Groupe de recherche en Agriculture Biologique.

Jama, B., & Getahun, A. (1992). Intercropping Acacia albida with maize (*Zea mays*) and green gram (*Phaseolus aureus*) at Mtwapa, Coast Province, Kenya. *Agroforest Systems, 14*, 193–205.

Jama, B., Getahun, A., & Ngugi, D. N. (1991). Shading effects of alley cropped *Leucaena leucocephala* on weed biomass and maize yield at Mtwapa, Coast Province, Kenya. *Agroforest Systems, 13*, 1–11.

Johansen, C., Haque, M. E., Bell, R. W., Thierfelder, C., & Esdaile, R. J. (2012). Conservation agriculture for small holder rainfed farming: Opportunities and constraints of new mechanized seeding systems. *Field Crops Research, 132*, 18–32.

Jones, L., & Clements, R. O. (1993). Development of a low input system for growing wheat (*Triticum vulgare*) in a permanent understorey of white clover (*Trifolium repens*). *Annals of Applied Biology, 123*, 109–119. https://doi.org/10.1111/j.1744-7348.1993.tb04078.x

Jordan, N. R., Zhang, J., & Huerd, S. (2000). Arbuscular-mycorrhizal fungi: Potential roles in weed management. *Weed Research, 40*, 397–410.

Jose, S. (2009). Agroforestry for ecosystem services and environmental benefits: An overview. *Agroforestry Systems, 76*, 1–10.

Kabir, Z., & Koide, R. T. (2002). Effect of autumn and winter mycorrhizal cover crops on soil properties, nutrient uptake and yield of sweet corn in Pennsylvania, USA. *Plant and Soil, 238*, 205–215.

Kamara, A. Y., Akobundu, I. O., Chikoye, D., & Jutzi, S. C. (2000). Selective control of weeds in an arable crop by mulches from some multipurpose trees in Southwestern Nigeria. *Agroforestry Systems, 50*, 17–26. https://doi.org/10.1023/A:1006425825597

Kaneko, M., Uozumi, S., Touno, E., & Deguchi, S. (2011). No-till, no-herbicide forage soybean (*Glycine max* (L.) Merrill) cropping system with an Italian ryegrass (*Lolium multiflorum* lam.) living mulch. *Grassland Science, 57*, 28–34. https://doi.org/10.1111/j.1744-697X.2010.00205.x

Kato-Noguchi, H., & Kurniadie, D. (2020). Allelopathy and allelopathic substances of mango (*Mangifera indica* L.). *Weed Biology and Management, 20*, 131–138. https://doi.org/10.1111/wbm.12212

Kaushal, R., & Verma, K. S. (2003). Tree-crop interaction studies in natural agroforestry system: A case study from western Himalayas in India. Paper presented at the XII World Forestry Congress, Quebec; 0234–B1.

Kelton, J. A., Price, A. J., van Santen, E., Balkcom, K. S., Arriaga, F. J., & Shaw, J. N. (2011). Weed seed bank density and composition in a tillage and landscape variability study. *Communications in Biometry and Crop Science, 6*, 21–30.

Khan, R., Khan, M. A., Sultan, S., Marwat, K. B., Khan, I., Hassan, G., & Shah, H. U. (2013). Nutritional quality of sixteen terrestrial weeds for the formulation of cost-effective animal feed. *JAPS., 23*, 75–79.

Koyejo, A. O., Okpara, D. A., & Agugo, B. A. C. (2021). Effect of alley cropping on soil, maize and mungbean grown under different maize spatial arrangements and mungbean spacings in south east Nigeria. *Agroforestry Systems, 97*(4), 1–10. https://doi.org/10.1007/s10457-021-00635-5

Kremer, R. J., & Li, J. (2003). Developing weed-suppressive soils through improved soil quality management. *Soil and Tillage Research, 72*, 193–202.

Kumar, H. D., & Aloke, P. (2020). Role of biostimulant formulations in crop production: An overview. *International Journal of Applied Research in Veterinary Medicine, 8*, 38–46.

Kumar, N., Nath, C. P., Hazra, K. K., Praharaj, C. S., Singh, S. S., & Singh, N. P. (2022). Long-term impact of zero-till residue management in post-rainy seasons after puddled rice and cropping intensification on weed seedbank, above-ground weed flora and crop productivity. *Ecological Engineering, 176*, 106540.

LaCanne, C. E., & Lundgren, J. G. (2018). Regenerative agriculture: Merging farming and natural resource conservation profitably. *PeerJ, 6*, e4428.

Lacerda, F., Miranda, I., Kato, O. R., Bispo, C. J. C., & Vale, I. D. (2013). Weed dynamics during the change of a degraded pasture to agroforestry system. *Agroforestry Systems, 87*(4). https://doi.org/10.1007/s10457-013-9607-z

Lal, R. (2020). Regenerative agriculture for food and climate. *Journal of Soil and Water Conservation, 75*, 123A–124A.

Larsen, F. V. (2023). The benefits of Lucerne as a perennial living mulch. *Agroganic*. Available from: https://agroganic.com/the-benefits-of-lucerne-as-a-perennial-living-mulch/

Larson, C. D., Menalled, F. D., Lehnhoff, E. A., & Seipel, T. (2021). Plant community responses to integrating livestock into a reduced-till organic cropping system. *Ecosphere, 12*, e03412.

Launchbaugh, K., & Walker, J. (2006). Targeted grazing—A new paradigm for livestock management. In K. Launchbaugh (Ed.), *Targeted grazing: A natural approach to vegetation management and landscape restoration* (pp. 2–9). ASI, A. Peischel and D.D. Henry, Jr.

Launchbaugh, K. L., Taylor, C. A., & Hohensee, S. D. (1997). Do different breeds of livestock have different dietary preferences? *Women and Natural Resources, 18*, 22–24.

Légère, A., Stevenson, F. C., & Benoit, D. L. (2005). Diversity and assembly of weed communities: Contrasting responses across cropping systems. *Weed Research, 45*, 303–315.

Lendzemo, V. W., Kuyper, T. W., Kropff, M. J., & van Ast, A. (2005). Field inoculation with arbuscular mycorrhizal fungi reduces *Striga hermonthica* performance on cereal crops and has the potential to contribute to integrated *Striga* management. *Field Crops Research, 91*, 51–61.

Li, M., Jordan, N. R., Koide, R. T., Yannarell, A. C., & Davis, A. S. (2016). Meta-analysis of crop and weed growth responses to arbuscular mycorrhizal fungi: Implications for integrated weed management. *Weed Science, 64*, 642–652.

Li, M., Jordan, N. R., Koide, R. T., Yannarell, A. C., & Davis, A. S. (2019). Interspecific variation in crop and weed responses to arbuscular mycorrhizal fungal community highlights opportunities for weed biocontrol. *Applied Soil Ecology, 142*, 34–42.

Li, H., Hill, N., & Wallace, J. (2023). A perennial living mulch system fosters a more diverse and balanced soil bacterial community. *PLoS One, 18*, e0290608. https://doi.org/10.1371/journal.pone.0290608

Liebman, M., & Staver, C. P. (2001). Crop diversification for weed management. In M. Liebman et al. (Eds.), *Ecological management of agricultural weeds* (pp. 322–374). Cambridge University Press.

Liu, C. L. C., Kucma, O., & Krutovsky, K. V. (2018). Mixed-species versus monocultures in plantation forestry: Development, benefits, ecosystem services and perspectives for the future. *Global Ecology and Conservation, 15*, e00419. https://doi.org/10.1016/j.gecco.2018.e00419

Lym, R. G., Sedivec, K. K., & Kirby, D. R. (1997). Leafy spurge control with angora goats and herbicides. *Journal of Range Management, 50*, 123–128.

Machado, V. D., Tuffi Santos, L. D., Santos, J. R., Mota, V. A., Padilha, S. V., & Santos, M. V. (2011). Fitossociologia de plantas daninhas em sistemas de integração de sorgo com braquiária sob diferentes formas de implantação da pastagem. *Planta Daninha., 29*, 85–94.

Maillard, E., Angers, D. A., Chantigny, M., Lafond, J., Pageau, D., Rochette, P., Lévesqueet, G., et al. (2016). Greater accumulation of soil organic carbon after liquid dairy manure application under cereal-forage rotation than cereal monoculture. *Agriculture, Ecosystems and Environment, 233*, 171–178.

Massenssini, A. M., Bonduki, V. H. A., Melo, C. A. D., Tótola, M. R., Ferreira, F. A., & Costa, M. D. (2014). Soil microorganisms and their role in the interactions between weeds and crops. *Planta Daninha, 32*, 873–884.

Mayor, J. P., & Maillard, A. (1995). Resultats d'un essai de culture sans labour depuis plus de 20 ans a changins. IV. Stock semencier et maitrise de la flore adventice. *Revue Suisse d'agric, 27*, 229–236.

McCravy, K. W., & Lundgren, J. G. (2011). Carabid beetles (*Coleoptera: carabidae*) of the Midwestern United States: A review and synthesis of recent research. *Terrestrial Arthropod Reviews, 4*, 63–94.

McCurdy, J. D., Small, Z. D., Tseng, T.-M., Brosnan, J. T., & Reasor, E. H. (2022). Effects of soil compaction and moisture on the growth of *Juncus tenuis*. *International Turfgrass Society Research Journal, 14*, 776–782.

McPheeters, D., Bruns, M. A., Karsten, H. D., & Dell, C. J. (2022). Integrated weed management with strategic tillage can maintain soil quality in continuous living cover systems. *Frontiers in Sustainable Food Systems, 6*, 907590.

Mellado, M., Rodríguez, A., Villarreal, J. A., Rodríguez, R., Salinas, J., & López, R. (2005). Gender and tooth wear effects on diets of grazing goats. *Small Ruminant Research, 57*, 105–114. https://doi.org/10.1016/J.SMALLRUMRES.2004.03.009

Menalled FD, Schonbeck M. Manage the weed seed Bank-minimize "deposits" and maximize "withdrawals". 2009. Accessed from: https://eorganic.org/node/2806

Menalled, F. D., Gross, K. L., & Hammond, M. (2001). Weed aboveground and seedbank community responses to agricultural management systems. *Ecological Applications, 11*, 1586–1601.

Menalled, F. D., Liebman, M., Renner, K. A., Singh, H., Batish, D., & Kohli, R. (2006). The ecology of weed seed predation in herbaceous crop systems. In H. P. Singh, D. R. Batish, & R. K. Kohli (Eds.), *Handbook of sustainable weed management* (pp. 297–327).

Menalled, F. D., Smith, R. G., Dauer, J. T., & Fox, T. B. (2007). Impact of agricultural management on carabid communities and weed seed predation. *Agriculture, Ecosystems and Environment, 118*, 49–54.

Mhlanga, B., Cheesman, S., Maasdorp, B., Muoni, T., Mabasa, S., Mangosho, E., & Thierfelder, C. (2015). Weed community responses to rotations with cover crops in maize-based conservation agriculture systems of Zimbabwe. *Crop Protection, 69*, 1–8.

Mia, M. S., Azam, G., Nouraei, S., & Borger, C. (2023). Strategic tillage in Australian conservation agricultural systems to address soil constraints: How does it impact weed management? *Weed Research, 63*, 12–26.

Michalsky, S., Neville, M., & Miller, A. J. (2022). *Targeted grazing: Plant and animal interactions*. Alberta EcoTrust Foundation.

Mickelson, J. A., & Grey, W. E. (2006). Effect of soil water content on wild oat (*Avena fatua* L.) seed mortality and seedling emergence. *Weed Science, 54*, 255–262.

Midega, C. A., Pittchar, J., Salifu, D., Pickett, J. A., & Khan, Z. R. (2013). Effects of mulching, N-fertilization and intercropping with *Desmodium uncinatum* on *Striga hermonthica* infestation in maize. *Crop Protection, 44*, 44–49.

Minhas, W. A., Mumtaz, N., Ur-Rehman, H., Farooq, S., Farooq, M., Ali, H. M., & Hussain, M. (2023). Weed infestation and productivity of wheat crop sown in various cropping systems under conventional and conservation tillage. *Frontiers in Plant Science, 14*, 1176738.

Mishra, J. S., & Singh, V. P. (2012). Tillage and weed control effects on productivity of a dry seeded rice–wheat system on a Vertisol in Central India. *Soil and Tillage Research, 123*, 11–20.

Mohammadi, G. R. (2010). Weed control in irrigated corn by hairy vetch interseeded at different rates and times. *Weed Biology and Management, 10*, 25–32. https://doi.org/10.1111/j.1445-6664.2010.00363.x

Moonen, A., & Barberi, P. (2004). Size and composition of the weed seedbank after 7 years of different cover-crop-maize management systems. *Weed Research, 44*, 163–177.

Moore, K. J., Anex, R. P., Elobeid, A. E., Fei, S., Flora, C. B., Goggi, A. S., Jacobs, K. L., et al. (2019). Regenerating agricultural landscapes with perennial groundcover for intensive crop production. *Agronomy, 9*, 458. https://doi.org/10.3390/agronomy9080458

Moyer, J., Roman, E., Lindwall, C., & Blackshaw, R. E. (1994). Weed management in conservation tillage systems for wheat production in North and South America. *Crop Protection, 13*, 243–259.

Mudge, K., & Gabriel, S. (2014). *Farming the woods: An integrated permaculture approach to growing food and medicinals in temperate forests*. Chelsea Green Publishing. ISBN 978-1-60-358507-1.

Murphy, W. M., Mena Barreto, A. D., Silman, J. P. & Dindal D.L. (1995). Cattle and sheep grazing effects on soil organisms, fertility, and compaction in a smooth-stalked meadowgrass-dominate white clover sward. Grass and Forage Science 50, 183-190.

Naeem, M., Hussain, M., Farooq, M., & Farooq, S. (2021a). Weed flora composition of different barley-based cropping systems under conventional and conservation tillage practices. *Phytoparasitica, 49*, 1–19.

Naeem, M., Mehboob, N., Farooq, M., Farooq, S., Hussain, S., Ali, H. M., & Hussain, M. (2021b). Impact of different barley-based cropping systems on soil physicochemical properties and barley growth under conventional and conservation tillage systems. *Agronomy, 11*, 8.

Nestel, D., & Altieri, M. A. (1992). The weed community of Mexican coffee agroecosystems: Effect of management upon plant biomass and species composition. *Acta Oecologica, 13*, 715–726.

Nichols, V., Verhulst, N., Cox, R., & Govaerts, B. (2015). Weed dynamics and conservation agriculture principles: A review. *Field Crops Research, 183*, 56–68.

Nikolić, L., Šeremesic, S., Milosev, D., Dalovic, I., & Latkovic, D. (2018). Weed infestation and biodiversity of winter wheat under the effect of long-term crop rotation. *Applied Ecology and Environmental Research, 16*, 1413–1426.

Nikolić, N., Squartini, A., Concheri, G., Stevanato, P., Zanin, G., & Masin, R. (2020). Weed seed decay in no-till field and planted riparian buffer zone. *Plants, 9*(3), 293.

Norsworthy, J. K., Ward, S. M., Shaw, D. R., & Llewellyn, R. S. (2012). Reducing the risks of herbicide resistance: Best management practices and recommendations. *Weed Science, 60*, 31–62.

O'Rourke, M. E., Heggenstaller, A. H., Liebman, M., & Rice, M. E. (2006). Post-dispersal weed seed predation by invertebrates in conventional and low-external-input crop rotation systems. *Agriculture, Ecosystems and Environment, 116*, 280–288.

Ohletz, J. L. (2022). Regenerative agriculture. In P. Scharf (Ed.), *Encyclopedia of smart agriculture technologies*. Springer. https://doi.org/10.1007/978-3-030-89123-7_252-1

Olson, B., & Launchbaugh, K. (2006). Managing herbaceous broadleaf weeds with targeted grazing. In K. Launchbaugh (Ed.), *Targeted grazing: A natural approach to vegetation management and landscape restoration* (pp. 57–66). ASI, A. Peischel and D.D. Henry, Jr..

Oluwatobi, A. S. (2020). Weed species composition and diversity of 2-year old oil palm tree/fruit vegetables intercrop in rainforest zone of Nigeria. *Preprint*. https://doi.org/10.1101/2020.05.31.126722

Oluwatobi, A. S., & Olorunmaiye, K. S. (2014). Weed species distribution of juvenile oil palm tree (*Elaeis guineensis*) intercropped with maize (*Zea mays*), okra (*Abelmoshus esculentus*) and pepper (*Capsicum anuum var. abbreviatum*). *Notulae Scientia Biologicae, 6*, 483–490. https://doi.org/10.1583/nsb649358

Ong, C. K., & Rao, M. R. (2001). Management of complex interactions for growth resources and of biotic stresses in agroforestry. In J. Nosberger, H. H. Geiger, & P. C. Struik (Eds.), *Crop science: Progress and prospects* (pp. 175–190). CABI International.

Orwa C, Mutua A, Kindt R, Jamnadass R, Simons A. Agroforestree database: A tree reference and selection guide. Version 4.0. 2022., Available from: https://apps.worldagroforestry.org/treedb2/speciesprofile.php?Spid=18109.

Parr, F. J., Papendick, I. R., Hornick, B. S., & Meyer, E. R. (1992). Soil quality: Attributes and relationship to alternative and sustainable agriculture. *American Journal of Alternative Agriculture, 7*, 5–11.

Peachey, R. E., & Mallory-Smith, C. (2007). Influence of winter seed position and recovery date on hairy nightshade (*Solanum sarrachoides*) recruitment and seed germination, dormancy, and mortality. *Weed Science, 55*, 49–59.

Peachey, R. E., & Mallory-Smith, C. (2011). Effect of fall tillage and cover crop strategies on wild-proso millet (*Panicum miliaceum*) emergence and interference in snap beans. *Weed Technology, 25*, 119–126.

Peixoto, D. S., da Silva, L. D. C. M., de LBB, M., et al. (2020). Occasional tillage in no-tillage systems: A global meta-analysis. *Science of the Total Environment, 745*, 140887.

Place, G., Bowman, D., Burton, M. P., & Rufty, T. (2008). Root penetration through a high bulk density soil layer: Differential response of a crop and weed species. *Plant and Soil, 307*, 179–190.

Popay, I., & Field, R. (1996). Grazing animals as weed control agents. *Weed Technology, 10*, 217–231.

Pouryousef, M., Yousefi, A. R., Oveisi, M., & Asadi, F. (2015). Intercropping of fenugreek as living mulch at different densities for weed suppression in coriander. *Crop Protection, 69*, 60–64. https://doi.org/10.1016/j.cropro.2014.12.004

Pritz, R. K., Launchbaugh, K. L., & Taylor, C. A. (1997). Effects of breed and dietary experience on juniperconsumption by goats. *Journal of Range Management, 50*, 600–606. https://doi.org/10.2307/4003454

Pumariño, L., Sileshi, G. W., Gripenberg, S., Kaartinen, R., Barrios, E., Muchane, M. N., & Midega, C. (2015). Effects of agroforestry on pest, disease and weed control: A meta-analysis. *Basic and Applied Ecology, 16*, 573–582. https://doi.org/10.1016/j.baae.2015.08.006

Rai, V., Pramanik, P., Aggarwal, P., Krishnan, P., & Bhattacharyya, R. (2018). Effect of conservation agriculture on soil physical health. *International Journal of Current Microbiology and Applied Sciences, 7*, 373–389.

Ralphs, M. H., & Provenza, F. D. (1999). Conditioned food aversions: Principles and practices, with special reference to social facilitation. *The Proceedings of the Nutrition Society, 58*, 813–820. https://doi.org/10.1017/S002966519900110X

Rao, M. R., Nair, P. K. R., & Ong, C. K. (1998). Biophysical interactions in tropical agroforestry systems. *Agroforestry Systems, 38*, 3–50.

Recasens, J., Valencia-Gredilla, F., Cabrera-Pérez, C., Baraibar, B., & Royo-Esnal, A. (2023). Dynamics of *Cynodon dactylon* and weed community composition in different cover crops in a vineyard. *Weed Research, 63*, 261–269.

Refsell, D. E., & Hartzler, R. G. (2009). Effect of tillage on common waterhemp (*Amaranthus rudis*) emergence and vertical distribution of seed in the soil. *Weed Technology, 23*, 129–133.

Reintam, E., & Kuht, J. (2012). Weed responses to soil compaction and crop management. In A. J. Price (Ed.), *Weed control* (pp. 243–264). InTech.

Rhodes, C. J. (2017). The imperative for regenerative agriculture. *Science Progress, 100*, 80–129.

Riaz Marral, M. W., Khan, M. B., Ahmad, F., Farooq, S., & Hussain, M. (2020). The influence of transgenic (Bt) and non-transgenic (non-Bt) cotton mulches on weed dynamics, soil properties and productivity of different winter crops. *PloS ONE, 15*, e0238716.

Rinaudo, V., Bàrberi, P., Giovannetti, M., & van der Heijden, M. G. A. (2010). Mycorrhizal fungi suppress aggressive agricultural weeds. *Plant and Soil, 333*, 7–20.

Rippin, M., Haggar, J. P., Kass, D., & Kopke, U. (1994). Alley cropping and mulch with *Erythrina poeppigiana* (Walp.) O.F. Cook and *Gliricidia sepium* (Jacq.) Walp.: Effects on maize=weed competition. *Agroforestry Systems, 25*, 119–134.

Rowell, M. J., & Florence, L. Z. (1993). Characteristics associated with differences between undisturbed and industrially-disturbed soils. *Soil Biology and Biochemistry, 25*, 1499–1511.

Schipanski, M. E., Barbercheck, M., Douglas, M. R., Finney, D. M., Haider, K., Kaye, J. P., Kemanian, A. R., et al. (2014). A framework for evaluating ecosystem services provided by cover crops in agroecosystems. *Agricultural Systems, 125*, 12–22.

Schreefel, L., Schulte, R. P. O., de Boer, I. J. M., Schrijver, A. P., & van Zanten, H. H. E. (2020). Regenerative agriculture-the soil is the base. *Global Food Security, 26*, 100404.

Scott, D., & Freckleton, R. P. (2022). Crop diversification and parasitic weed abundance: A global meta-analysis. *Scientific Reports, 12*, 19413. https://doi.org/10.1038/s41598-022-24047-2

Server, M. (2021). *Strategic tillage has its place in no-till agriculture. Crops and Soils Magazine*. https://doi.org/10.1002/crso.20150

Shahzad, M., Farooq, M., Jabran, K., & Hussain, M. (2016). Impact of different crop rotations and tillage systems on weed infestation and productivity of bread wheat. *Crop Protection, 89*, 161–169.

Sharma, P, & Verma, K. (2019). Weed management through agroforestry systems. 3rd International Conference Global Initiatives in Agricultural and Applied Sciences for Eco Friendly Environment (GIASE-2019).

Sharma, T., Das, T. K., Maity, P. P., Biswas, S., Sudhishri, S., Govindasamy, P., & Raj, R. (2023). Long-term conservation agriculture influences weed diversity, water productivity, grain yield, and energy budgeting of wheat in North-Western Indo-Gangetic Plains. *Sustainablity, 15*, 7290.

Shearin, A., Reberg-Horton, S., & Gallandt, E. (2007). Direct effects of tillage on the activity and density of ground beetle (Coleoptera: Carabidae) weed seed predators. *Environmental Entomology, 36*, 1140–1146.

Shekoofa, A., Brosnan, J. T., Vargas, J. J., Tuck, D. P., & Elmore, M. T. (2020). Environmental effects on efficacy of herbicides for postemergence goosegrass (*Eleusine indica*) control. *Scientific Reports, 10*, 20579.

Shirtliffe, S. J., & Benaragama, D. (2014). Sometimes you need a big hammer: Evaluating and appraising selected non-herbicidal weed control methods in an integrated weed management system. In R. C. Martin & R. MacRae (Eds.), Managing energy, nutrients, and pests in organic field crops (1st ed., Chapter 7). CRC Press.

Shukla, A., Kumar, A., Jha, A., Dhyani, S. K., & Vyas, D. (2012). Cumulative effects of tree-based intercropping on arbuscular mycorrhizal fungi. *Biology and Fertility of Soils, 48*, 899–909. https://doi.org/10.1007/s00374-012-0682-5

Sileshi, G., & Mafongoya, P. L. (2003). Effect of rotational fallows on abundance of soil insects and weeds in maize crops in eastern Zambia. *Applied Soil Ecology, 23*, 211–222.

Sileshi, G., Mafongoya, P. L., Chintu, R., & Akinnifesi, F. K. (2008). Mixed-species legume fallows affect faunal abundance and richness and N cycling compared to single species in maize fallow rotation. *Soil Biology and Biochemistry, 40*, 3065–3075.

Simard, S. W., Beiler, K. J., Bingham, M. A., Deslippe, J. R., Philip, L. J., & Teste, F. P. (2012). Mycorrhizal networks: Mechanisms, ecology and modelling. *Fungal Biology Reviews, 26*, 39–60.

Simon, A., Chetan, F. & Chetan C. (2019) Influence of tillage systems on the degree of weeding on agricultural crops at ARDS Turda. *Agriculture, 76*(2), 99–106.

Singh, V., Johar, V., Kumar, R., & Chaudhary, M. (2021). Socio-economic and environmental assets sustainability by agroforestry systems: A review. *IJAEB, 14*, 521–533. https://doi.org/10.30954/0974-1712.04.2021.6

Soussana, J. F., & Lemaire, G. (2014). Coupling carbon and nitrogen cycles in European grasslands through multi-functionality: Towards an integrated ecosystem approach. *European Journal of Agronomy, 57*, 2–14.

Sportelli, M., Frasconi, C., Fontanelli, M., Pirchio, M., Gagliardi, L., Raffaelli, M., Peruzzi, A., et al. (2022). Innovative living mulch management strategies for organic conservation field vegetables: evaluation of continuous mowing, flaming, and tillage performances. *Agronomy Journal, 12*, 622. https://doi.org/10.3390/agronomy12030622

Stampe, E. D., & Daehler, C. C. (2003). Mycorrhizal species identity affects plant community structure and invasion: A microcosm study. *Oikos, 100*, 362–372.

Steponavičienė, V., Marcinkeviciene, A., Butkeviciene, L. M., Skinulienė, L., & Bogužas, V. (2021). The effect of different soil tillage systems and crop residues on the composition of weed communities. *Agron., 11*, 1276.

Steppler, H. A., & Nair, P. K. R. (1987). *Agroforestry a decade of development*. International Council for Research in Agroforestry, ICRAF.

Summers, H., Karsten, H. D., Curran, W., & Malcolm, G. M. (2021). Integrated weed management with reduced herbicides in a no-till dairy rotation. *Agronomy Journal, 113*, 3418–3433.

Swanton, C. J., & Weise, S. F. (1991). Integrated weed management: The rationale and approach. *Weed Technology, 5*, 657–663.

Swanton, C. J., Clements, D. R., & Derksen, D. A. (1993). Weed succession under conservation tillage: A hierarchical framework for research and management. *Weed Technology, 7*, 286–297.

Swanton, C. J., Shrestha, A., Roy, R. C., Ball-Coelho, B. R., & Knezevic, S. Z. (1999). Effect of tillage systems, N, and cover crop on the composition of weed flora. *Weed Science, 47*, 454–461.

Synnestvedt, P. (2022). The economic and environmental benefits of managed grazing. Accessed from: https://www.nofence.no/en/blog/the-economic-and-environmental-benefits-of-managed-grazing.

Taa, A., Tanner, D., & Bennie, A. T. (2004). Effects of stubble management, tillage andcropping sequence on wheat production in the south-eastern highlands of Ethiopia. *Soil and Tillage Research, 76*, 69–82.

Teague, W. R., & Barnes, M. K. (2017). Grazing management that regenerates ecosystem function and grazingland livelihoods. *African Journal of Range and Forage Science, 34*, 77–86.

Teasdale, J. R. (2018). The use of rotations and cover crops to manage weeds. In R. Zimdahl (Ed.), *Integrated weed management for sustainable agriculture* (pp. 227–260). Burleigh Dodds Science.

Tillmann, M., von Tiedemann, A., & Winter, M. (2017). Crop rotation effects on incidence and diversity of Fusarium species colonizing stem bases and grains of winter wheat. *Journal of Plant Diseases and Protection, 124*, 121–130.

Torralba, M., Fagerholm, N., Burgess, P. J., Moreno, G., & Plieninger, T. (2016). Do European agroforestry systems enhance biodiversity and ecosystem services? A meta-analysis. *Agriculture, Ecosystems & Environment, 230*, 150–161.

Tørresen, K. S., Skuterud, R., Tandsaether, H. J., & Hagemo, M. B. (2003). Long-term experiments with reduced tillage in spring cereals. I. Effects on weed flora, weed seedbank and grain yield. *Crop Protection, 22*, 185–200.

Travlos, I. S., Cheimona, N., Roussis, I., & Bilalis, D. J. (2018). Weed-species abundance and diversity indices in relation to tillage systems and fertilization. *Frontiers in Environmental Science, 6*, 11.

Trinchera, A., & Warren Raffa, D. (2023). Weeds: An insidious enemy or a tool to boost mycorrhization in cropping systems? *Microorganisms., 11*, 334. https://doi.org/10.3390/microorganisms11020334

Tscharntke, T., Clough, Y., Bhagwat, S. A., Buchori, D., Faust, H., Hertel, D., Holscher, D., et al. (2011). Multifunctional shade-tree management in tropical agroforestry landscapes—A review. *Journal of Applied Ecology, 48*, 619–629.

Tuesca, D., Puricelli, E., & Papa, J. (2001). A long-term study of weed flora shifts in different tillage systems. *Weed Research, 41*, 369–382.

Turlin, J. P. (2016). Agriculture de conservation des sols: semis direct sous couvert permanent de trefle blanc. *Fourrages., 228*, 265–270.

Udawatta, R. P., Rankoth, L. M., & Jose, S. (2019). *Agroforestry and biodiversity. Sustain., 11*, 2879. https://doi.org/10.3390/su11102879

Umar, Y., Syakir, M. I., Yusuff, S., Azhar, B., & Tohiran, K. A. (2023). The integration of cattle grazing activities as potential best sustainable practices for weeding operations in oil palm plantations. *IOP ConferenceSeries: Earth and Environmental Science, 1167*, 012014. https://doi.org/10.1088/1755-1315/1167/1/012014

Van der Heijden, M. G. A., & Horton, T. R. (2009). The role of mycorrhizal networks in plant communities. *Ecology Letters, 12*, 1152–1162.

Vatovec, C., Jordan, N., & Huerd, S. (2005). Responsiveness of certain agronomic weed species to arbuscular mycorrhizal fungi. *Renewable Agriculture and Food Systems, 20*, 181–189.

Vickers L, Bondaroff K, Burton S. Nutritional value of thistle. Forage Fact # 104, 2017, Published by Peace River forage association of BC, Canada.

Voth K. Educated cows eat weeds. NODPA, 2024. Accessed from: https://nodpa.com/n/496/Educated-Cows-Eat-Weeds. (25 Aug. 2024).

Walker, J. W., Coffey, L., & Faller, T. (2006). Improving grazing lands with multi-species grazing. In K. Launchbaugh (Ed.), *Targeted grazing: A natural approach to vegetation management and landscape restoration* (pp. 50–55). ASI, A. Peischel and D.D. Henry, Jr..

Weisberger, D., Nichols, V., & Liebman, M. (2019). Does diversifying crop rotations suppress weeds? A meta-analysis. *PLoS ONE., 14*, e0219847.

Weisberger, D., Bastos, L. M., Sykes, V. R., & Basinger, N. T. (2023). Do cover crops suppress weeds in the U.S. Southeast? A meta-analysis. *Weed Science, 71*, 244–254.

Westerman, P., Luijendijk, C., Wevers, J., & Van der Werf, W. (2011). Weed seed predation in a phenologically late crop. *Weed Research, 51*, 157–164.

Wezel, A., Soboksa, G., McClelland, S., Delespesse, F., & Boissau, A. (2015). The blurred boundaries of ecological, sustainable, and agroecological intensification: A review. *Agronomy for Sustainable Development, 35*, 1283–1295.

White, S. S., Renner, K. A., Menalled, F. D., & Landis, D. A. (2007). Feeding preferences of weed seed predators and effect on weed emergence. *Weed Science, 55*, 606–612.

Wiggans, D. R., Singer, J. W., Moore, K. J., & Lamkey, K. R. (2012). Response of continuous maize with Stover removal to living mulches. *Agronomy Journal, 104*, 917–925.

Wozniacka, G. (2019). With regenerative agriculture booming, the question of pesticide use looms large. Accessed from on 20 May 2024: https://civileats.com/2019/09/05/with-regenerative-agriculture-booming-the-question-of-pesticide-use-looms-large/.

Woźniak, A. (2018). Effect of tillage system on the structure of weed infestation of winter wheat. *Spanish Journal of Agricultural Research, 16*, e1009.

Woźniak, A. (2020). Effect of cereal monoculture and tillage systems on grain yield and weed infestation of winter durum wheat. *International Journal of Plant Production, 14*, 1–8.

Woźniak, A., & Soroka, M. (2015). Biodiversity of weeds in pea cultivated in various tillage system. *Romanian Agricultural Research, 32*, 231–237.

Woźniak, A., & Soroka, M. (2018). Effect of crop rotation and tillage system on the weed infestation and yield of spring wheat and on soil properties. *Applied Ecology and Environmental Research, 16*, 3087–3096.

Yasin, M., Rosenqvist, E., & Andreasen, C. (2017). The effect of reduced light intensity on grass weeds. *Weed Science, 65*, 603–613.

Yu, Q., Wang, H., Wen, P., Wang, S., Li, J., Wang, R., & Wang, S. (2020). A suitable rotational conservation tillage system ameliorates soil physical properties and wheat yield: An 11-year in-situ study in a semi-arid agroecosystem. *Soil and Tillage Research, 199*, 104600.

Zhang J. (1993). Biology of *Harpalus rufipes* DeGeer (Coleoptera: Carabidae) in Maine and dynamics of seed predation. [dissertation] University of Maine.

Zhang, C., & Fu, S. (2010). Allelopathic effects of leaf litter and live roots exudates of Eucalyptus species on crops. *Allelopathy Journal, 26*, 91–99.

Zhang, J., & Wu, L. F. (2021). Impact of tillage and crop residue management on the weed community and wheat yield in a wheat–maize double cropping system. *Agriculture, 11*, 265.

Chapter 7
Aquatic Plant Management

Tobias O. Bickel

Abstract While aquatic plants are an integral component of freshwater ecosystems, invasive aquatic plants can cause severe environmental and socioeconomic impacts. Managing aquatic plants is highly complex for a range of reasons. Tools developed for terrestrial weed management are frequently unsuitable for freshwater environments. Also, unfamiliarity with the effective control of invasive aquatic plants often leads to a haphazard approach, lacking clear management goals, or tools are chosen based on hearsay or personal preferences. As a result, the management of aquatic plants is often inefficient, and at worst, causes significant environmental damage to sensitive aquatic environments.

The management of aquatic plants is highly situation-dependent, and control tools should be selected based on management goals, the scale of the problem, the stage of the invasion process, the availability of funds, social acceptability, the growth form of the target species, and the specific settings of the aquatic environment (e.g., bathymetry, water movement, water physicochemical parameters, and human use). There is a wide variety of available control tools that fit into five categories: (1) surveillance and prevention, (2) physical control that uses various machinery and manual techniques, (3) chemical control with herbicides, (4) biological control, and (5) habitat modification and restoration activities. None of these control tools is inherently better than others, as all have benefits and limitations in terms of direct and indirect costs, control efficacy, and situation-specific suitability. It is important to intelligently choose and integrate tools for situation-specific management goals.

This chapter provides an overview of the ecology of aquatic plants, discusses the benefits and limitations of various control tools, and lastly discusses how to apply these tools for the effective management of aquatic plants.

T. O. Bickel (✉)
Invasive Plants and Animals Research, Biosecurity Queensland, Department of Primary Industries, Brisbane, QLD, Australia
e-mail: tobias.bickel@daf.qld.gov.au

B. S. Chauhan, G. Mahajan (eds.), *Recent Advances in Weed Science*,
https://doi.org/10.1007/978-3-031-91014-2_7

Keywords Invasive aquatic plants · Environmental impacts · aquatic weeds control tools · IWM

1 Introduction

Aquatic plants are most commonly defined as plants that spend at least part of their lifecycle in water (Cook et al., 1974). Overall, aquatic plants are not a particularly diverse group (Murphy et al., 2019). However, they contain economically, culturally, and environmentally important species, such as rice. On the other hand, aquatic plants also include some of the world's worst alien invasive species, such as *Eichhornia crassipes* (water hyacinth) (Lowe et al., 2000).

This chapter will focus on freshwater macrophytes, including complex macroalgae (charophytes). Hence, from here on, the term "aquatic plants" will be used to describe macrophytes. This work will not discuss the management of planktonic or attached algae, such as filamentous algae or cyanobacteria.

2 Aquatic Plant Ecology and Impacts

Aquatic plants play crucial environmental roles in many freshwater ecosystems and deliver important ecosystem services, such as providing fish and invertebrate habitat, influencing key system components in the oxygen/carbon and nutrient cycles, stabilizing substrate and increase the structural diversity of the littoral (Carpenter & Lodge, 1986; Jeppesen et al., 1998).

However, macrophytes can grow at undesirable densities and impact on socioeconomic and environmental aspects of freshwater systems. Foremost, alien invasive macrophytes can cause severe impacts to aquatic ecosystems (e.g., a reduction in biodiversity) or act as ecosystem engineers that alter the functioning of aquatic habitats (Crooks, 2002; Emery-Butcher et al., 2020; Kovalenko & Dibble, 2011; Kuehne et al., 2016). Aquatic weeds pose a serious threat to agricultural, amenity, and environmental aspects of the world's limited and precious freshwater resources. These impacts include: (1) reduction in the quality and quantity of irrigation and drinking water, (2) interference with recreational activities on lakes and rivers (boating, fishing, and swimming), (3) a lower aesthetic appeal of water bodies (reduction in property values), (4) ongoing maintenance costs and damage to irrigation and water infrastructure, and (5) profound environmental effects on sensitive freshwater ecosystems (loss of biodiversity).

While most weedy aquatic plants are alien invasive species, native macrophytes can require management as well. Therefore, water or aquatic weeds can be alien invasive species or native macrophytes that grow at nuisance levels. The focus of this work is the management of alien invasive species in water bodies. The definition of a "weed" is loose and depends on individual views, in particular regarding aquatic plants. In this text, "aquatic" or "water weeds" means aquatic plants that are either alien invasive or native water plants that have become noxious.

An overabundance of aquatic plants is often a symptom of anthropogenic impacts on the catchment, such as eutrophication and changed hydrological processes, that will lead to a proliferation of plants (Crooks et al., 2011). Degraded ecosystems are also more prone to invasion by alien species. Lastly, poorly designed constructed wetlands in urban developments are frequently plagued by ongoing water weed problems due to conflicts between public safety (shallow water depth), amenity values, aesthetics (cleared shorelines that lack nutrient retention), and ecological functioning.

The overall small area of the aquatic system belies their functional significance in the wider landscape. Therefore, maintenance and protection of freshwater resources is frequently seriously underfunded compared to terrestrial areas. Despite the profound impacts of aquatic plants, there is often a lack of efficient management strategies that control macrophytes affecting lakes, reservoirs, wetlands and irrigation systems. Consequently, aquatic weed management is often conducted haphazardly and results in poor control.

Management of aquatic weeds is notoriously difficult because of the uniqueness of the aquatic environment. Equipment and herbicides for weed control are primarily designed for terrestrial applications and are not readily transferable to aquatic areas. For example, the use of herbicides is complicated through the movement and dilution of the chemical in the water column, which can severely reduce efficacy and also carries a high risk of non-target damage in the sensitive aquatic environment, including flora and fauna, irrigated crops, and human use. Therefore, inappropriate management actions can lead to far-reaching negative outcomes.

This chapter aims to assist with a goal-oriented approach to aquatic plant management that is situation-dependent, uses available control tools smartly and aims to ameliorate real (as opposed to perceived) impacts.

3 Growth Forms

Macrophytes occur in four distinct growth forms that are defined by the position of leaves with respect to their position on the water surface and if roots are anchored in the substrate or not (Fig. 7.1). In still water, there is typically a change in the predominant growth form in plant communities from the shoreline to deeper water: emergent plants on the terrestrial- hydrological interface, floating attached, and then submersed plants in deeper water (Figs. 7.1 and 7.4). Some macrophytes can be polymorphic and change between different growth forms that are morphologically distinct. This can be linked to their life cycle or environmental conditions, foremost sexual reproduction, season and water depth. For example, when *Sagittaria platyphylla* (sagittaria, slender arrowhead or delta arrowhead) germinates from seeds, it

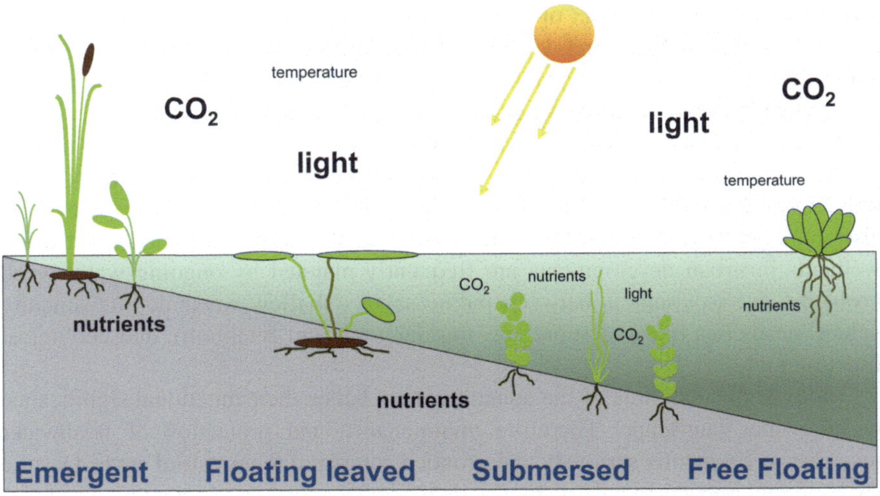

Fig. 7.1 The transition of aquatic plant communities dominated by different growth forms from the shoreline to deeper water. Availability of resources is indicated by font size; large abundance is indicated by large lettering

grows into fully submersed rosettes with linear leaves, which can persist in this form for extended periods of time. If the water level is shallow enough, *Sagittaria platyphylla* starts to grow lanceolate leaves that emerge above the water surface. Once fully emerged, the plant will start to flower. Most plants that are fully submersed through their entire lifecycle still produce flowers that emerge past the water surface to allow pollination. The growth form of aquatic plants has important implications for their management, as it will determine the effective use of different control tools and the ease of monitoring and surveillance. Additionally, the growth form will determine limiting environmental conditions and how water plants will respond to anthropogenic disturbances.

3.1 Submersed

The shoots of submersed plants are completely submerged below the water surface and are generally rooted in the substrate (Fig. 7.2), but some can also be detached and free-floating in the water column, e.g., *Ceratophyllum demersum* (hornwort, coon tail). As submersed plants are supported by the water, they are often structurally weak and can easily fragment. Some submersed plants produce flowers that extend above the water surface, but the predominant mode of reproduction is asexual through stem fragmentation or adventitious runners or shoots that grow along the substrate.

Submersed plants are protected from adverse climatic conditions (e.g., frost) but their growth can be limited by light extinction and the quantity and quality (chemical form) of carbon available for photosynthesis due to the slow diffusion of CO_2 in

Fig. 7.2 *Cabomba caroliniana* (fanwort) anchored in the substrate at 2 m depth in Ewen Maddock Dam, Australia

water and its reaction with ions depending on the pH of the water (Fig. 7.1). In tropical and subtropical areas, submersed plants can thrive throughout the year, but low light reduces growth in temperate areas in winter. They can access nutrients from the water column or the substrate. Therefore, submersed plants can grow luxuriously even in nutrient-poor water as they can access nutrients stored in the sediment. Submerged aquatic plants are important ecological habitats for fish and invertebrates and have an important function in maintaining water quality.

Hidden below the water surface, they can elude detection until growth is sufficient to reach the water surface and the problem becomes apparent. This affects the chance of early detection in surveillance programs for invasive aquatic plants and interferes with monitoring after control efforts. Submersed plants can be monitored through mechanical samplers, diving, underwater video if the water clarity is sufficient, or hydro-acoustics (sonar). The submersed growth form can also counter management actions, e.g., plants can be out of reach of mechanical cutters, or the application of herbicides can be complicated.

Submerged plants are popular as ornamental plants in aquariums, resulting in a large number of submerged alien invasive plants that either were unintentionally released into the environment or are intentionally cultivated for the nursery trade.

3.2 Emergent

Emergent plants are rooted in the substrate, with their leaves extending above the water surface (Fig. 7.3). Typical examples are reeds and sedges that grow on the margins of the waterbody as emergent plants are limited to shallow water. Some emergent aquatic plants, such as alligator weed, grow from the terrestrial toward the aquatic habitat and then extend as floating mats into the water. Emergent plants

Fig. 7.3 Top: Community of emergent aquatic plants *Myriophyllum aquaticum* (parrot's feather), *Heteranthera reniformis* (kidney-leaf mud plantain), *Persicaria* sp. (knotweeds) and *Typha* sp. (cattail, bulrush). Bottom: Emergent growth form of *Sagittaria platyphylla* in the background; the submersed rosettes are just visible in the deeper water in the front. Both photos from urban creeks in Logan, Australia

create important ecological buffers along shorelines that can prevent run-off of nutrients into water bodies and are frequently used for habitat regeneration or waste and storm water treatment.

Sexual reproduction is more common in emergent plants, and the plants have often a high potential for dispersal as seeds can be transported by water currents and wind. Nevertheless, asexual reproduction through fragments, tubers, and adventitious growth is still prevalent.

The emergent growth form allows these plants to satisfy their nutrient needs from the substrate, while having full access to CO_2 and light for photosynthesis (Fig. 7.1). Emergent plants are highly seasonal in temperate climates due to their sensitivity to frost (Burnett et al., 2007).

Most emergent weeds can be managed by several control tools as access is generally easier and they are highly visible, but the production of seeds and below ground asexual propagules can complicate long term control. Some species are also resistant to herbicides.

3.3 Floating-Attached

Floating attached plants are rooted in the substrate and have elongated petioles, with their leaves floating on the water surface (Fig. 7.4). Water lilies (Nymphaeaceae) are the most common example of this growth form. These plants often grow persistent rhizomes but also produce large numbers of seeds for reproduction. Floating leaved plants are an important ecological component in the shallow water of many still waters. They can grow deeper than emergent plants but are limited to about 2 m water depth.

Nymphaea species have been traded for their ornamental flowers for a long time. Therefore, there are many species and hybrids that are cultured worldwide. Despite this, there are few highly invasive species in this group. However, the few species that are invasive can monopolize shallow water bodies and exclude other plant species. Due to their considerable rhizome and large seed bank, these species can be difficult to manage.

3.4 Free Floating

As the name implies, free-floating plants float freely on the water surface, with their roots extending into the water column (Fig. 7.5). As they are not rooted, they are highly mobile and affected by wind and water currents. The free-floating habit gives them access to the entire water surface, often outcompeting submerged vegetation. However, the lack of access to the substrate means they are limited by nutrients available in the water (Fig. 7.1). As the leaves extend past the water surface, they are also limited by water temperature. Therefore, most of the larger free-floating plants are found in sub-tropical and tropical climates. Free-floating plants rapidly reproduce asexually and can double their extent in a very short time under favorable conditions. Seeds are also produced for long-term survival by some species (e.g., *Pistia stratiotes*).

While smaller free-floating aquatic plants, especially Lemnoideae species (duckweeds), can be environmentally beneficial, larger free-floating tropical species are environmentally the most impactful species (Perna et al., 2011). Water hyacinth is

Fig. 7.4 Floating leaved *Nymphoides indica* (water snowflake) covering the water surface along the shoreline in a suburban lake at the Gold Coast, Australia (top). Note the transition from emergent (*Persicaria* sp.) to floating leaved (*N. indica*) and then submersed plants (*C. caroliniana*) with increasing water depth (compare with Fig. 7.1). Close-up of *N. indica* floating leaves of and flowers (bottom)

considered the world's worst weed (Lowe et al., 2000). Free-floating water plants can grow in thick mats that cover the entire water surface, interfering with boating and fishing. They also are a suitable habitat for mosquito proliferation; therefore, they cause public health issues (Ferreira et al., 2003; Russell, 1999). Their thick cover also completely transforms the underlying aquatic ecosystem (Figs. 7.5 and 7.6). Gas exchange through the water surface is prevented and the blocking of light prevents submerged photosynthesis by phytoplankton and macrophytes. The extensive root system acidifies the water column through respiration, and the plants produce large amounts of organic matter. Overall, free-floating plants can transform a healthy freshwater system fueled by photosynthesis into a detrital system dominated by decomposition. As a result, the water becomes anoxic and highly acidic and is an unsuitable habitat for most aquatic organisms (Perna & Burrows, 2005).

Fig. 7.5 A small farm dam covered by a thick layer of free-floating *Limnobium laevigatum* (Amazon frogbit). The water below the dense plant mat is 2 m deep. The water below the plant canopy is hypoxic and highly acidic (dissolved oxygen 0 mg L^{-1}, pH 4; T O Bickel pers. observation)

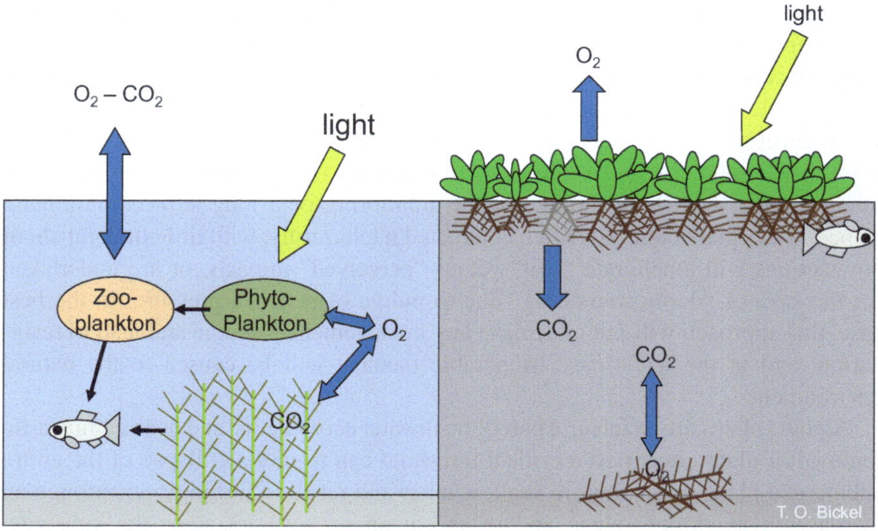

Fig. 7.6 Simplified functioning of a healthy aquatic ecosystem driven by primary production and free gas exchange through the water surface (left), and the same system after invasion by free floating plants that is heavily shaded and dominated by detrital decomposition; note the lack of oxygen production and increase in CO_2 that makes the water acidic

Due to their reliance on warm temperatures and high nutrients in the water, free-floating water plants benefit particularly from anthropogenic impacts on the environment, namely global warming and eutrophication. Climate change scenarios predict an increase in problematic invasive floating water plants and a range extension into temperate areas (Feuchtmayr et al., 2009; Scheffer et al., 2003).

Free-floating macrophytes are highly visible but can be challenging to manage when they infest large water bodies. The high biomass makes mechanical removal costly. Most species are sensitive to a range of herbicides, but application can be challenging due to the difficult access. The mobility of the plants also interferes with monitoring as plants treated with herbicides today might be in a different location tomorrow. Lastly, free-floating plants can "hide" between shoreline vegetation, and some produce long-lived seeds. Unless great care is taken during control efforts and monitored regularly, management can be frustrated as the plants regrow so quickly.

4 Management

The terms management and control are often used interchangeably in the aquatic plant literature. In the context of this text, management of aquatic plants means the use of suitable control tools to achieve a desired management outcome.

The management of aquatic plants in natural water bodies is highly complex due to the environmental heterogeneity, the high risk of nontarget damage, social resistance, and the unsuitability of weed control tools that were originally designed for terrestrial situations. However, one of the biggest challenges to efficient and long-term management of aquatic plants is the lack of clearly defined or unrealistic goals. Most water bodies have multiple user groups that often have conflicting ideas about the outcome of management actions. Additionally, control tools are chosen for social acceptability or "hear say" and not to suit the specific situation, degree of plant density and species, or the planned outcome. Lastly, aquatic plant control is often initiated with little thought about monitoring or a long-term commitment. Hence, water plant control is often conducted haphazardly, with little thought about how actions will ameliorate "real" versus "perceived" impacts, or are undertaken for the sake of "doing something" due to public or political pressure. In the best case, this approach will fail to achieve any management outcome and lead to resignation, and in the worst case irreparable damage will be caused to the natural environment.

Aquatic plants are an integral part of freshwater ecosystems, and over-enthusiastic removal of plant cover past a critical threshold can trigger a collapse of the entire submerged plant community in shallow lakes and result in turbid water through an increase in planktonic algae and resuspension of particles from the substrate. Shallow lakes exist in either of two stable states: a macrophyte dominated clear water state or a turbid, phytoplankton-dominated state (Scheffer et al., 1993; van Nes & Scheffer, 2005). Once a system is shifted from the environmentally desirable clear water state, it is very difficult to reverse this process. Therefore, care must be taken not to remove too much vegetation. A third state dominated by free-floating macrophytes can also be triggered by the invasion of large free-floating species, e.g., water hyacinth. The probability of the free-floating plant dominated state increases with eutrophication and increasing temperatures associated with climate change (Feuchtmayr et al., 2009; Scheffer et al., 2003).

Basically, there are five different management goals that differ in the intensity of management effort from high to low: (1) prevention and surveillance; (2) eradication, or the complete removal of aquatic vegetation to manage incursions of invasive alien species (IAS) or to suit operational use (e.g., irrigation); (3) containment, where there is little scope to achieve long term reduction, but further spread is prevented; (4) maintenance, where plants are controlled in strategic areas or around infrastructure (asset protection) such as boat ramps, swimming areas, and water intakes for recreational and economical use of water bodies; and (5) a "do nothing" approach or enhancement, where water bodies are restored with native species or water plants are considered as beneficial, e.g. for recreational fishing.

In terms of IAS management, the first four goals presented would align with the invasion status from prevention to new incursion (eradication), containment during the establishment phase to prevent further dispersal, and asset-based protection (maintenance) once species have fully established and are widespread in its new environment (Fig. 7.7). The "do nothing" approach is not an option if IAS management is a priority.

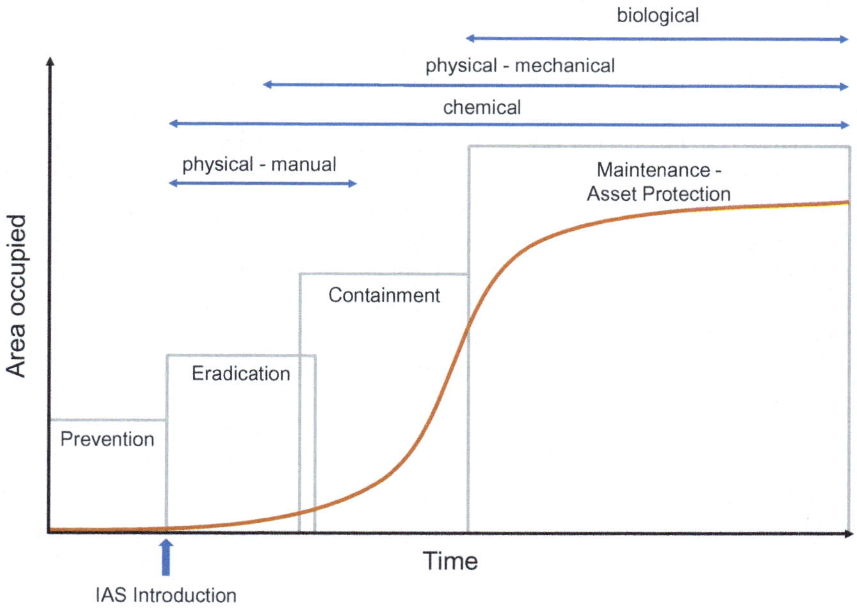

Fig. 7.7 Change in management goals (grey boxes) with an increase in the area occupied by an Invasive Alien Species (IAS, orange curve). Blue arrows indicate appropriate control tools used during the different stages of species invasion

5 Control Tools

There is a wide variety of control tools available for aquatic plant management. All these tools have benefits and limitations in terms of direct and indirect costs, control efficacy, and situation specific suitability. Hence, there are no control tools that are superior or inferior per se, but the use of appropriate tools will always depend on the scale of the problem, the availability of funds, social acceptability, the stage of the invasion process for the management of IAS, and lastly the management goals. It is important to intelligently choose and integrate tools for situation-specific management goals.

Basically, aquatic plant control tools can be separated into five groups: (1) surveillance and prevention, (2) physical control through mechanical tools that use various machinery and manual techniques, (3) chemical control with herbicides, (4) biological control through the introduction of natural enemies or pathogens (classical biocontrol) or the use of generalist grazers (fish), and (5) habitat modification and restoration activities that directly or indirectly interfere with or affect aquatic plant physiology. Different control tools are appropriate during the stages of the invasion process, with physical and chemical tools for eradication, mechanical and chemical tools for containment, and a combination of all for asset protection (Fig. 7.7).

5.1 *Surveillance and Prevention*

Due to the high management cost, once an aquatic plant becomes problematic, prevention of arrival through movement/border control and early detection through surveillance is highly cost effective. The risk of establishment of an aquatic invader in a new habitat is a probabilistic process that consists of several discreet steps (e.g., uptake, transport, survival, regeneration, colonization) (Colautti et al., 2006; Floerl et al., 2009). Each step has its own statistical probability. As the probabilities are multiplicative, even a reasonable decrease in any of the risks can have a profound effect on the overall probability of the establishment (Buchan & Padilla, 1999; Clayton & Champion, 2006). Furthermore, as the chance of eradication declines and management costs rapidly escalate with increasing spread of an IAS, early detection through surveillance is critical (Anderson, 2005; Hussner et al., 2017). Border control is largely legislative and will not be described in detail here, but surveillance is highly relevant for aquatic plant managers at different levels of establishment.

The ease of detection depends in part on the specific growth form of aquatic plants. Free-floating, floating-attached, and emergent plants are readily observable, through visual observations or remote sensing, as they grow on or beyond the water surface. However, submersed invasive aquatic plants, hidden below the water surface, often elude detection and only become apparent when the infestation reaches a large extent with a visible canopy formation at the water surface. Therefore, effective prevention can only be achieved through a structured monitoring program of high-risk areas by expert divers that will have a high chance of early detection, but this can only be applied on a small scale. The probability of detection depends highly on water clarity and the experience of monitoring staff. High-confidence detection can only be achieved by implementing active surveys undertaken by experienced divers or systematic surveys with submersed surveying equipment (underwater drones, cameras, or similar technology), and requires a high effort.

Given the difficulty in detecting submersed aquatic plants below the water surface at an early stage of infestation, future use of eDNA (environmental DNA) techniques will be invaluable for early detection programs (Edmunds et al., 2019; Ghahramanzadeh et al., 2013; Scriver et al., 2015).

Once invasive aquatic plants cross the border and establish in the environment, "hitchhiking" of aquatic plant fragments on boats and fishing equipment becomes a major dispersal pathway for invasive aquatic plants (Barnes et al., 2013; Bickel, 2015; Jacobs & Macisaac, 2009; Johnson et al., 2001; Johnstone et al., 1985). Therefore, human-mediated dispersal between water bodies (containment) can be controlled through water body access management or removal of water plants from watercraft launch areas to prevent uptake and transport (Rothlisberger et al., 2010). This is achieved by creating water plant-free corridors around access points with sufficient spatial extent and depth to prevent watercraft and launching equipment (trailers) from picking up potential fragments. Additionally, the installation of wash-down facilities near boat ramps can be beneficial, as they allow users to clean off aquatic weeds before transport (Hippolite & Kurapa, 2018). The efficacy of wash-down facilities, general weed hygiene, and other biosecurity measures is strongly dependent on public awareness, so these measures should be coupled with strong public awareness campaigns.

5.2 *Physical Control: Manual Removal*

Physical control tools include hand removal/weeding and a large number of mechanical tools using various machinery.

Manual removal of plants can be highly effective in controlling small infestations in the early stage of incursion or as a follow-up to other control measures if all remaining plant material is effectively removed. Emergent and free-floating plants can be hand-weeded in shallow water, but submersed plants will have to be removed by trained divers. As most submersed macrophytes readily regenerate from fragments (Barrat-Segretain & Bornette, 2000; Bickel & Perrett, 2014; Combroux et al., 2001; Riis et al., 2010), great care needs to be taken to remove all material, and efficacy is highly dependent on the divers' skills and visibility (water clarity). Due to the high labor effort, manual removal is only efficient for small areas or low-density scattered populations that remain after other control techniques were carried out.

Hand weeding can be conducted in conjunction with routine monitoring (see previous section) and as a follow-up to previous control efforts with other techniques, that is, to “mop up” remaining plant material (Clayton, 1996, 2006; Oosterhout, 2009). For example, hand weeding was used to remove outlier stands of *Lagarosiphon major* (curly water weed) during routine monitoring efforts in a large lake (Lake Wanaka, New Zealand) spanning many kilometers of the lake shore (Clayton, 1996). Hand weeding was also successfully employed to remove remaining *Salvinia molesta* plants from a wetland (area in the range of ha) after the majority of plants had been killed with herbicides (Honey Dam, North Queensland (A. Petroeschevsky, pers. com.) and Myall Lakes catchment (Oosterhout, 2009).

Hand removal is unlikely to cause measurable environmental, social, or economic side effects, mainly due to the small scale to which it is applied. As plants are removed specifically by divers, the impact on nontarget species is minimal. Since no chemicals or heavy machinery are used, this method is highly acceptable to most stakeholders and has high public acceptability.

Hand weeding and removal by trained divers is a suitable method to cost-efficiently and rapidly eradicate early incursions of invasive aquatic plants. It is an invaluable control tool for removing outliers and remaining material after large scale application of other control techniques.

5.3 Physical Control: Mechanical

There is a plethora of different machinery used for the mechanical control of aquatic plants that either remove plants from the substrate, mow the plants to a certain depth, or harvest plants from the surface (or a combination of all three options). Only the most commonly used machinery will be described here.

Mechanical devices are often used to remove or dislodge aquatic weeds from the substrate, such as suction dredging, rototilling, and hydro-venturi. These methods are useful for infestations that are too large for manual removal, but the costs can be high due to the considerable initial outlay for the machinery and ongoing staff costs (Clayton, 1996; Cooke et al., 2005; Gettys et al., 2020; Pieterse & Murphy, 1990). The environmental impacts are also higher, mainly due to the disturbance of the natural substrate, creation of fragments, or nutrient release (James et al., 2002). The duration of control depends on the prevention of recolonization from remaining stem fragments or untreated areas (van Valkenburg et al., 2011).

Suction dredging uses machines originating from the mining industry (in-stream alluvial gold mining), which are basically large aquatic vacuum machines that suck up weeds and the upper substrate layer, then pump the material onto a barge for later disposal on land. The suction device is manipulated by divers and can be targeted to remove all plants from a designated area. The removal efficiency can be very high, as plant and embedding substrate are removed, and nontarget damage can be relatively low as the suction hose is guided by divers. Suction dredging is more suitable for small and strategic areas and, in some cases, can also be suitable for the eradication of early incursions. In Australia, management of *C. caroliniana* with suction dredging in Ewen Maddock Dam and Lake Macdonald (monthly treatment) provided short term control in strategic areas for containment (Oosterhout, 2009).

Rototilling uses an underwater rotary hoeing machine that tills the underwater substrate and thereby uproots aquatic plants. Rototilling gives good medium- to long-term control, depending on substrate penetration. For example, experimental rototilling of *Lagarosiphon major* resulted in 0.5–2 years of control in Lake Wanaka, NZ (Clayton et al., 2000). However, rototilling efficacy depends on the contour of the lakebed and the presence of obstacles. Hydro-venturi, a water jet, is used to dislodge aquatic plants from the substrate. The plant material can be collected afterward. *C. caroliniana* was experimentally removed by hydro-venturi from a canal in the Netherlands, which resulted in the efficient but short term control (van Valkenburg et al., 2011). Lastly, excavation is frequently used to remove nuisance aquatic plant vegetation from flowing waters, such as drains and canals. It can be used to clear long stretches of drains on a scale of 100 m to km. However, excavation is limited to shallow areas and narrow canals, and by the reaching ability of the excavator equipment (Clayton, 1996; Hussner et al., 2010).

The last group of mechanical control tools are different types of weed-cutting machinery that cut submersed aquatic weeds down to a certain depth (~2 m) and remove the cut plants via conveyer belts and barges for later disposal (harvesters). The efficiency of plant removal after mowing is critical. For example, inappropriate mowing produced plant fragments that aided in the spread of *C. caroliniana* in the Netherlands (Matthews et al., 2013). As this technique is practically similar to mowing a lawn, it has to be applied repeatedly, often multiple times per growing season and should be part of an ongoing maintenance program (Clayton, 1996; Cooke et al., 2005; Gettys et al., 2020).

Overall, the efficacy of mechanical control with machinery will depend on the water depth and the contour (bathymetry), the presence of underwater obstacles, and access to the water body to launch the machinery (Clayton, 1996; Hussner et al., 2017). Control will be medium-term, requiring a regular maintenance program for long-term plant management. Mechanical control, such as suction dredging, can be used on a small scale for early eradication or control around structures such as jetties and boat ramps), but the high initial costs would rarely justify the use of these machines on a small geographical and temporal scale (Clayton, 1996; Cooke et al., 2005; Hussner et al., 2017). The area that can be cleared in a day depends on the density of the plants but would be on the scale of less than a hectare per day. Frequently, these machines are used for ongoing maintenance of lakes (up to several hundred ha) or rotated between lakes to clear high-use recreational areas, such as swimming areas, on a regular basis (Clayton, 1996; Oosterhout, 2009; van Valkenburg et al., 2011).

Mechanical control methods rely on the use of heavy machinery, which could negatively impact public perception or amenity values. There are also negative environmental impacts due to the disturbance of the aquatic substrate and possible temporary increase of turbidity. However, mechanical devices for aquatic weed control are usually more acceptable to the public than the use of herbicides, mainly due to the perceived lower environmental impacts. However, this depends on the local urgency and interest in getting the plants removed quickly and cheaply (using herbicides) versus slowly and costly (via mechanical means) (pers. obs.). While mechanical control options usually incur fairly high management costs, the long-term costs of inaction potentially far outweigh the initial outlay for mechanical control.

5.4 Chemical

Herbicides are widely used for chemical control of aquatic plants and work by physically damaging target plants (e.g., desiccants) or by interfering with their physiological processes (e.g., growth hormone mimics). Herbicides have a long history of use for the control of aquatic weeds (Clayton, 1996; Cooke et al., 2005; Gettys et al., 2020; Pieterse & Murphy, 1990). This chapter will not provide a list of herbicide actives used for aquatic areas as the availability of products will depend on local legislation, new products become available regularly, and lastly, herbicides have to be chosen specifically for target weeds that differ between countries. The user should study labels of locally available aquatic herbicides to choose suitable products. Instead, this text will discuss the general application of herbicides in water bodies and highlight the benefits and limitations of herbicides compared to other control tools.

Depending on the target weed, herbicides can either be applied as a foliar spray (free-floating and emergent plants; Fig. 7.8) or injected into the water column (submersed macrophytes) or as a combination of these two approaches. Either application technique poses challenges. For foliar sprays, safe access to water plants can be challenging due to the limited reach of spray equipment, which can result in poor coverage and subsequent need for reapplication of herbicide. Mixing herbicides with adjuvants that assist the transfer of the herbicide through cuticles and stabilize the spray mixture (e.g., through acidification) can be advantageous for foliar applications. Adjuvants can be beneficial when treating plants that are covered by hairs or have a waxy cuticle, such as *Salvinia molesta* (giant salvinia) or *Myriophyllum aquaticum* (parrot's feather). Only adjuvants approved for aquatic use should be applied, as terrestrial formulations can pose considerable environmental risks.

Effective injection of herbicides into the water column is affected by environmental factors, resulting in dilution, displacement through currents, deactivation by suspended particles in the water column, and increased risk of nontarget damage (Gettys et al., 2020) (Fig. 7.9). As the entire water column needs to be dosed for effective treatment, herbicide injection can be costly in deeper water. Some of these difficulties with injection can be overcome using suitable carriers, such as gelling agents or diatomaceous earth, that allow more targeted delivery of herbicide (Barrett & Logan, 1982; Clayton, 1996; Tanner & Clayton, 1984).

Herbicidal control of aquatic plants can be achieved with one or multiple applications, depending on the herbicide used and situation-specific conditions, such as

Fig. 7.8 Foliar application of florpyrauxifen-benzyl to an infestation of the emergent *Myriophyllum aquaticum* in a farm dam in Queensland, Australia. The herbicide is sprayed from a vehicle mounted tank with an extendable hose. Note the use of personal protective equipment

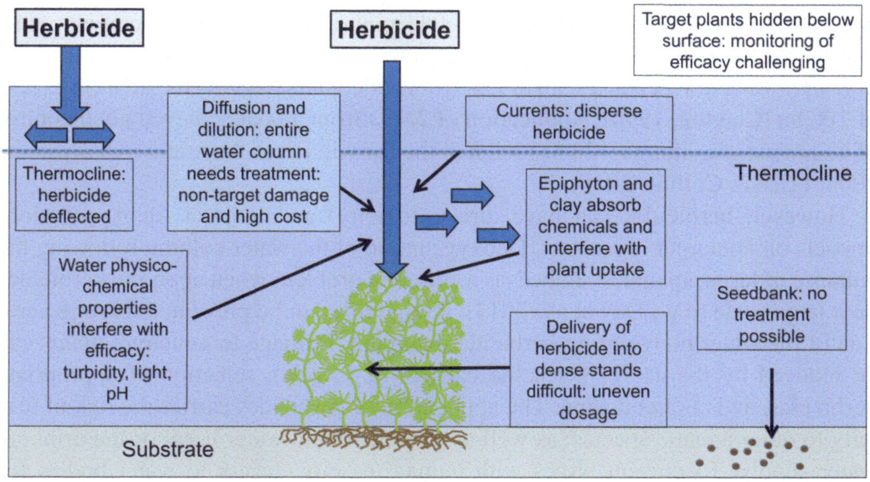

Fig. 7.9 Injecting herbicides into the water column to control submersed aquatic plants meets many environmental challenges. Situation dependent application is critical for control efficacy

the extent of the infestation and plant density (biomass), bathymetry, water physico-chemical parameters, water movement, and application technique. Control is usually achieved within days after application, but long-term monitoring after treatment is necessary to prevent re-establishment. The most critical factor for effective submersed aquatic plant management is the concentration and exposure time (CET) relationship, which depends on the uptake rate of a target macrophyte and the specific herbicide (Netherland et al., 1991). Response to dosage or specific herbicides can vary widely between species. Therefore, good data is necessary to predict control efficacy and choose the correct herbicide. The speed of herbicide uptake is also highly variable between herbicides. Some herbicides can provide effective control within hours of application, while others need exposure times of several days. Therefore, to achieve effective control, both an appropriate dose and exposure time must be achieved. Environmental factors such as water currents, stratification (prevention of mixing), and water chemistry will affect both dose and exposure time through dilution, movement of target sites, and herbicide hydrolysis or deactivation (binding to organic or clay particles). Taken all this together, successful control of submersed aquatic plants through injection is complex and requires considerable expertise. Despite these technical and operational difficulties, herbicides can be a highly effective tool for controlling a wide range of aquatic plants and are particularly useful when integrated with other control techniques.

One of the advantages of herbicides is that they are frequently more economical than mechanical tools and can be applied on a wide range of scales, from spot spraying with a backpack sprayer to large-scale application from drones or aircraft. Therefore, herbicides can be used to control aquatic plant infestations in small to large water bodies. For example, excellent herbicidal control of *C. caroliniana* was achieved in wetlands and farm dams measuring up to 8 ha in the Northern Territory and New South Wales, Australia, with single applications of 2,4-D or carfentrazone (Anonymous, 2008; Inkson et al., 2014). On the other end of the scale, management of *L. major* in New Zealand with diquat from helicopters has covered areas in excess of 100 ha (Clayton, 1996). Application of 2,4-D from a boat achieved good control of *C. caroliniana* along a 12 km stretch of the Darwin River, Australia (Anonymous, 2008; Price & Collins, 2016).

However, herbicides can have direct (toxicity) and indirect (deoxygenation) impacts on nontarget organisms. Deoxygenation of the water column following the collapse of large aquatic weed beds is a common problem when applying herbicides on a large scale (e.g., Day et al., 2014), though this can be prevented through aeration of the water body during herbicide treatment. Damage to nontarget plants can be reduced by the use of herbicide carriers (e.g., gels), selection of appropriate herbicides, and correct dosing. The application of herbicides carries the risk of toxicity to other aquatic species, as well as to recreational water users or the drinking water supply. To prevent issues with human toxicity, access to water bodies for recreational use can be temporarily restricted, which can be unpopular with users. Depending on specific withholding periods, the use of herbicides can temporarily restrict the use of water for irrigation or livestock watering.

Acceptability of the use of herbicides in water bodies depends on the specific local community. While some communities reject the use of herbicides in water outright for their perceived health and environmental risks (chemophobia), other communities are highly supportive, as herbicides provide a rapid and visible reduction in plant infestations. While herbicides can cause nontarget damage when used inappropriately, they can be highly cost-efficient in removing invasive species or enabling eradication and subsequent restoration of water bodies. The long-term environmental impacts of invasive species can far outweigh any potential short-term nontarget damage caused by herbicides.

5.5 Biological Control

Two types of biological control exist for the control of aquatic weeds: (1) "classical" biological control, which relies on host-specific invertebrate herbivores or pathogens that consume and/or damage the target plant and (2) generalist herbivores (grass carp).

The principle of classical biological control is based on the "enemy release" hypothesis, which suggests that invasive organisms do so well outside their native range because their natural predators are missing (Keane & Crawley, 2002; McFayden, 1998; Muniappan et al., 2009; Murdoch et al., 1985; Williamson, 1996). Thus, the logical step is to introduce predators from their natural range to provide natural control. Once the natural predators are present, the target plant and its herbivores will establish a new equilibrium, resulting in a reduced and acceptable density of the target plant.

Potential biological control agents have to go through a rigorous host testing regime (i.e., they need to be host-specific) before they can be deemed safe for release. Biological control agents initially require a high effort in scientific personnel and funds and are time consuming (multiple years). This is due to the extensive field work required in the natural host range, the rigorous testing for specificity, and the need for specialized scientific infrastructure (e.g., quarantine facilities). Furthermore, there is no guarantee that a specific agent will be identified. Nevertheless, biological control can be highly cost effective in the long term and there are examples of spectacular successes with several aquatic agents that have provided good control of aquatic weeds. The most efficient biological control agents available are for free-floating aquatic weeds, such as *S. molesta*, *Pistia stratiotes*, and *E. crassipes* (Gassmann et al., 2006). Under suitable environmental conditions, spectacular successes have been achieved within a few years, such as the large-scale reduction of *E. crassipes* in Lake Victoria or the Sepik River in Papua New Guinea (Williams et al., 2005).

There are fewer efficient biological control agents for submersed aquatic plants (Coetzee et al., 2011), with so far limited control success achieved for species such as *L. major* and *Hydrilla verticillata* (hydrilla or waterthyme) (Baars et al., 2010; Wheeler & Center, 2001). However, a promising biological control agent for *C. caroliniana* has recently been released in Australia (Cabrera-Walsh et al., 2011; Schooler et al., 2006).

Once biological control agents are established, they continuously control the target aquatic plant infestation at the scale of an entire water body with little further input. Some biological control agents are able to naturally disperse and thus control the target plant at a landscape level (Muniappan et al., 2009; Williamson, 1996). Biological control agents are usually less suitable for smaller water bodies (e.g., ponds, farm dams), as plant infestations in these can be more efficiently managed with other tools (herbicides, physical methods and habitat manipulation). Because of the long time for development and because agents don't entirely remove the target plant, biological control is not a suitable control tool for eradication targets.

Apart from target specific agents, the grass carp (*Ctenopharyngodon idella*), a large herbivorous cyprinid fish, has been used for aquatic plant control for several decades in the EU, USA, and New Zealand (Bain, 1993; Chilton & Muoneke, 1992; Clayton et al., 1992; Pípalová, 2006; Rowe & Champion, 1994). Grass carp are generalist feeders and, while showing preference for certain plant species (Chilton & Muoneke, 1992; Hanlon et al., 2000; Mitchell, 1980), they will, at the correct stocking density, consume all aquatic vegetation in a water body (Hanlon et al., 2000; Pípalová, 2006). Once the desired level of control is achieved (complete removal of the target species), the fish can be removed by netting or with the use of rotenone (Rowe & Champion, 1994). Grass carp have been successfully used for control and site-specific eradication of a wide range of submersed aquatic plants, such as *L. major*, *H. verticillata*, and *Egeria densa* (dense water weed) (Clayton et al., 1992; Mitchell, 1980; Rowe & Champion, 1994).

Grass carp control aquatic plants within any stocked water body, independently of the area. Efficient control depends only on appropriate stocking density (Chilton & Muoneke, 1992; Fowler & Robson, 1978; Noble et al., 1986). However, the feasibility of stocking grass carp in large lakes can be limited by the large number of fish required and subsequent difficulty of removal, if required. Stocking of large lakes can also result in unpredictable macrophyte control due to fish movement (Noble et al., 1986). Large systems are also often part of a wider catchment, so grass carp stocking can become problematic in terms of the containment of the fish and prevention of non-target damage in case of escape (Bain, 1993).

Compared to other control tools, biological control is widely accepted by the public, as it has no direct environmental impact and does not interfere with recreational activities. When using biological control for aquatic plant management, public education can be highly beneficial as sufficient control is often achieved only after a few years. Therefore, the public can become impatient and interfere through political channels with management because the biological control "doesn't work" (pers. observation). This can result in a waste of management effort, for example, using expensive mechanical harvesters to remove plants so water body managers can show instant "success" and demonstrate that "something is done" when plants would have been largely eliminated by biological control in the long term without any further effort.

Grass carp have highly specific breeding requirements and are thus unlikely to reproduce in stocked water bodies. The risk can be further ameliorated by using triploid fish (Bain, 1993; Pípalová, 2006). In case of escape (e.g., after floods), the fish can potentially cause non-target damage off-site, but as they would be dispersed at low densities, large scale damage is unlikely (Clayton & Wells, 1999). Compared to classical biocontrol, grass carp are a more "radical" approach because they can remove all aquatic plants in stocked systems and have been shown to have other ecological impacts (Bain, 1993; Dibble & Kovalenko, 2009).

5.6 Habitat Manipulation (Shading, Dyes, Benthic Blankets, and Water Drawdown)

Habitat manipulation to control aquatic weeds usually involves management actions that directly interfere with aquatic plant growth, such as the use of benthic blankets, shading with plastic sheets or artificial dyes, and the removal of water through a drawdown (Hussner et al., 2017). The suitability of each of these methods depends on the specific circumstances, for example, shading with plastic or dyes is only suitable for small water bodies, while benthic blankets can be strategically placed in large water bodies to support recreational users. Water levels can only be manipulated in lakes and dams that have the necessary infrastructure. There are also more large-scale approaches that modify the environment to make it less amenable to aquatic plant growth, such as restoring shoreline vegetation to create buffers that prevent nutrient runoff into water bodies and managing entire catchments to reduce nutrient loading.

A reduction of light availability for photosynthesis, and subsequent plant death, can be achieved through artificial shading of water bodies, either through dyes that specifically filter out light necessary for photosynthesis, or through the use of shading material (e.g., plastic sheeting). For example, the shading of a small farm dam with dark plastic sheets that covered the entire water surface provided several months of *C. caroliniana* control (Schooler, 2008; Oosterhout, 2009). Field trials in the Netherlands also used shading and resulted in the removal of *C. caroliniana* from a 100 m stretch of a canal (van Valkenburg et al., 2011). Dyes are used mainly in small artificial water bodies, such as golf courses.

The installation of benthic blankets, either made of geotextile or of natural fiber material, is a popular aquatic weed control tool. These blankets are either installed by divers or during lake drawdowns. The blankets suffocate and shade plant material trapped underneath and also reduce colonization on top of the blanket. The efficacy of benthic blankets relies on the absence of wave action or currents, and on the contour of the bottom of the water body. The choice of benthic blanket material depends on the situation and management goals. Natural fibers can break down over time and help native plants to re-establish after invasive species are removed. Blankets made out of artificial material will potentially provide a longer-term control, because they break down much slower. However, long-term efficacy to prevent plant colonization will depend on maintenance and preventing the build-up of organic material which will render benthic blankets ineffective. Plastic sheets are probably the most inexpensive option, but they can create problems when large amounts of plant biomass are covered, as the subsequent decomposition will accumulate gas that can lift the blankets of the lake bottom.

A trial with benthic blankets to control *C. caroliniana* in Lake Benalla, Australia, was carried out as part of a lake drawdown experiment (Dugdale et al., 2013). Geotextile blankets were installed on the exposed lakebed and provided good control for 2 years after filling (Tony Dugdale, pers. comm.). Benthic blankets were successfully used to control *Lagarosiphon major* (Caffrey et al., 2010). Benthic blankets are particularly useful for long-term control around boat ramps and swimming areas.

Success with drawdowns relies on sufficient drying out of the substrate, susceptibility of the target species, and occurrence of temperature extremes (heat or cold) (Cooke, 1980). Drawdowns must be long enough to allow complete desiccation of aquatic plants (in summer) or death through freezing (in winter). For example, several draw downs (4–6 weeks) were trialed in Lake Benalla, Victoria, Australia, to control *C. caroliniana*. Despite the duration of the drawdowns, the mud remained wet and some of the Cabomba material was viable and resulted in quick regrowth once the lake was refilled (Dugdale et al., 2013).

While habitat manipulation control techniques might be more acceptable to the public than the use of herbicides, due to the lower perceived environmental impact, some of the methods have a high impact on the recreational and economic use of water resources, and are thus less acceptable (e.g., to irrigators, drinking water producers). Shading and benthic blankets will affect all aquatic plants and impact water quality due to the lack of photosynthesis. The level of environmental impact will strongly depend on the area that is treated (complete water body vs. strategic areas). A drawdown also has significant environmental impacts. In addition, the loss of water for irrigation and power generation (in hydro-power lakes) is often prohibitive. The emptying of a water body also carries a high risk of spreading macrophytes elsewhere. While these methods are socially acceptable, most of them will detrimentally affect the recreational use of water bodies.

5.7 Integrated Control

As all individual control tools have their benefits and limitations, successful management of aquatic plants often benefits from the integration of multiple control tools. The choice of which tools to integrate will be situation and management goal specific. Integration of tools will be most useful with increase in infested area along the invasion process. For example, to contain or manage large infestations (asset protection), biological control (if available for target) could be used to reduce the overall density of the infestation to benefit recreational use of water bodies. At the same time, herbicides or suction dredging is used around access points, such as boat ramps (prevent dispersal) or critical infrastructure. Additionally, benthic blankets could be installed to keep certain areas clear, such as swimming areas.

Integration does not imply that tools will always be used simultaneously but could be used successively to improve overall management. For example, biological control (if available for target) could be employed to remove the bulk of aquatic plant biomass, followed up by hand weeding or herbicide spot spraying to achieve eradication in a site. Similarly, harvesters can be used to remove the bulk of free-floating aquatic plants, followed up by spot spraying remaining vegetation in refuges in between shoreline plants with herbicides. This approach will greatly improve time till reinfestation or enable site-specific eradication.

References

Anderson, L. W. J. (2005). California's reaction to Caulerpa taxifolia: A model for invasive species rapid response. *Biological Invasions, 7*(6), 1003–1016. https://doi.org/10.1007/s10530-004-3123-z

Anonymous. (2008). *Northern territory cabomba eradication program 2005/06*. N. T. Government.

Baars, J. R., Coetzee, J. A., Martin, G., Hill, M. P., & Caffrey, J. M. (2010). Natural enemies from South Africa for biological control of Lagarosiphon major (Ridl.) Moss ex wager (Hydrocharitaceae) in Europe. *Hydrobiologia, 656*(1), 149–158. https://doi.org/10.1007/s10750-010-0427-0

Bain, M. B. (1993). Assessing impacts of introduced aquatic species: Grass carp in large systems. *Environmental Management, 17*(2), 211–224. https://doi.org/10.1007/bf02394691

Barnes, M. A., Jerde, C. L., Keller, D., Chadderton, W. L., Howeth, J. G., & Lodge, D. M. (2013). Viability of aquatic plant fragments following desiccation. *Invasive Plant Science and Management, 6*(2), 320–325. https://doi.org/10.1614/IPSM-D-12-00060.1

Barrat-Segretain, M.-H., & Bornette, G. (2000). Regeneration and colonization abilities of aquatic plant fragments: Effect of disturbance seasonality. *Hydrobiologia, 421*(1), 31–39. https://doi.org/10.1023/A:1003980927853

Barrett, P., & Logan, P. (1982). The localised control of submerged aquatic weeds in lakes with diquat-alginate. In *Proceedings 6th EWRS international symposium on aquatic weeds*. European Weed Research Society.

Bickel, T. O. (2015). A boat hitchhiker's guide to survival: Cabomba caroliniana desiccation resistance and survival ability. *Madoqua, 746*(1), 123–134. https://doi.org/10.1007/s10750-014-1979-1

Bickel, T. O., & Perrett, C. (2014). Competitive performance of Cabomba caroliniana. In *19th Australasian weeds conference*. Council of Australasian Weed Societies.

Buchan, L. A. J., & Padilla, D. K. (1999). Estimating the probability of long-distance overland dispersal of invading aquatic species. *Ecological Applications, 9*(1), 254–265.

Burnett, D. A., Champion, P. D., Clayton, J. S., & Ogden, J. (2007). A system for investigation of the temperature responses of emergent aquatic plants. *Aquatic Botany, 86*(2), 187–190. https://doi.org/10.1016/j.aquabot.2006.09.015

Cabrera-Walsh, G., Schooler, S., & Julien, M. (2011). Biology and preliminary host range of Hydrotimetes natans Kolbe (Coleoptera: Curculionidae), a natural enemy candidate for biological control of Cabomba caroliniana gray (Cabombaceae) in Australia. *Australian Journal of Entomology, 50*, 200–206. https://doi.org/10.1111/j.1440-6055.2010.00793.x

Caffrey, J. M., Millane, M., Evers, S., Moran, H., & Butler, M. (2010). A novel approach to aquatic weed control and habitat restoration using biodegradable jute matting. *Aquatic Invasions, 5*(2), 123–129. https://doi.org/10.3391/ai.2010.5.2.01

Carpenter, S. R., & Lodge, D. M. (1986). Effects of submerged macrophytes on ecosystem processes. *Aquatic Botany, 26*, 341–370.

Chilton, E. W., & Muoneke, M. I. (1992). Biology and management of grass carp (Ctenopharyngodon idella, Cyprinidae) for vegetation control: A north American perspective. *Reviews in Fish Biology and Fisheries, 2*(4), 283–320. https://doi.org/10.1007/BF00043520

Clayton, J., & Champion, P. (2006). Risk assessment method for submerged weeds in New Zealand hydroelectric lakes. In J. M. Caffrey, A. Dutartre, J. Haury, K. J. Murphy, & P. M. Wade (Eds.), *Macrophytes in aquatic ecosystems: From biology to management* (pp. 183–188). Springer. https://doi.org/10.1007/978-1-4020-5390-0_26

Clayton, J. S. (1996). Aquatic weeds and their control in New Zealand lakes. *Lake and Reservoir Management, 12*(4), 477–486.

Clayton, J. S. (2006). Assessment of the 2005/06 Lagarosiphon major control programme in Lake Wanaka (HAM2006–081), National Institute of Water and Atmospheric Research, Hamilton, New Zealand.

Clayton, J. S., Champion, P. D., & McCarter, N. H. (1992). *Control of Hydrilla verticillata in a New Zealand lake using triploid grass carp*. In Proceedings of the eighth international symposium on biological control of weeds, February 1992, Lincoln, New Zealand.

Clayton, J. S., Wells, R., Champion, P., & Blair, N. (2000). Rototilling and alternative options for control of Lagarosiphon major in Paddock Bay, Lake Wanaka (NIWA report 78), Hamilton, New Zealand.

Clayton, J. S., & Wells, R. D. S. (1999). Some issues in risk assessment reports on grass carp and silver carp (conservation advisory science notes 257, Department of Conservation, Wellington, New Zealand.

Coetzee, J. A., Bownes, A., & Martin, G. D. (2011). Prospects for the biological control of submerged macrophytes in South Africa. *African Entomology, 19*(2), 469–487.

Colautti, R., Grigorovich, I., & MacIsaac, H. (2006). Propagule pressure: A null model for biological invasions. *Biological Invasions, 8*(5), 1023–1037. https://doi.org/10.1007/s10530-005-3735-y

Combroux, I., Bornette, G., Willby, N., & Amoros, C. (2001). Regenerative strategies of aquatic plants in disturbed habitats: The role of the propagule bank. *Archiv für Hydrobiologie, 152*(2), 215–232.

Cook, C. D. K., Gut, B. J., Rix, E. M., & Scheller, J. (1974). *Water plants of the world* (1st ed.). Springer.

Cooke, G. D. (1980). Lake level drawdown as macrophyte control technique. *Water Resources Bulletin, 16*(2), 317–322.

Cooke, G. D., Welch, E. B., Peterson, S., & Nichols, S. A. (2005). *Restoration and management of lakes and reservoirs*. CRC Press.

Crooks, J., Chang, A., & Ruiz, G. (2011). Aquatic pollution increases the relative success of invasive species. *Biological Invasions, 13*(1), 165–176. https://doi.org/10.1007/s10530-010-9799-3

Crooks, J. A. (2002). Characterizing ecosystem-level consequences of biological invasions: The role of ecosystem engineers. *Oikos, 97*(2), 153–166. https://doi.org/10.1034/j.1600-0706.2002.970201.x

Day, C. (2014). Managing a priority outlier infestation of Cabomba caroliniana in a natural wetland in the Blue Mountains, NSW, Australia: Could this be eradication? 19th Australasian weeds conference, Hobart, Australia.

Day, C., Petroeschevsky, A., Pellow, B., Bevan, J., O'Dwyer, T., Lawrence, A. S., & Smith, G. (2014). Managing a priority outlier infestation of Cabomba caroliniana in a natural wetland in the Blue Mountains, NSW, Australia: Could this be eradication? Hobart. Retrieved from https://www.cabdirect.org/cabdirect/abstract/20163345217

Dibble, E. D., & Kovalenko, K. (2009). Ecological impact of grass carp: A review of the available data. *Journal of Aquatic Plant Management, 47*, 1–15.

Dugdale, T. M., Butler, K. L., Clements, D., & Hunt, T. D. (2013). Survival of cabomba (Cabomba caroliniana) during lake drawdown within mounds of stranded vegetation. *Lake and Reservoir Management, 29*(1), 61–67. https://doi.org/10.1080/10402381.2013.775198

Edmunds, R. C., Villacorta-Rath, C., Collins, C. J., & Burrows, D. (2019). Monitoring for invasive Cabomba caroliniana eDNA in two Darwin River billabongs following eradication attempts. Report 19/25. Centre for Tropical Water and Aquatic Ecosystem Research (TropWATER), James Cook University, Townsville, Australia.

Emery-Butcher, H. E., Beatty, S. J., & Robson, B. J. (2020). The impacts of invasive ecosystem engineers in freshwaters: A review. *Freshwater Biology*. https://doi.org/10.1111/fwb.13479

Ferreira, R. L. M., Pereira, E. D. S., Har, N. T. F., & Hamada, N. (2003). Mansonia spp. (Diptera: Culicidae) associated with two species of macrophytes in a Varzea lake, Amazonas, Brazil. *Entomotropica, 18*(1), 21–25.

Feuchtmayr, H., Moran, R., Hatton, K., Connor, L., Heyes, T., Moss, B., Harvey, I., & Atkinson, D. (2009). Global warming and eutrophication: Effects on water chemistry and autrophic communities in experimental hypertrophic shallow lake mesocosms. *Journal of Applied Ecology, 46*, 713–723. https://doi.org/10.1111/j.1365-2664.2009.01644.x

Floerl, O., Inglis, G. J., Dey, K., & Smith, A. (2009). The importance of transport hubs in stepping-stone invasions. *Journal of Applied Ecology, 46*(1), 37–45. https://doi.org/10.1111/j.1365-2664.2008.01540.x

Fowler, M. C., & Robson, T. O. (1978). The effects of the food preferences and stocking rates of grass carp (Ctenopharyngodon idella Val.) on mixed plant communities. *Aquatic Botany, 5*, 261–276. https://doi.org/10.1016/0304-3770(78)90069-4

Gassmann, A., Cock, M., Shaw, R., & Evans, H. (2006). The potential for biological control of invasive alien aquatic weeds in Europe: A review. *Hydrobiologia, 570*(1), 217–222. https://doi.org/10.1007/s10750-006-0182-4

Gettys, L. A., Haller, W. T., & Petty, D. G. (2020). *Biology and control of aquatic plants: A best management practices handbook* (4th ed.). Aquatic Ecosystem Restoration Foundation.

Ghahramanzadeh, R., Esselink, G., Kodde, L. P., Duistermaat, H., van Valkenburg, J. L. C. H., Marashi, S. H., Smulders, M. J. M., & van de Wiel, C. C. M. (2013). Efficient distinction of invasive aquatic plant species from non-invasive related species using DNA barcoding. *Molecular Ecology Resources, 13*(1), 21–31. https://doi.org/10.1111/1755-0998.12020

Hanlon, S. G., Hoyer, M. V., Cichra, C. E., & Canfield, J. D. E. (2000). Evaluation of macrophyte control in 38 Florida lakes using triploid grass carp. *Journal of Aquatic Plant Management, 38*, 48–54.

Hippolite, J., & Kurapa, T. W. M.-T. (2018). Aquatic pest report 2018. Bay of Plenty Regional Council, Whakatane, New Zealand.

Hussner, A., Stiers, I., Verhofstad, M. J. J. M., Bakker, E. S., Grutters, B. M. C., Haury, J., van Valkenburg, J. L. C. H., Brundu, G., Newman, J., Clayton, J. S., Anderson, L. W. J., & Hofstra, D. (2017). Management and control methods of invasive alien freshwater aquatic plants: A review. *Aquatic Botany, 136*(Supplement), 112–137. https://doi.org/10.1016/j.aquabot.2016.08.002

Hussner, A., Van de Weyer, K., Gross, E. M., & Hilt, S. (2010). Comments on increasing number and abundance of non-indigenous aquatic macrophyte species in Germany. *Weed Research, 50*(6), 519–526. https://doi.org/10.1111/j.1365-3180.2010.00812.x

Inkson, T., Petroeschevsky, A., Officer, D., & Schmitzer, T. (2014). Cabomba control for the protection of the Ramsar listed Myall Lakes. 19th Australasian weeds conference, Hobart.

Jacobs, M. J., & Macisaac, H. J. (2009). Modelling spread of the invasive macrophyte Cabomba caroliniana. *Freshwater Biology, 54*(2), 296–305. https://doi.org/10.1111/j.1365-2427.2008.02108.x

James, W. F., Barko, J. W., & Eakin, H. L. (2002). Water quality impacts of mechanical shredding of aquatic macrophytes. *Journal of Aquatic Plant Management, 40*, 36–42.

Jeppesen, E., Søndergaard, M., Søndergaard, M., & Christoffersen, K. (1998). *The structuring role of submerged macrophytes in lakes* (Vol. 131). Springer. https://doi.org/10.1007/978-1-4612-0695-8

Johnson, L. E., Ricciardi, A., & Carlton, J. T. (2001). Overland dispersal of aquatic invasive species: A risk assessment of transient recreational boating. *Ecological Applications, 11*(6), 1789–1799. https://doi.org/10.1890/1051-0761(2001)011[1789:ODOAIS]2.0.CO;2

Johnstone, I. M., Coffey, B. T., & Howard-Williams, C. (1985). The role of recreational boat traffic in interlake dispersal of macrophytes: A New Zealand case study. *Journal of Environmental Management, 20*, 263–279.

Keane, R. M., & Crawley, M. J. (2002). Exotic plant invasions and the enemy release hypothesis. *Trends in Ecology & Evolution, 17*(4), 164–170. https://doi.org/10.1016/S0169-5347(02)02499-0

Kovalenko, K., & Dibble, E. (2011). Effects of invasive macrophyte on trophic diversity and position of secondary consumers. *Hydrobiologia, 663*(1), 167–173. https://doi.org/10.1007/s10750-010-0570-7

Kuehne, L. M., Olden, J. D., & Rubenson, E. S. (2016). Multi-trophic impacts of an invasive aquatic plant. *Freshwater Biology*. https://doi.org/10.1111/fwb.12820

Lowe, S., Browne, M., Boudjelas, S., & M., D. P. (2000). *100 of the worlds worst invasive alien species: A selection from the global invasive species database*. Invasive Species Specialist Group. Retrieved from www.issg.org/booklet.pdf

Matthews, J., Beringen, R., Lamers, L. P. M., Odé, B., Pot, R., van der Velde, G., van Valkenburg, J. L. C. H., Verbrugge, L. N. H., & Leuven, R. S. E. W. (2013). *Risk analysis of the non-native fanwort (Cabomba caroliniana) in the Netherlands*. Netherlands Food and Consumer Product Safety Authority.

McFayden, R. E. C. (1998). Biological control of weeds. *Annual Review of Entomology, 43*, 369–393.

Mitchell, C. P. (1980). Control of water weeds by grass carp in two small lakes. *New Zealand Journal of Marine and Freshwater Research, 14*, 381–390.

Muniappan, R., Reddy, G. V. P., & Raman, A. (2009). *Biological control of tropical weeds using arthropods*. Cambridge University Press.

Murdoch, W. W., Chesson, J., & Chesson, P. L. (1985). Biological control in theory and practice. *The American Naturalist, 125*(3), 344–366. http://www.jstor.org/stable/2461520

Murphy, K., Efremov, A., Davidson, T. A., Molina-Navarro, E., Fidanza, K., Crivelari Betiol, T. C., Chambers, P., Tapia Grimaldo, J., Varandas Martins, S., Springuel, I., Kennedy, M., Mormul, R. P., Dibble, E., Hofstra, D., Lukács, B. A., Gebler, D., Baastrup-Spohr, L., & Urrutia-Estrada, J. (2019). World distribution, diversity and endemism of aquatic macrophytes. *Aquatic Botany, 158*, 103127. https://doi.org/10.1016/j.aquabot.2019.06.006

Netherland, M. D., Green, W. R., & Getsinger, K. D. (1991). Endothall concentration and exposure time relationships for the control of Eurasian watermilfoil and hydrilla. *Journal of Aquatic Plant Management, 29*, 61–67.

Noble, R. L., Bettoli, P. W., & Betsill, R. K. (1986). Considerations for the use of grass carp in large, open systems. *Lake and Reservoir Management, 2*(1), 46–48. https://doi.org/10.1080/07438148609354599

Oosterhout, E. V. (2009). *Cabomba control manual: Current management and control options for cabomba (Cabomba caroliniana) in Australia*. NSW Department of Primary Industries. Retrieved from https://www.cabdirect.org/cabdirect/abstract/20093341841

Perna, C., & Burrows, D. (2005). Improved dissolved oxygen status following removal of exotic weed mats in important fish habitat lagoons of the tropical Burdekin River floodplain, Australia. *Marine Pollution Bulletin, 51*(1-4), 138–148. https://doi.org/10.1016/j.marpolbul.2004.10.050

Perna, C. N., Cappo, M., Pusey, B. J., Burrows, D. W., & Pearson, R. G. (2011). Removal of aquatic weeds greatly enhances fish community richness and diversity: An example from the Burdekin River floodplain, tropical Australia. *River Research and Applications*. https://doi.org/10.1002/rra.1505

Pieterse, A. H., & Murphy, K. J. (Eds.). (1990). *Aquatic weeds: The ecology and management of nuisance aquatic plants*. Oxford University Press.

Pípalová, I. (2006). A review of grass carp use for aquatic weed control and its impact on water bodies. *Journal of Aquatic Plant Management, 44*, 1–12.

Price, T., & Collins, C. (2016). Silt, snags and snapping handbags - Managing Cabomba caroliniana in the tropics of the Northern Territory. 20th Australasian weeds conference, Perth.

Riis, T., Lambertini, C., Olesen, B., Clayton, J. S., Brix, H., & Sorrell, B. K. (2010). Invasion strategies in clonal aquatic plants: Are phenotypic differences caused by phenotypic plasticity or local adaptation? *Annals of Botany, 106*(5), 813–822. https://doi.org/10.1093/aob/mcq176

Rothlisberger, J. D., Chadderton, W. L., McNulty, J., & Lodge, D. M. (2010). Aquatic invasive species transport via trailered boats: What is being moved, who is moving it, and what can be done. *Fisheries, 35*(3), 121–132. https://doi.org/10.1577/1548-8446-35.3.121

Rowe, D. K., & Champion, P. D. (1994). *Biomanipulation of plants and fish to restore Lake Parkinson: A case study and its implications*. New Zealand Limnological Society.

Russell, R. C. (1999). Constructed wetlands and mosquitoes: Health hazards and management options - an Australian perspective. *Ecological Engineering, 12*, 107–124.

Scheffer, M., Hosper, S. H., Meijer, M. L., Moss, B., & Jeppesen, E. (1993). Alternative equilibria in shallow lakes. *Trends in Ecology & Evolution, 8*(8), 275–279. https://doi.org/10.1016/0169-5347(93)90254-M

Scheffer, M., Szabó, S., Gragnani, A., van Nes, E. H., Rinalid, S., Kautsky, N., Norberg, J., Roijackers, R. M. M., & Franken, R. J. M. (2003). Floating plant dominance as a stable state. *Proceedings of the National Academy of Sciences of the United States of America, 100*(7), 4040–4045.

Schooler, S., Julien, M., & Walsh, G. C. (2006). Predicting the response of Cabomba caroliniana populations to biological control agent damage. *Australian Journal of Entomology, 45*(4), 327–330. https://doi.org/10.1111/j.1440-6055.2006.00559.x

Schooler, S. S. (2008). Shade as a management tool for the invasive submerged macrophyte, Cabomba caroliniana. *Journal of Aquatic Plant Management, 46*(2), 168.

Scriver, M., Marinich, A., Wilson, C., & Freeland, J. (2015). Development of species-specific environmental DNA (eDNA) markers for invasive aquatic plants. *Aquatic Botany, 122*, 27–31. https://doi.org/10.1016/j.aquabot.2015.01.003

Tanner, C. C., & Clayton, J. S. (1984). *Control of submerged weeds in flowing water using viscous gel diquat*. In Proceedings of the New Zealand weed and pest control conference. 46–49.

van Nes, E. H., & Scheffer, M. (2005). Implications of spatial heterogeneity for catastrophic regime shifts in ecosystems. *Ecology, 86*(7), 1797–1807. Retrieved from http://www.jstor.org/stable/3450623

van Valkenburg, J. L. C. H., Roijackers, R. M. M., & Leonard, R. (2011). *Cabomba caroliniana gray in The Netherlands*. In 3rd international symposium on weeds and invasive plants, Ascona, Switzerland.

Wheeler, G. S., & Center, T. D. (2001). Impact of the biological control agent Hydrellia pakistanae (Diptera: Ephydridae) on the submersed aquatic weed Hydrilla verticillata (Hydrocharitaceae). *Biological Control, 21*(2), 168–181. https://doi.org/10.1006/bcon.2001.0927

Williams, A. E., Duthie, H. C., & Hecky, R. E. (2005). Water hyacinth in Lake Victoria: Why did it vanish so quickly and will it return? *Aquatic Botany, 81*(4), 300–314. https://doi.org/10.1016/j.aquabot.2005.01.003

Williamson, M. H. (1996). *Biological invasions*. Chapman & Hall.

Chapter 8
Nanoherbicides: A New Frontier for Weed Management in Agriculture

Simerjeet Kaur, Gulshan Mahajan, and Bhagirath Singh Chauhan

Abstract There is significant potential for the application of nanotechnology to improve weed management. Nano-agrochemicals can act as fertilizers, pesticides, and plant growth-promoting rhizobacteria. Among nano-enabled pesticides, nanoherbicides offer desirable characteristics, including targeted delivery, chemical stability, controlled release, enhanced bioavailability and solubility, and reduced sorption and decomposition losses. These benefits can be realized through nanoformulations and nanocarriers, for example, the use of nanomaterials in formulations may significantly alter the environmental fate and effectiveness of active ingredients. The use of nanotechnology in herbicide formulations should be approached cautiously because of the potential alteration of risk to nontarget organisms. Therefore, a critical analysis of factors influencing the interaction of plants with nanoparticles, the fate of herbicides in the environment, and strategies for ecological risk assessment of nanoherbicides is warranted.

Keywords Herbicide efficacy · Nano-formulation · Nanoparticles · Precision agriculture

1 Introduction

Five decades after Feynman postulated that "there's plenty of room at the bottom" (Feynman, 1960), a novel technology called nanotechnology emerged. Nanotechnology is the fifth revolutionary technology of the century after

S. Kaur (✉)
Punjab Agricultural University, Ludhiana, India
e-mail: simer@pau.edu

G. Mahajan · B. S. Chauhan
Queensland Alliance for Agriculture and Food Innovation (QAAFI), The University of Queensland, Gatton, QLD, Australia

B. S. Chauhan, G. Mahajan (eds.), *Recent Advances in Weed Science*,
https://doi.org/10.1007/978-3-031-91014-2_8

biotechnology. This technology employs nanomaterials with at least one external dimension ranging from 1 to 100 nm, which have novel size-dependent properties. Clay minerals are the most abundant naturally occurring nanomaterials with active sites (clay–water surface and particle–particle interactions) at the nanoscale (Johnston, 2010). This nanoscale size of the particles plays a significant role in biological applications and imparts unique characteristics compared to their bulk counterparts. In general, nanoparticles can be metallic, polymeric, quantum dots, fullerene, carbon tubes, and nanorod nanoparticles (Khan et al., 2019). Polymeric nanostructures or nanocomposites are made up of natural or synthetic polymers such as protein-based polymers like albumin, gelatin, and collagen, as well as polysaccharides like chitosan, alginate, starch, lignin, and cyclodextrin. Synthetic polyesters like polyethylene glycol (PEG), polycaprolactone (PCL), poly-glycolic acid (PGA), poly-lactic acid (PLA), poly(lactic-co-glycolic acid) (PLGA), and poly-D,L-lactic acid have widespread use in nanopesticide synthesis (Rani et al., 2023). The size-structure-dependent physicochemical properties of nanomaterials are being used in material science, medicine, electronics, food, pharmacology, and agriculture. A range of nanoparticles like zinc, silver, silica, gold, aluminum, copper, titanium, nano clay, carbon nanotubes, chitin, and graphene, and bio-nano-hybrid agro particles (e.g., carboxymethyl cellulose) in the agricultural sector (Yin et al., 2018). There has been remarkable progress in the use of nanoparticles for the development of nanopesticides and nanofertilizers. Over the past decade, nanotechnology has been explored to enhance input use efficiency and ensure the controlled release and targeted delivery of agrochemicals to plants. The number of peer-reviewed research and review papers on this subject has shown exponential growth over the last decade.

2 Scope of Nanotechnology for Weed Control

Weeds compete for growth resources like nutrients, moisture, space, light, etc., with all field crops throughout life and, therefore, cause more damage to crops than any other pest. Worldwide, food loss due to weed competition accounts for 11.5% of total food production (Oerke, 2006). The yield loss varies in different crops (40–69% in rice, 34–43% in wheat, 50–56% in pulses, 52–75% in sugarcane, and 67–90% in vegetables), depending on the type of weeds, time of weed emergence, weed density, type of crop, duration of crop, season, etc. (Chauhan et al., 2017). In some cases, severe crop-weed competition may lead to complete crop failure. Weeds also reduce the quality of economic produce and can act as alternate hosts for insect pests and pathogens. Climate change may intensify weed competition with crops and pose challenges in enhancing and sustaining agricultural productivity.

The global agricultural production needs to increase by 63% relative to levels in 2005 to meet global food security challenges (FAO, 2024). Effective weed control is essential for achieving the potential yield of crops and global food security. Weeds can be controlled by chemical, physical, cultural, and biological methods. Physical measures of weed control, such as hand weeding, is a time-consuming, laborious, and costly operation. On a long-term basis, tillage may cause soil erosion due to wind and water, damage soil structure, and stimulate weed seedling recruitment from the seed bank buried in the soil. Conservation agriculture systems offer the benefits of reduced fuel and labor use along with significantly reduced soil erosion. However, the problem of perennial weeds persists in farming systems because of their below-ground plant parts—such as tubers (e.g., *Cyperus esculentus* L.), suckers (e.g., *Cynodon dactylon* (L.) Pers.), runners (e.g., *Equisetum hyemale* L.), bulbs (e.g., *Allium triquetrum* L.), rhizomes (e.g., *Sorghum halepense* (L.) Pers.)), etc., disperse and get established in the ensuing season.

Herbicides have become an essential tool of weed control in modern agriculture due to their ease, rapidity, and cost-effectiveness (Chauhan & Mahajan, 2014). Herbicide production has become a multi-billion-dollar industry, comprising nearly two-thirds of the total pesticide market in many countries, such as the USA and Australia. However, only a fraction of applied herbicides (less than 0.1%) is responsible for weed control. The rest of it is lost via leaching, microbial decomposition and hydrolysis in the environment (Caldwell & Mohler, 2001). Herbicides used at higher than recommended doses or inappropriate herbicide mixtures can leave residues in crop produce and contaminate soil or underground water, which may be detrimental to human and ecosystem health. The use of herbicides that persist for long periods in the soil and crop residues can prove hazardous to the succeeding crops. The volume of agrochemicals used, their low efficacy, off-site migration and contamination of food and environment are some of the concerns associated with chemical weed control. Chemical control of perennial weeds is difficult due to a limited number of herbicides available for their control. The challenges associated with conventional herbicides can potentially be addressed with the application of nanotechnology to develop more effective and environmentally benign herbicides. Designing nanoscale agrochemicals that combine optimal safety and functionality profiles will be helpful for sustainable weed control and agroecosystems (Zhang et al., 2021). With the help of emerging nanotechnology, it is possible to improve weed control while decreasing environmental risks.

3 Role of Nanotechnology in Weed Control

There is growing interest in the application of nanotechnology for developing new plant protection products. A nano-enabled pesticide is a product in which a nanomaterial has been used to enhance the functionality and/or alter the risk profile of a conventional active ingredient. Different types of herbicides based on colloidal solutions of nanoparticles like copper, zinc, silver, iron, and cupric oxide have been synthesized and used for weed management (Konotop et al., 2014). Nanoformulations may be prepared by combining nanometer-sized inorganic components, organic polymers, and active ingredients. Different techniques, such as encapsulation and entrapment of the active ingredient with polymers and/or dendrimers with attachments like surface ionic and weak bonds, may be utilized for devising nano herbicide formulations (Kah et al., 2013). The nanoencapsulation process comprises sealing active ingredients that have been used for decades in conventional formulations within a nano-sized thin-walled sac or shell as a protective coating. In addition, nano-coating of herbicides on seeds can be done to eliminate early crop-weed competition. A range of technologies, such as nanocomposites, nanoemulsions, and nanocapsules, etc., can be utilized to nano-enable herbicides for their enhanced efficacy and better environmental footprint. Polymeric nanoparticles (e.g., poly-ε-caprolactone for encapsulation of atrazine (Pereira et al., 2014) and alginate/chitosan for encapsulation of paraquat (Grillo et al., 2014)) and inorganic nanoparticles (such as silica dioxide nanoparticle for pH-sensitive herbicide) have been used to reduce the adverse environmental effects caused by conventional herbicides. However, the majority of nanoherbicides greatly exceed the 100 nm size boundary. Many products such as nanoemulsions and formulants such as clays and polymers have been on the market without being labeled or classified as "nano" (Kah, 2015; Kah et al., 2013; Walker et al., 2017). Therefore, establishing a size threshold to distinguish "nano" from "non-nano" is of limited value for agrochemicals.

The increasing use of nanotechnology in herbicide development, particularly through innovative formulations and encapsulation techniques, offers enhanced efficacy and reduced environmental impact. However, the absence of clear size thresholds and labeling for herbicides poses challenges to effective regulation and consumer awareness in the agricultural sector.

4 Functionality of Nano-Enabled Herbicides

Precision agriculture requires novel approaches to optimize the targeting, uptake, delivery, and long-term impacts of agrochemicals on soil microbial communities. Many reports have foreseen that nanotechnology will further result in the development of precision agriculture and will facilitate the optimum use of inputs, including pesticides, fertilizers, and water. It has been observed that herbicide efficacy is enhanced by 20–30% in nano-based formulations compared to their conventional analogs (Kah et al., 2018). Some of the functionalities that can be achieved with the application of nanotechnology include sustained release, targeted delivery, and improved bioavailability and efficacy, resulting in lower application rates and reduced loss to environmental compartments (de Oliveira et al., 2015; Hussein et al., 2005). Some examples of these are given in Table 8.1, elaborated in Fig. 8.1, and discussed in the following sections.

Table 8.1 Some examples of functionality of nanoformulations and the benefits associated with nano-enabled herbicides

Functionality	Formulation	Example	References
Controlled release of active ingredient (AI)	Nanocapsules, nanospheres, solid-lipid encapsulation	Pendimethalin, acetochlor, diuron, nanocomposite 2,4-D and 2,5-T, simazine, atrazine, paraquat	Kanimozhi & Chinnamuthu (2012), Walker et al. (2017), Guo et al. (2014), Yu et al. (2015), Sarijo et al. (2010), Choy et al. (2006), Hussein et al. (2005), de Oliveira et al. (2015), Sousa et al. (2018), Grillo et al. (2014), Silva et al. (2010)
Targeted delivery of active ingredient	Nanocapsules	Nanoencapsulated glyphosate or sulfonylurea herbicide	Pérez-de-Luque & Rubiales (2009)
Increased rainfastness/ improved efficacy	Dendrimer technology	Glyphosate	Carpenter (2016)
Lowered dose and improved efficacy of AI	Polymer encapsulation	Metsulfuron methyl loaded on pectin	Choy et al. (2006), Kumar et al. (2017)
Increased solubility	Dendrimer technology	Trifluralin	Carpenter (2016)
Lowered toxicity of AI to nontarget organism (lower dose)	Solid-lipid encapsulation, polymer	Simazine, atrazine, imazapyr, imazapic	De Oliveira et al. (2015), Pereira et al. (2014), Silva et al. (2010), Grillo et al. (2014), Maruyama et al. (2016)
Protected AI against premature degradation	Nanocapsules with catalyst AI conjugate	Pendimethalin	Walker et al. (2017)
Enhanced decomposition of AI in soil and/or plant	Nanocatalyst conjugated AI in microcapsules	2,4-D	Luo et al. (2014)
Increased weed control spectrum	Nanoencapsulation	Metsulfuron methyl for parasitic weed control	Pérez-de-Luque & Rubiales (2009)

4.1 Controlled Release

Herbicide transport processes, mainly soil leaching and surface runoffs, can be avoided using controlled-release formulations. The release rate of the active ingredient can be slowed down by the use of nano-sized clay-based adsorbent or carrier. Nano-enabled herbicide formulations may increase herbicide activity and improve the delivery of herbicide either due to better adhesion of the nano-formulation to the leaves or by increased uptake by leaf stomata. The formulation of active ingredients

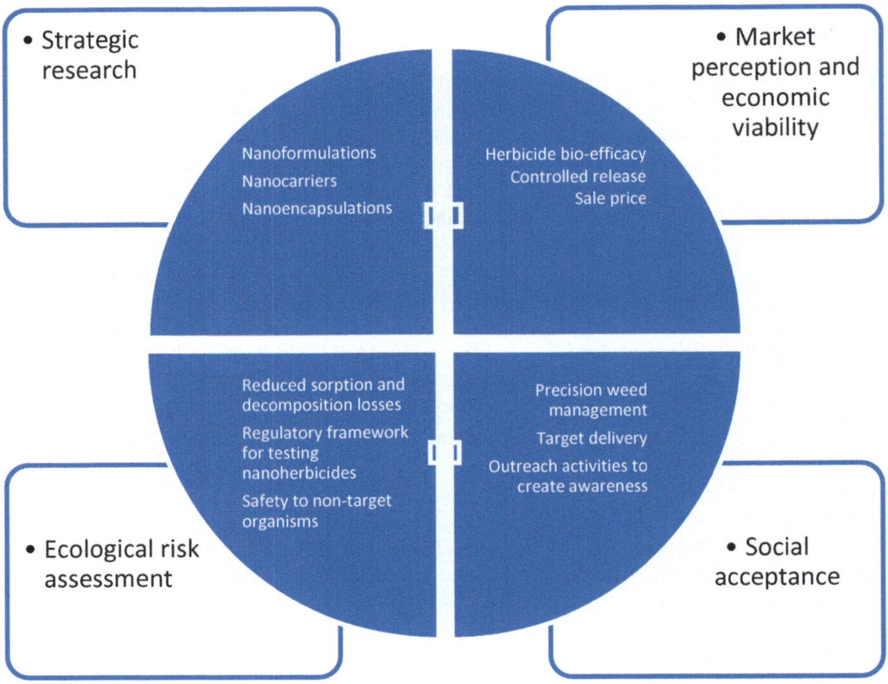

Fig. 8.1 Functionalities of nanoherbicides and strategies for precision management

with nanoparticles helps improve herbicide efficacy by modifying the release of the active ingredient over time. The restricted release of the herbicide over time in a formulation containing nanoparticles as a carrier may protect the active ingredient from degradation and improve the effectiveness of the formulation. Effective and long-term weed control with a preemergent herbicide can only be achieved if the herbicide is persistent for time long enough to control weeds during the critical period of crop-weed competition. The slower release of pendimethalin was made possible using manganese carbonate core-shell nanoparticles (Kanimozhi & Chinnamuthu, 2012; Walker et al., 2017). Nano-encapsulation of herbicides with poly-ε-caprolactone, chitosan, or carboxymethyl chitosan improved herbicidal activity by modified-release of the active ingredient over time. The controlled release of acetochlor was achieved with the use of poly-butyl methacrylate-diacetone acrylamide (Guo et al., 2014) and of diuron by cross-linking with glutathione-responsive carboxymethyl chitosan nanoparticles (Yu et al., 2015). Hybridization of phenoxy herbicides like 2,4-D and 2,4,5-T into zinc-aluminum double hydroxide nanoparticles can be done for the controlled release of these herbicides over time and improved weed control efficacy (Sarijo et al., 2010). For the sustained release of pesticides, the use of low-cost carriers like clay nanotubes resulted in 70-80% less use of pesticides and better contact of pesticides with plants (Choy et al., 2006). These types of clay nanotubes can also be developed to enhance herbicides' efficacy.

Both commercial formulations and poly-ε-caprolactone nanocapsules containing atrazine (at 2000 g ha^{-1}) reduced the shoot and root growth of *Amaranthus viridis* and *Bidens pilosa* plants, leading to a loss of plant biomass (Sousa et al., 2018). Controlled-release formulations can reduce the leaching and migration of herbicides to surface and groundwater. Radian and Mishael (2008) demonstrated that a polycation-clay nanocomposite imazapyr formulation resulted in nearly twofold reduced leaching while maintaining good herbicidal activity. Nano-capsules encapsulated by chitosan and tripolyphosphate (TPP) containing formulations are known for their photo-responsive sustained release behavior. Nitrobenzyl-carboxymethyl cellulose-encapsulated diuron demonstrated a 97% release rate over 8 h at pH 7.0 under solar irradiation, and no release was noted in the absence of light (Ye et al., 2015). Thus, controlled-release herbicide formulations using nano-sized carriers can greatly improve herbicide effectiveness at lower rates while reducing environmental contamination with reduced leaching and runoff.

4.2 Targeted Delivery

Encapsulating herbicide molecules with nanoparticles for target-specific delivery is another revolutionary step to avoid harm to nontarget organisms. Target weed plants have a specific receptor on their roots, and encapsulated herbicides target those receptors, which may result in the death of weed plants. Encapsulation of atrazine with poly-ε-caprolactone nanoparticle (Pereira et al., 2014), clomazone with chitosan/alginate nanoparticle (Silva et al., 2010), and paraquat with silver nanoparticles-chitosan/tripolyphosphate nanoparticle (Grillo et al., 2014) was aimed to minimize environmental contamination and reduce toxicity to nontarget organisms. Further, there was reduced toxicity of nano herbicides in nontarget plants such as *Allium cepa* and organisms, such as algae and human cells, compared with free atrazine (Clemente et al., 2014; Pereira et al., 2014). For instance, imazapic and imazapyr, when encapsulated in chitosan-TPP nanoparticles, showed less cytotoxicity and genotoxicity to bacteria linked to the nitrogen cycle (Maruyama et al., 2016). Oliveira et al. (2015) demonstrated that atrazine-containing poly-ε-caprolactone nanocapsules led to oxidative stress and caused little adverse effects on photosynthetic parameters in maize leaves after application, but these effects were transient and shoot growth remained unaffected. The innovative use of nanotechnology can detect and use precisely the right amount of herbicide, which provides effective weed control and enhances productivity while ensuring environmental safety and high efficacy. Nanoencapsulation also helps in alleviating environmental problems due to reductions in chemical runoff.

Overall, nanoencapsulation of herbicides enhances weed control efficiency through target-specific delivery in order to reduce the negative impact on non-target organisms and contamination of the environment through precise application.

4.3 Increased Availability and Durability of Herbicides

Nanoencapsulation of herbicides may increase the durability of herbicide formulations (persistence) in a diverse environment. The formulation of active ingredients with nanoparticles helps to increase herbicide shelf-life by shielding the active ingredient against unfavorable environmental conditions. Nano-emulsions of active ingredients may be prepared to improve the physicochemical properties like particle size, particle ageing rate and thermo-stability. Sousa et al. (2018) demonstrated that physiological parameters and growth of *Bidens pilosa* L. and *Amaranthus viridis* L. were more affected by atrazine-loaded poly-ε-caprolactone nanocapsules than by commercial atrazine. This increased activity of atrazine-loaded nanocapsules on these weeds was not due to increased toxicity, but rather was due to the modified release pattern of atrazine by the nanocapsules that improved the delivery to the target organism through better adhesion to leaves and by enhancing the uptake by leaf stomata. For pendimethalin (a preemergent herbicide that does not move much into the soil, prone to volatilization losses, and requires moisture for high efficacy), a carrier like hydrogel may deliver the chemical deep into the soil to such a level that it increases its efficacy and is also attractive from an environmental perspective (Walker et al., 2017). Similarly, the activity of metolachlor-loaded poly(lactide-co-glycolides) was also higher than the conventional formulation of metolachlor (Tong et al., 2017). Agro-industrial waste-based nanoparticles, such as a nano-sized rice husk, were used as nano-carriers for 2,4-D, which resulted in enhanced and reversible sorption of 2,4-D with similar herbicide activity (Abigail et al., 2016). Schnoor et al. (2018) synthesized poly(lactic-co-glycolic acid) nanoherbicides with 50% encapsulation efficiency for atrazine as the active ingredient, which produced a decrement of 40% in root length with less phytotoxicity to nontarget plants. Because nanotechnology may be used to increase the persistence and residual effect of herbicides in the soil, this may in turn, decrease the amount of herbicides required for effective weed control over an area and lessen associated adverse environmental effects. In short, by extending shelf-life, improving delivery mechanisms, and moderating persistence in the environment, nanoencapsulation makes herbicides more durable and effective and potentially reduces chemical use and environmental impact.

4.4 Lower Application Rates

Product-containing nanoparticles can provide similar effectiveness and duration of control as classical herbicide formulations, but using 10–15 times less product. It was observed that metsulfuron-methyl loaded onto pectin (a polysaccharide) nanoparticles resulted in improved efficacy on *Chenopodium album* L. with the reduced application (Kumar et al., 2017). SThe use of 10-fold diluted atrazine-loaded PCL nanocapsules (200 g ha^{-1}) resulted in the same inhibitory effect in root and shoot growth of *Amaranthus viridis* and *Bidens pilosa* plants as the commercial formulation at the standard atrazine dose at 2000 g ha^{-1} (Sousa et al., 2018). With the help of nanotechnology, herbicides can be applied as required in a localized region, to prevent wastage. Nanocapsules of herbicides may aid in targeted delivery, thus, ensuring the effective control of parasitic weeds, which otherwise is not possible (Pérez-de-Luque & Rubiales, 2009). As soon as nanocapsules reach the parasitic weed (due to the strong sink), the herbicide will be released or activated, resulting in efficient weed control. Overall, nanotechnology will allow us to apply herbicides in a more precise, targeted manner with greatly reduced application rates and still gain the same level of control of hard-to-kill weeds.

4.5 Increased Retention on Leaves and Enhanced Rainfastness

Nanoscale herbicide carriers may help in increasing herbicide retention on leaves. Herbicide activity can be increased with the help of nanoparticles as observed in a glyphosate formulation having nanosurfactant (Soysoap) prepared from soybean micelles (Biobased USA, 2012). Soysoap helps to reduce the chelation of glyphosate with salts present in the carrier water, such as manganese, zinc and other trace elements. The nanoscale carriers for soil-applied herbicides may be designed to anchor plant roots to the surrounding soil colloids and organic matter. The use of nanotechnology in this direction will help in reducing the soil uptake of active ingredients, thus reducing the amount to be used and the residual effects. The improved rainfastness of glyphosate was demonstrated in glasshouse trials using Starpharma's Pro Star® dendrimers and got license for increasing solubility of active ingredients to achieve an improved 2,4-D formulation in the United States (Carpenter, 2016).

5 Other Potential Applications of Nanotechnology

The large surface-to-volume ratio of nanoparticles leads to efficient rates of chemical reactivity.

1. The problem of herbicide residues restricts the choice of crops to be sown in an area and thus warrants mitigation strategies. The use of nanoparticles may lead to more rapid degradation of toxic compounds and disinfection of soil due to photocatalysis (Mueller & Nowack, 2008). This phenomenon may be employed for the decomposition of herbicides having higher residual levels, which otherwise take months to degrade under normal conditions.
2. This application of nanotechnology may help in reducing selection pressure caused by persistent herbicides, and therefore useful in delaying the evolution of herbicide resistance (Chauhan et al., 2017). The intelligent delivery, lower appli-

cation rates, and high penetration efficiency of nanoherbicides may help in eliminating the weeds before resistance could evolve at the early stage (Abigail & Chidambaram, 2017).

3. Ground and surface water contamination may be reduced by remediation (removal and/or immobilization). Clay minerals, naturally occurring and synthetic nanomaterials, can be used as adsorbents for herbicides. Celis et al. (2012) successfully demonstrated the use of montmorillonite-chitosan bionanocomposites as an adsorbent of clopyralid in soil and water under mildly acidic conditions.
4. The negative effect of herbicides on living organisms in ponds infested with aquatic weeds is an emerging issue, and nanoparticles may be employed for the bioremediation of aquatic bodies. The use of photocatalytic degradation for wastewater treatment (for purification and decontamination) has gained popularity, and this may be useful in aquatic weed control.
5. Photoluminescence using nanoparticles may detect the presence of herbicide residues in water. Based on chemiluminescence, aryloxy-phenoxy-propionic acid herbicides can be detected in water with the help of carbon nanoparticles, triazine herbicides with the use of graphene nanoparticles, and 2,4-D can be detected with the help of gold or silver nanoparticles (Luo et al., 2014).

6 Challenges Associated with Nanoherbicides

Sustained and increased efforts are needed to strengthen the research and development of suitable nanotechnology for herbicides with desired properties both in terms of improved weed control and minimal environmental impact.

1. Major concerns are the usage of toxic elements and the generation of heat in the conventional method of synthesis of nanoparticles (Kavitha et al., 2013). Bioconjugation and encapsulation with biogenic sources like plants, fungi, bacteria, actinomycetes, algae, or their products can be employed in the synthesis of different types of nanoparticles.
2. The ultra-small size of nanoparticles is a technological marvel, but due to indiscriminate use, the same property may cause significant hazards to target and non-target users including humans, plants, animals and the environment. Airborne nanoparticles as herbicides (nanoparticles of 200–400 nm size) may be deposited on above-ground parts of nontarget plants and clog stomata or prevent pollen tube penetration on the stigma by creating a fine physical barrier layer (Jampílek & Kráľová, 2018). These may enter the vascular tissue and hinder the translocation of water, minerals, and photosynthates. Airborne nanoparticles may also enter the human body through the respiratory system or through skin pores, which may cause health hazards like inflammation, fibrillation, or genotoxicity.
3. The agricultural potential across diverse production systems can be increased by using nanotechnology in eco-friendly ways. The use of chemical herbicides for

weed control is very limited in crops grown in water-deficit (in rainfed and dry farming) regions and weeds can damage the whole crop in cases of severe crop-weed competition. For example, a soil-applied encapsulated nanoherbicide that is designed and formulated to become active only when there has been rainfall or sufficient water is available could provide highly effective weed control.

4. Nanotechnology must be used with caution and utmost care with a life-cycle assessment approach. A risk assessment framework is very important when developing nanoherbicides. There is a need to examine the agronomic relevance of nano-formulations, target and nontarget impact on the desired plants, natural enemies of agricultural pests, beneficial micro-flora and micro-fauna, and weeds. The basic aim of a risk assessment framework for nanoherbicides is to address those questions that are relevant to the product (Kookana et al., 2014). There are three important elements to be considered for the ecological risk assessment of a nanoherbicide: active ingredient, nanocarrier, and durability of the nanocarrier-active ingredient complex. If the durability of the nanocarrier-active ingredient complex of the given nanoherbicide is comparable or more than the persistence of the active ingredient, then ecotoxicology studies would have greater importance when evaluating risks. Novel nano-Quantitative Structure-Activity Relationship/Quantitative Structure-Property Relationship (nano-QSAR/QSPR) tools offer viable solutions to address experimental issues observed during pesticide risk assessment (Villaverde et al., 2018).
5. Candidate nanoparticles must have the potential for weed control along with other ideal characteristics, including biocompatibility, cost effectiveness, and minimal risk with maximum activity. Jin et al. (2014) demonstrated that single-walled carbon nanotubes have a significant negative relationship with the biomass of a broad range of microbial groups. Therefore, to reduce the risk to nontarget users, the choice of suitable nanocarrier is very important.
6. Protocols for seed coating with nanoherbicides must be designed for producing customized seeds for effective weed control to eliminate crop-weed competition in the initial period and reduce weed seed banks.
7. The profitability of the use of nanotechnology in weed control in small land holdings (small and marginal growers) must be worked out by conducting pilot field trials and participatory field demonstrations.
8. Comprehensive field studies on the efficacy of nanoherbicides and their impact on nontarget organisms are lacking. Laboratory results on nanoherbicides in terms of their increased efficacy or lower toxicity to nontarget organisms cannot be directly extrapolated to field conditions. Ideally, nanoherbicides need to be evaluated under field conditions (in vivo toxicological studies) for their efficacy and environmental fate for a period that should be long enough to assess both aspects. Jacques et al. (2017) assessed the safety of solid lipid, polymeric, and chitosan/tripolyphosphate (CS/TPP) nanoparticles loaded or not with the atrazine, simazine, and paraquat herbicides. They observed that both loaded and unloaded formulations of polymeric nanoparticles, such as NC_PCL, were the most toxic to the developmental rate of roundworms, even at lower concentra-

tions. In contrast, both loaded and unloaded CS/TPP formulations were the least toxic, even at higher concentrations.
9. The claims about the increased efficacy of nanoherbicides over the conventional formulation need to be comprehensively assessed using peer-reviewed field trials. Until such time, the broader community may remain less than convinced about the true potential of nanotechnology applications for herbicides.

7 Way Forward

Nanotechnology is increasingly being embraced across the globe for its potential use in the agriculture and food sectors. Globally, out of 242 agricultural nano-based products available in the market, 106 (44%) are nanofertilizers or nanopesticides. India is producing the largest number of nano-based agricultural products, followed by the UK and Japan (StatNano, 2023). The use of this novel technology in weed management is still in the nascent stage, and there is still a long way to achieve the goal. It is expected that it will lead to greater uptake of precision farming, and open up new avenues for sustainability and profitability of the agri-sector. The site-specific target-delivery, reduced product doses, controlled release, and longer shelf life of herbicides with nanoparticles can usher in more productive, sustainable, and environment friendly agriculture (Liu & Lal, 2015; Sekhon, 2014). However, a pre-requisite to this is to prove that nanoherbicides will not create a greater risk to the environment and human health. Hence, strategic research and policy for the safe use of nanotechnology in agriculture is crucial. There is a need to develop eco-friendly nanoparticles (biological entities, bioconjugates, and encapsulation with nanoparticles with bioactive molecules) for use in agriculture. Moving to a broader concept of nano-enabled technology will be more valuable in supporting the development of more sustainable agrochemicals (Kah, 2015). An efficacy evaluation is required before registering a nanotechnology-enabled plant protection product. This information must include the products' economic impact, agronomic sustainability, and direct efficacy against the targeted pests (FAO, 2006).

Current challenges to the realization of the full potential of nanoherbicides include a lack of adequate field evaluation, frameworks, and capability ecological risk assessment of nanomaterials (Walker et al., 2017) as well as the perception of the community (Kah, 2015), the market, and the cost of production (Kah, 2015; Pérez-de-Luque, 2017; Walker et al., 2017). A detailed investigation into the interaction of plants with nanoparticles under controlled environment will be the first step in this direction (Pérez-de-Luque, 2017). Suitable protocols for the evaluation and use of nanoherbicides need to be developed by a multidisciplinary team. Very few studies have provided evidence of how well nano-formulations behave in real-world conditions. The release and degradation behavior of nano-pesticides in the environment could be altered by many factors like nature of the active ingredient, the nature of the carrier, the association between the carrier and active ingredient, biotic (microbial degradation) and abiotic (photodegradation and chemical breakdown) processes, soil types (moisture, pH, organic carbon, soil mineral phase), and other climatic factors (such as temperature, humidity, sunlight). Therefore, it is very difficult to determine their unpredictable behavior in the environment and to address health risks and other environmental issues. The fate and behavior of nanoforms in the environment may be more complex compared to their conventional counterparts (Kah et al., 2014, 2016). Precision agriculture requires nanoinformatics approaches to optimize targeting, uptake, delivery, nutrient capture, and long-term impacts on soil microbial communities, which will enable the design of nanoscale agrochemicals that combine optimal safety and functionality profiles (Zhang et al., 2021).

There is a need to work out the economics of using nanoherbicides by formulating statistically sound pilot projects and demonstrating the technology in identified niche areas through a public-private partnership. To ensure human and environmental health in accordance with Food and Agriculture Organization (FAO) and World Health Organization (WHO) standards, clear guidelines need to be developed for proper monitoring, evaluation of nanoherbicides for intelligent delivery in the crop-weed environment complex and ecotoxicological studies. These studies are important for public awareness and to avoid being rejected by the public as has happened with many genetically modified crops. Further, there is a need for long-term evaluation of nanoherbicides under various agroclimatic conditions and crop ecologies for precision agriculture solutions in the context of climate change.

References

Abigail, E. A., & Chidambaram, R. (2017). Nanotechnology in herbicide resistance. In M. S. Seehra (Ed.), *Nanostructured materials - Fabrication to applications* (pp. 207–212). IntechOpen.

Abigail, M. E. A., Samuel, M. S., & Chidambaram, R. (2016). Application of rice husk nanosorbents containing 2,4-dichlorophenoxyacetic acid herbicide to control weeds and reduce leaching from soil. *Journal of the Taiwan Institute of Chemical Engineers, 63*, 318–326.

Biobased USA. (2012). *Soysoap research report*. Retrieved March 10, 2024, from www.biobased.us/2012-Soysoap-Research-Report.pdf

Caldwell, B., & Mohler, C. L. (2001). Stale seedbed practices for vegetable production. *Horticultural Science, 36*, 703–705.

Carpenter S. (2016). *A star performer - Priostar® dendrimers*. Grainews. Retrieved March 8, 2024, from http://news.agropages.com/News/NewsDetail-19235.html

Celis, R., Adelino, M. A., Hermosin, M. C., & Cornejo, J. (2012). Montmorillonite-chitosan bionanocompisites as adsorbents of the herbicide clopyralid in aqueous solution and soil/water suspensions. *Journal of Hazardous Materials, 209-210*, 67–76.

Chauhan, B. S., & Mahajan, G. (Eds.). (2014). *Recent advances in weed management* (pp. 1–407). Springer. https://doi.org/10.1007/978-1-4939-1019-9

Chauhan, B. S., Matloob, A., Mahajan, G., Aslam, F., Florentine, S. K., & Jha, P. (2017). Emerging challenges and opportunities for education and research in weed science. *Frontiers in Plant Science, 8*, 1537. https://doi.org/10.3389/fpls.2017.01537

Choy, J. H., Choi, S. J., Oh, J. M., & Park, T. (2006). Clay minerals and double layered hydroxides for novel biological applications. *Applied Clay Science, 36*, 122–132.

Clemente, Z., Grillo, R., Jonsson, M., Santos, N. Z. P., Feitosa, L. O., Lima, R., & Fraceto, L. F. (2014). Ecotoxicological evaluation of poly(ϙ-caprolactone) nanocapsules containing triazine herbicides. *Journal of Nanoscience and Nanotechnology, 14*, 4911–4917. https://doi.org/10.1166/jnn.2014.8681

de Oliveira, J. L., Campos, E. V., Gonçalves da Silva, C. M., Pasquoto, T., Lima, R., & Fraceto, L. F. (2015). Solid lipid nanoparticles co-loaded with simazine and atrazine: Preparation, characterization, and evaluation of herbicidal activity. *Journal of Agricultural and Food Chemistry, 63*, 422–432.

FAO. (2006). *International code of conduct on the distribution and use of pesticides guidelines on efficacy evaluation for the registration of plant protection products*. FAO. Retrieved from https://www.fao.org/publications/card/en/c/f693e5df-bbde-443d-97c8-c91d7b06ae5b/

FAO. (2024). *The future of food and agriculture—Trends and challenges* (pp. 1–163). FAO. Retrieved March 12, 2024, from http://www.fao.org/3/a-i6583e.pdf

Feynman, R. P. (1960). There's plenty of room at the bottom. *Engineering and Science, 23*, 22–36.

Grillo, R., Pereira, A. E. S., Nishisaka, C. S., Lima, R. D., Oehlke, K., Greiner, R., Leonardo, F., & Fraceto, L. F. (2014). Chitosan/tripolyphosphate nanoparticles loaded with paraquat herbicide. An environmentally safer alternative for weed control. *Journal of Hazardous Materials, 278*, 163–171.

Guo, Y. Z., Yang, Q., Yan, W., Li, B., Qian, K., Li, T., Xiao, W., & He, L. (2014). Controlled release of acetochlor from poly (butyl methacrylate-diacetone acrylamide) based formulation prepared by nanoemulsionpolymerisation method and evaluation of the efficacy. *International Journal of Environmental Analytical Chemistry, 94*, 1001–1012.

Hussein, M. Z., Yahaya, A. H., Zainal, Z., & Kian, L. H. (2005). Nanocomposite-based controlled release formulation of an herbicide, 2,4-dichlorophenoxyacetate incapsulated in zinc-aluminium-layered double hydroxide. *Science and Technology of Advanced Materials, 6*, 956–962. https://doi.org/10.1016/j.stam.2005.09.004

Jacques, M. T., Oliveira, J. L., Campos, E. V., Fraceto, L. F., & Avila, D. S. (2017). Safety assessment of nanopesticides using the roundworm Caenorhabditis elegans. *Ecotoxicology and Environmental Safety, 139*, 245–253.

Jampílek, J., & Kráľová, K. (2018). Benefits and potential risks of nanotechnology applications in crop protection. In K. Abd-Elsalam & R. Prasad (Eds.), *Nanobiotechnology applications in plant protection. Nanotechnology in the life sciences* (pp. 189–246). Springer.

Jin, L., Son, Y., De Forest, J. L., Kang, Y. J., Kim, W., & Chung, H. (2014). Single-walled carbon nanotubes alter soil microbial community composition. *Science of the Total Environment, 466*, 533–538.

Johnston, C. T. (2010). Probing the nanoscale architecture of clay minerals. *Clay Minerals, 45*, 245–279.

Kah, M. (2015). Nanopesticides and nanofertilizers: Emerging contaminants or opportunities for risk mitigation? *Frontiers in Chemistry, 3*, 64. https://doi.org/10.3389/fchem.2015.00064

Kah, M., Beulke, S., Tiede, K., & Hofmann, T. (2013). Nanopesticides: State of knowledge, environmental fate, and exposure modeling. *Critical Reviews in Environmental Science and Technology, 43*, 1823–1867.

Kah, M., Kookana, R. S., Gogos, A., & Bucheli, T. D. (2018). A critical evaluation of nanopesticides and nanofertilizers against their conventional analogues. *Nature Nanotechnology, 13*(8), 677–684. https://doi.org/10.1038/s41565-018-0131-1

Kah, M., Machinski, P., Koerner, P., Tiede, K., Grillo, R., Fraceto, L. F., & Hofmann, T. (2014). Analyzing the fate of nanopesticides in soil and the applicability of regulatory protocols using a polymer-based nanoformulation of atrazine. *Environmental Science and Pollution Research, 21*, 11699–11707.

Kah, M., Weniger, A. K., & Hofmann, T. (2016). Impacts of (nano) formulations on the fate of an insecticide in soil and consequences for environmental exposure assessment. *Environmental Science & Technology, 50*, 10960–10967.

Kanimozhi, V., & Chinnamuthu, C. R. (2012). Engineering core/hallow shell nanomaterials to load herbicide active ingredient for controlled release. *Research Journal of Nanoscience and Nanotechnology, 2*, 58–69.

Kavitha, K. S., Baker, S., Rakshith, D., Kavitha, H. U., Rao, Y. H. C., Harini, B. P., & Satish, S. (2013). Plants as green source towards synthesis of nanoparticles. *International Research Journal of Biological Sciences, 2*, 66–76.

Khan, I., Saeed, K., & Khan, I. (2019). Nanoparticles: Properties, applications and toxicities. *Arabian Journal of Chemistry, 12*, 908–931.

Konotop, Y. O., Kovalenko, M. S., Ulynets, V. Z., Meleshko, A. O., Batsmanova, L. M., & Taran, N. Y. (2014). Phytotoxicity of colloidal solutions of metal-containing nanoparticles. *Cytology and Genetics, 48*, 99–102.

Kookana, R. S., Boxall, A. B. A., Reeves, P. T., Ashauer, R., Beulke, S., Chaudhry, Q., Cornelis, G., Fernandes, T. F., Gan, J., Kah, M., Lynch, I., Ranville, J., Sinclair, C., Spurgeon, D., Tiede, K., & Brink, P. J. (2014). Nanopesticides: Guiding principles for regulatory evaluation of environmental risks. *Journal of Agricultural and Food Chemistry, 62*, 4227–4240.

Kumar, S., Bhanjana, G., Sharma, A., Dilbaghi, N., Sidhu, M. C., & Kim, K. H. (2017). Development of nanoformulation approaches for the control of weeds. *Science of the Total Environment, 586*, 1272–1278.

Liu, R., & Lal, R. (2015). Potentials of engineered nanoparticles as fertilizers for increasing agronomic productions. *Science of the Total Environment, 514*, 131–139.

Luo, M., Liu, D., Zhao, L., Han, J., Liang, Y., Wang, P., & Zhou, Z. (2014). A novel magneticionic liquid modified carbon nanotube for the simultaneous determination of aryloxyphenoxy-propionate herbicides and their metabolites in water. *Analytica Chimica Acta, 852*, 88–96.

Maruyama, C. R., Guilger, M., Pascoli, M., Bileshy-Jose, N., Abhilash, P. C., Fraceto, L. F., & de Lima, R. (2016). Nanoparticles based on chitosan as carriers for the combined herbicides imazapic and imazapyr. *Scientific Reports, 6*, 1–15.

Mueller, N. C., & Nowack, B. (2008). Exposure modeling of engineered nanoparticles in the environment. *Environmental Science & Technology, 42*, 4447–4453.

Oerke, E. C. (2006). Crop losses to pests. *The Journal of Agricultural Science, 144*, 31–43.

Oliveira, H. C., Stolf-Moreira, R., Martinez, C. B., Sousa, G. F., Grillo, R., de Jesus, M. B., & Fraceto, L. F. (2015). Evaluation of the side effects of poly(epsilon-caprolactone) nanocapsules containing atrazine toward maize plants. *Frontiers in Chemistry, 3*, 61. https://doi.org/10.3389/fchem.2015.00061

Pereira, A. E., Grillo, R., Mello, N. F., Rosa, A. H., & Fraceto, L. F. (2014). Application of poly (epsilon-caprolactone) nanoparticles containing atrazine herbicide as an alternative technique to control weeds and reduce damage to the environment. *Journal of Hazardous Materials, 268*, 207–215.

Pérez-de-Luque, A. (2017). Interaction of nanomaterials with plants: What do we need for real applications in agriculture? *Frontiers in Environmental Science, 6*, 12. https://doi.org/10.3389/fenvs.2017.00012

Pérez-de-Luque, A., & Rubiales, D. (2009). Nanotechnology for parasitic plant control. *Pest Management Science, 65*, 540–545. https://doi.org/10.1002/ps.1732

Radian, A., & Mishael, Y. G. (2008). Characterizing and designing polycation-clay nanocomposites as a basis for imazapyr controlled release formulations. *Environmental Science & Technology, 42*, 1511–1516.

Rani, N., Duhan, A., Pal, A., Kumar, P., Beniwal, R. K., Verma, D., Goyat, A., & Singh, R. (2023). Are nano-pesticides really meant for cleaner production? An overview on recent developments, benefits, environmental hazards and future prospectives. *Journal of Cleaner Production, 411*, 137232.

Sarijo, S. H., Hussein, M. Z., Yahaya, A. H. J., & Zainal, Z. (2010). Effect of incoming and outgoing exchangeable anions on the release kinetics of phenoxy herbicides nanohybrids. *Journal of Hazardous Materials, 182*, 563–569.

Schnoor, B., Elhendawy, A., Joseph, S., Putman, M., Chacon-Cerdas, R., Flores-Mora, D., Bravo-Moraga, F., Gonzalez-Nilo, F., & Salvador-Morales, C. (2018). Engineering atrazine loaded poly (lactic-co-glycolic acid) nanoparticles to ameliorate environmental challenges. *Journal of Agricultural and Food Chemistry, 66*, 7889–7898.

Sekhon, B. S. (2014). Nanotechnology in agri-food production: An overview. *Nanotechnology, Science and Applications, 7*, 31–53.

Silva, M. D., Cocenza, D. S., de Melo, N. F. S., Grillo, R., Rosa, A. H., & Fraceto, L. F. (2010). Nanopartículas de alginatocomosistema de liberação para o herbicidaclomazone. *Quim Nova, 33*, 1868–1873.

Sousa, G. F., Gomes, D. G., Campos, E. V., Oliveira, J. L., Fraceto, L. F., Stolf-Moreira, R., & Oliveira, H. C. (2018). Post-emergence herbicidal activity of nanoatrazine against susceptible weeds. *Frontiers in Environmental Science, 6*, 12. https://doi.org/10.3389/fenvs.2018.00012

StatNano. (2023). *Nanotechnology products database*. Retrieved February 28, 2023, from https://product.statnano.com/

Tong, Y., Wu, Y., Zhao, C., Xu, Y., Lu, J., Xiang, S., Zong, F., & Wu, X. (2017). Polymeric nanoparticles as a metolachlor carrier: Water-based formulation for hydrophobic pesticides and absorption by plants. *Journal of Agricultural and Food Chemistry, 65*, 7371–7378.

Villaverde, J. J., Sevilla-Moran, B., Lopez-Goti, C., Alonso-Prados, J. L., & Sandin-Espana, P. (2018). Considerations of nano-QSAR/QSPR models for nanopesticide risk assessment within European legislative framework. *Science of the Total Environment, 634*, 1530–1539.

Walker, G. W., Kookana, R. S., Smith, N. E., Kah, M., Doolette, C. L., Reeves, P. T., Lovell, W., Anderson, D. J., Turney, T. W., & Navarro, D. A. (2017). Ecological risk assessment of nano-enabled pesticides: A perspective on problem formulation. *Journal of Agricultural and Food Chemistry, 66*, 6480–6486.

Ye, Z., Guo, J., Wu, D., Tan, M., Xiong, X., Yin, Y., & He, G. (2015). Photo-responsive shell cross-linked micelles based on carboxymethyl chitosan and their application in controlled release of pesticide. *Carbohydrate Polymers, 132*, 520–528.

Yin, J., Wang, Y., & Gilbertson, L. M. (2018). Opportunities to advance sustainable design of nano-enabled agriculture identified through a literature review. *Environmental Science, 5*(1), 11–26. https://doi.org/10.1039/c7en00766c

Yu, Z., Sun, X., Song, H., Wang, W., Ye, Z., Shi, L., & Ding, K. (2015). Glutathione-responsive carboxymethyl chitosan nanoparticles for controlled release of herbicides. *Materials Sciences and Applications, 6*, 591–604.

Zhang, P., Guo, Z., Ullah, S., Melagraki, G., Afantitis, A., & Lynch, I. (2021). Nanotechnology and artificial intelligence to enable sustainable and precision agriculture. *Nature Plants, 7*, 864–876. https://doi.org/10.1038/s41477-021-00946-6

Chapter 9
Molecular Biology in Weed Management

Sachin Kumar, Priyanka, Jyoti Kumari, and Prashant Vikram

Abstract Globally, the rising number of herbicide-resistant weeds is a consequence of the dissipation and overuse of herbicides, as well as insufficient unique active ingredients. Furthermore, the challenges in managing weeds are made worse by weed characteristics that support the persistence of weed seed banks. Therefore, molecular control of weeds is an applied science that includes constant and continuous development of molecular genetic techniques that reduce reliance on herbicides while managing stubborn weed species. Weed management–related technologies developed through molecular biology include molecular genetic analysis, genetic transformation, finding and isolating genes of interest, and introducing "suicide" genes into plants of interest. Moreover, molecular biology has been utilized to understand more about how weeds endure adverse environmental conditions, engage in competition and interaction with nearby plants, and develop resistance to the herbicides employed to control them. Whether through increasing crop-weed competitiveness for space, light, nutrients, and water or via allelochemicals, molecular biology should be taken into consideration for weed management without the use of or with limited herbicide application. The possibility of controlling weeds by direct genetic manipulation has become prominent with the introduction of CRISPR/Cas9 gene-editing tools, as well as recent breakthroughs in "omics" and less expensive sequencing technology, which could be accomplished either permanently or temporarily. The first generation of transgenic products has proven successful, and the massive commercialization effort to use chemical and molecular tools to solve weed control challenges has substantially benefitted crop production. Furthermore,

S. Kumar
Department of Agronomy, Faculty of Agricultural Sciences, SGT University, Gurugram, Haryana, India

Priyanka (✉) · P. Vikram
Department of Crop Improvement, Faculty of Agricultural Sciences, SGT University, Gurugram, Haryana, India

J. Kumari
Deparment of Genetics and Plant Breeding, CSK Himachal Pradesh Agriculture University, Palampur, India

B. S. Chauhan, G. Mahajan (eds.), *Recent Advances in Weed Science*,
https://doi.org/10.1007/978-3-031-91014-2_9

these strategies comprise an array of techniques that could potentially be utilized to alter the crop to boost its competitiveness or resistance to herbicides or to alter the expression of significant genes in weeds to lessen their degree of fitness and effectiveness. Despite knowing that applying these promising molecular technologies to weed control will present considerable practical, legal, and technological difficulties.

Keywords CRISPR · Gene drive · Genomics · Herbicides · Omics · Weed

1 Introduction

As weeds are plants that flourish in areas where they are not wanted, it implies that such plants have existed for as long as humans have, even though they are unwanted. In traditional agriculture, weed management consisted primarily of hand uprooting or the use of tools like hoes to disturb plant roots. Although plant-killing chemicals like salt and olive oil have been used for hundreds of years to kill plants (Smith & Secoy, 1975), along with the discovery of synthetic selective herbicides like 2,4-dichlorophenoxyacetic acid (2,4-D) in the mid-1900s opened a new field of scientific study. However, a significant interaction between plants and chemical molecules and their action on the physiological processes are the major concerns of modern weed science. Comprehensive goals in the domain of weed science are to lessen the adverse impact of weeds on agricultural productivity, quality, and the sustainability of natural ecosystems. However, the long-term objective is to minimize weed seed reservoirs on and below the soil surface to control weed populations. Since the late 1960s, the primary source for controlling weeds worldwide has been the application of herbicides. Still, eventually, resistant weed populations have been noticed due to the application of the same herbicide or herbicides with a similar mode of action. However, the rising challenge in modern farming is the rise in herbicide-resistant weeds, which generally thrive and procreate despite exposure to a concentration of herbicide.

The resistance evolution process is generally categorized into two classes: (1) The agronomic techniques and human interventions, which include the application of herbicides and their dosages, varying the number of treatments over time, and tillage operations, and (2) biological components comprise genetics, ecology, and the life cycle of each type of weed. As a consequence, weed populations developed resistance to the herbicides of a similar mode of action, usually shown as a shift in plant metabolism or a gene mutation. Therefore, weed management necessitates novel approaches. Molecular biology has been found to have numerous significant roles related to agricultural biotechnology and plant breeding. The main goals of molecular crop enhancement strategies have been to increase crop resilience, adaptation, and yield and resistance to weeds, pest insects, and diseases. Worldwide agricultural production is being seriously eroded by the presence of weeds in developed and developing nations (Chauhan, 2020). The invention of herbicide-resistant crops and, to some extent, weed population genetics have been the most significant scientific advances in the field of weed management. Herbicide application has been combined with some molecular approaches to lessen the spread of weeds on agricultural fields. Among them, glyphosate is a nonselective herbicide that can be used to eradicate undesirable weeds in the field at different times of application, which, as a result, improves the degree of weed control. Nevertheless, the rise of genetically modified (GM) crops resistant to herbicides promotes the use of already-approved herbicides (Duke, 2012). Nevertheless, there are several obstacles to overcoming molecular technology to control weeds, which include the application and control of genetically modified (GM) technology for controlling noncrop plants and wild weeds, as well as the possible ecological consequences associated with the intentional dispersion of GM materials into the environment (Barrett et al., 2019).

Apart from GM techniques, transient methods utilize nontransformative applications of the RNA interference (RNAi) mechanism, offering effective molecular strategies for weed control as an alternative to herbicides. Therefore, other than using traditional pesticides, there is a need for innovative ways to control weeds. Consequently, weed control technology needs to advance constantly to keep up with weed adaptation and evolution; therefore, molecular approaches may be further investigated. This chapter examines the challenges of managing herbicide resistance in weeds and explores potential molecular techniques that could enhance current weed management strategies in the future.

2 Complexities in Modern Weed Control Approaches

The main causes of the difficulties in managing weeds are the diversity of crop and weed species, together with management strategies and various climate conditions that make it a challenge for agricultural productivity. However, weeds cannot be sufficiently controlled by cultural, mechanical, or chemical control measures whenever applied alone. Thus, integrating mechanical, chemical, and biological control methods can help reduce the weed seed bank effectively and offer advantages for the environment and agricultural profitability (Jabran & Chauhan, 2018; Alagbo et al., 2022). Although climate change has made many agricultural systems more complex, managing weeds has also become more difficult (Varanasi et al., 2016). Weeds possess defense mechanisms that enable them to withstand, particularly in the dry regions, due to their potential to gather quick mutations and adaptability compared to the crops that depend on breeding programs to introgress desirable characteristics in a steady manner under changing climate scenarios (Ramesh et al., 2017; Wong et al., 2022). However, under changing climatic scenarios, it is necessary to develop new weed control techniques, as the present herbicides will become more unstable. Therefore, herbicides with higher heat tolerances will need to be used in unexpectedly warmer regions, or spray techniques will need to be changed to account for uneven rainfall patterns.

2.1 Weeds and Herbicide Resistance

One major obstacle to efficient weed control is the rise of herbicide-resistant weeds, which are mostly caused by certain agricultural practices or the negligent and repeated application of similar herbicides.

Herbicide resistance mechanisms in weeds have also evolved due to a variety of biological variables, including their genetic makeup, life cycle, and habitat. Moreover, it has been more than 30 years since a new mode of action, herbicide, was introduced to the market (Duke, 2012), leading to the pressure to control the growing number of herbicide-resistant weeds. While new modes of action for herbicides have been developed, namely cinmethylin, tetflupyrolimet, and cyclopyrimorate; however, weeds may evolve resistance to these innovative herbicides; thus, their use cannot be the sole approach adopted to restrict weed population. However, the number of globally recorded herbicide-resistant weed instances grew from 2 to 507 between 1957 and 2020 (Heap, 2022). Herbicide resistance mechanisms fall into two principal categories: target-site resistance (TSR) and nontarget site resistance (NTSR; Gaines et al., 2020). Commonly, TSR is caused by specific site alterations in the target enzyme that inhibit the target enzyme's ability to bind herbicide. However, an enzyme can develop mutations in its binding sites or other regions, changing the enzyme's structure and making the herbicide less effective at inhibiting the enzyme's activity. Target-site resistance also encompasses target gene amplification, where the number of target gene copies rises, as well as increased expression of the target gene. Gene amplification of the herbicide target gene 5-enolpyruvylsh ikimate-3-phosphate synthase (EPSPS) has been observed in *Amaranthus palmeri* and *Hordeum glaucum*, whereas *Digitaria sanguinalis* has been shown to amplify the ACCase gene (Laforest et al., 2017). Conversely, nontarget site resistance is driven by the plant's physiological traits, including how it absorbs, detoxifies, or stores herbicides; however, nontarget site resistance is more complex to detect, as it often involves various gene families that regulate key processes like herbicide metabolism, movement, and sequestration. Proteins such as glutathione-s-transferases (GSTs) and glycosyltransferases (GTs), which play roles in herbicide detoxification at various stages, are linked to an accelerated metabolism of herbicide molecules (Perotti et al., 2020). Conversely, they participate in the early phase of herbicide metabolism by modifying the chemical functional groups of herbicide molecules, enabling their conjugation with thiol groups or glucose through GSTs or GTs. Thereafter, transporter proteins, such as the ATP-binding cassette proteins and cation amino acid transporter, subsequently carry conjugated herbicide molecules to vacuoles for segregation and depletion. Several weed species have shown reduced translocation of 2,4-D (Goggin et al., 2016), paraquat, and glyphosate (Moretti & Hanson, 2017).

2.2 Weed Seed Bank Persistence

Most weed species exhibit both resilience and resistance, producing thousands of seeds that can endure a variety of unfavorable environmental circumstances while remaining dormant in the soil for a prolonged period (Chauhan & Manalil, 2022); however, under optimal conditions, weed seeds germinate and compete with crops on the same plot of land. However, numerous factors, including dormancy processes, temperature and moisture levels, burial depth, and predator movement, can influence the persistence and weed seed duration in the soil. Furthermore, weed seed banks in agricultural farmlands may include seeds from several distinct weed species. Although each weed species has its own unique biological traits, life cycle, and ecological characteristics, therefore, various integrated weed management (IWM) techniques would be needed. Considering seed dormancy is a highly variable characteristic in weed species and is the primary cause of persistent weed seed banks globally, therefore, controlling weeds becomes challenging due to adaptation to different cropping systems (Schwartz-Lazaro & Copes, 2019). External conditions that induce dormancy affect the internal physiological and biochemical processes of seeds, including enzyme activity, nutrient transfer to the embryo, and metabolic functions. However, certain weed species possess seasonal dormancy, allowing them to evade risks or adverse environmental conditions which are observed in annual, biennial, and perennial weeds. However, annual weeds that grow in summer will not germinate in winter due to unfavorable growth conditions for their germination, development, and survival. Similarly, weeds that grow in winter will not germinate during the summer season. In perennial weeds, *Prosopis farcta* Macbride enters a state of dormancy by late fall and winter, shedding all its leaves and remaining inactive throughout the winter months. The weed releases bud dormancy at the beginning of spring and reproduces vegetatively during spring and summer. To enhance the effectiveness of sustainable weed management, integrated weed management (IWM) combines various approaches, including chemical, biological, and mechanical weeding, and specific crop management practices, which reduce the dependency on herbicides alone. Understanding weed population dynamics, their interactions with the environment, and how weeds contribute to ecosystem services are the main goals of ecologically based weed management. Similarly, farmers can precisely map and track weed infestations in real time with precision agricultural tools like satellite images, drones, and sensors (Zhang et al., 2022). However, these technologies provide crucial information for decision-making, enabling farmers to implement precise interventions in targeted areas. Furthermore, substitution for herbicide-resistant crops and weeds is provided by the application of beneficial insects, developments in RNAi technology, and microbial bioherbicides (Zabala-Pardo et al., 2022). Therefore, the development of appropriate and long-lasting IWM tactics can be facilitated by knowledge of the impact of seed depth and seed persistence duration.

2.3 Challenges Due to Inadequate Genomic Resources in Weeds

A major barrier to the effective implementation of molecular techniques for weed management is the insufficient genetic knowledge of various important weeds, which includes a notable absence of genomic and molecular research compared to that of crops. To assist in understanding the study of herbicide resistance mechanisms, genetic data on weeds should be obtained using tools from molecular biology and genomics. However, recently, a comprehensive sequence of several significant weed species, including *Conyza sumatrensis* and *Lolium rigidum*, has been done (Manning, 2021). Additionally, this effort and forthcoming research to improve genomic resources for several major weed species will enhance our understanding of weed biology, weedy traits, and the evolving landscape of herbicide resistance. The genomic resources of these important weed species could also uncover potential genetic resources for future crop breeding, allowing for the integration of beneficial traits from weeds into crops. Furthermore, acquiring insight into the biotic and abiotic tolerance of weeds is also made feasible through using the genetic characteristics of weed species. Herbicide selection and rotation are two instances of management decisions that may benefit from additional genomic data, such as population genetics (Perotti et al., 2020). There are many sedges like *Cyperus rotundus* and *Scirpus ancistrochaetus* found globally, and they have distinct properties, but there is very little genomic information accessible. Even well-known weeds, *Digitaria sanguinalis*, that damage croplands are not fully understood or described. Furthermore, higher-quality genomes are needed to provide insight into relevant biological concerns. We anticipate that there must be additional research on the population genetics of weeds, which will expand our knowledge of their evolutionary ecology and tactics while also providing more alternatives for weed control. Therefore, the discovery of genetic material has greatly enhanced our understanding of how various weed species develop herbicide resistance, particularly nontarget site resistance, to the same class of herbicides.

3 Solving Problems in Weed Management with Molecular Approaches

Global food security is seriously threatened by the rapidly spreading herbicide-resistant weeds, which can drastically lower agricultural yield and result in significant losses. Cultivating herbicide-resistant crops widens the herbicidal spectrum and reduces crop phytotoxicity, making it an efficient weed control technique when combined with a dearth of innovative herbicides. This section discusses potential applications of molecular methods and technologies in weed management.

3.1 Herbicide Efficiency

Molecular biology advancements have led to a better understanding of plant biology, which should make it easier to produce new herbicides that are safe, selective, and do not remain in the environment longer than desirable. With the early days of the invention of synthetic herbicides, weeds' resistance to herbicides has emerged as a significant contributing factor to issues with food security. Herbicide resistance has evolved in several economically significant weedy species, including careless weed (*Amaranthus palmeri*) (Manicardi et al., 2023), common wind grass (*Apera spica-venti*) (Papapanagiotou et al., 2022), black grass (*Alopecurus myosuroides*) (Lan et al., 2022), barren brome (*Bromus sterilis*) (Sen et al., 2021), and *Lolium* spp. (Zhu et al., 2023). Because of the selective pressure that results from repeatedly applying comparable herbicidal modes of action (MOAs), some species have evolved herbicide resistance. Numerous weeds were shown to be generating multiple resistances against herbicides with multiple modes of action, based on recent studies (Papapanagiotou et al., 2022; Zhu et al., 2023).

Herbicide resistance can generally be attained by several methods, including metabolism, enhanced herbicide detoxification, target site mutation, and target site gene amplification (Sammons & Gaines, 2014). The main resistance mechanism is targeting site mutation, and target genes like *ALS*, *ACCase*, and *EPSPS* have offered a multitude of possible genes to generate crops resistant to herbicides (Dong et al., 2021). Herbicide-resistant crop improvement was previously aided by classical mutagenesis breeding, though this method is labor- and time-intensive. Over the past few decades, transgenic techniques have been successfully applied for crop enhancement since they became available (Ogawa et al., 2008). Recent findings suggest that glyphosate-resistant soybeans, canola, rice, maize, wheat, cotton, and sugar beets are mostly responsible for the higher yield observed in transgenic herbicide-resistant crops (Shaner, 2000). Unfortunately, because alien genes are transferred, it is not possible to promote or employ genetically modified products.

Herbicide-resistant crops created using traditional mutagenesis breeding are viewed as non-GM, in contrast to the transgenic method. The creation and propagation of herbicide-resistant crops by mutagenesis has advanced quickly since 1992. Mutagenesis breeding was used to create a series of ALS inhibitor-resistant oilseed rape, sunflower, wheat, rice, and maize, which are marketed under the Clearfield® brand (Sudianto et al., 2013). Mutagenesis breeding has not been widely used because of its low mutation frequency and random mutation production, which makes it practically hard to induce several mutations for simultaneous herbicide resistance (Endo & Toki, 2013). As a result, a more exact breeding method is necessary. Advances in molecular biology have made genome editing techniques-like the recently discovered CRISPR/Cas-mediated genome editing techniques—a viable substitute for genetic modification in a variety of crops (Zhou et al., 2014). Additionally, these methods hold a lot of potential for speeding up the creation of superior crop varieties (Dong et al., 2021). Therefore, genome editing technology may be used to ameliorate certain DNA sequence variations linked to herbicide resistance.

The identification of gene families implicated in herbicide resistance, particularly the NTSR (non-target-site-based resistance) gene family, has also made considerable use of RNA-seq transcriptome sequencing (Chen et al., 2021). There are currently several instances of the use of transcriptomics and metabolomics for herbicide-resistance research. Even though transcriptomics can find the genes that are differentially expressed, the metabolome can be considered the basis for the system's apparent phenotype because the metabolites are the result of regulated gene transcription. However, problems in integrated transcriptomic–metabolomic investigations include a lack of association between some genes and metabolites, a large species variety, and a lack of direct connection between metabolite and transcript (Sen et al., 2023). While many cutting-edge technologies and multi-omics analyses are currently underutilized in weed science, their growing utilization could transform weed control techniques in the future (Chen et al., 2024). To create long-term weed control strategies, weed scientists, extension professionals, and farmers must work together in a collaborative, holistic manner using an integrated multidisciplinary approach. Additionally, the complexity and amount of multi-omics data can be efficiently handled by combining multi-omics with artificial intelligence (AI) and machine learning (ML) approaches, yielding insightful interpretations and predictions.

3.2 Increased Competitiveness

A competitive edge over weeds can be obtained by cultivating crops with higher seed rates and rearranging crop rows, all of which contribute to increased crop competition, which is particularly desirable. By making crop plants more competitive against weed species, weed management might be improved. The use of allelopathy, the exploitation of microbes, and enhancing the plant's availability of water, nutrients, and light (either through altered morphology or improved plant vigor) were the three main strategies previously found that could do this. Reducing the degree of "fitness" in weed populations is another strategy that is covered.

3.2.1 Improving Access to Water, Nutrients, and Light

Increasing crop plants' capacity to compete with weeds can be achieved, of course, by improving their availability of light, water, and nutrients. Studies have indicated that genotypes exhibiting big embryos and relatively thin leaves produce more robust seedlings with larger initial leaves and roots. These seedlings can take advantage of moisture and nutrients at an early stage of development, making them more likely to outcompete weeds. For mapping purposes, a population that is segregated based on both leaf thickness and embryo size is being produced. Linked markers could then be used to help choose these qualities once they have been identified. However, another characteristic related to early vigor and competitiveness is coleoptile length (Liang & Richards, 1994). Once more, the possibility of marker-assisted selection is presented by the identification of a DNA marker for coleoptile length. The linked marker is anticipated to be very helpful for the pyramiding of various genes encoding high seedling vigor, such as for the development of genotypes with both large embryos and long coleoptiles, even though selection for the trait may be made easily using a visual score.

Plant height, high tiller density, early canopy cover, high leaf area (leaf area index), vertical leaf orientation, large specific leaf area during vegetative growth, rapid biomass accumulation at the early crop growth stage, large root biomass, high shoot dry weight, and root volume are all associated with weed competitiveness (Saito et al., 2010). High yield potential and enhanced weed-suppressive capacity are compatible (Mahajan et al., 2014, 2015). As a result, breeding for both traits should result in more robust integrated crop management techniques. These studies' correlation analyses showed a favorable relationship between quick early development and the genotype-dependent weed-suppressive capacity. However, correlation analysis showed a favorable relationship between quick early development and the genotype-dependent weed-suppressive ability. By carefully choosing a competitive crop and cultivar and adjusting the sowing period, seed rate, crop row orientation, crop geometry, and intercropping, crop competitiveness against weeds can be increased. The capacity of crop plants to make use of the resources (water, light, nutrients, and space) in the presence of competing weeds is influenced by the morphological, physiological, and metabolic characteristics of the plant. By affecting the canopy architecture attributes (plant height, leaf area index, canopy density, rate of leaf area development, and leaf dispersion), cultural activities can also enhance the crop's capacity to suppress weeds (Sardana 2016).

Effective weed management employing crop competitiveness would undoubtedly benefit from a deeper comprehension of the biology and ecology of weeds. Different cultural methods that promote soil surface cover through rapid crop development and extended crop canopy can improve the competitiveness of crops. Nowadays, it is thought that using competitive crops or cultivars might improve weed control and maximize crop output, particularly for cereals (Blackshaw et al., 2007). This approach can also be economical. A crop cultivar's capacity to compete against weeds is influenced by how well it intercepts light (Holt, 1995). Slow-growing crop cultivars are less effective in inhibiting weed growth than crop cultivars that quickly shade the soil surface (Holt, 1995). According to Kumar et al., (2013), competitive cultivars can inhibit weed growth in standing crops, limit the production of weed seeds, stop further weed infestation, and potentially be an inexpensive weed management technique. Barley (*Hordeum vulgare L.*) is more resistant to weed competition than wheat (*Triticum aestivum L.*), rye (*Secale cereale L.*), and oats (*Avena sativa L.*) because of its more extensive root system at an early stage. Crops may be more able to compete with weeds if they are planted in narrow rows, with less spacing between plants, or with more seed. In direct-seeded rice, narrow row spacing reduced biomass and weed density. It also improved wheat yields across several areas and increased wheat's competitiveness against a range of weeds (Sardana 2016). The potential benefits of tight row spacing for controlling a range of weed species and increasing soybean yields because of superior competitive skills have been documented in numerous research projects conducted worldwide.

3.2.2 Allelopathy

The full potential of allelopathy in crops must be explored; nevertheless, it may be better leveraged by modifying the management of currently grown crops (e.g., by retaining stubble or altering rotations). These tactics might yield better outcomes than more costly genetic fixes. It has already been shown that genes producing allelochemicals were transferred from crop progenitors to modern crop cultivars (Wu et al., 1999). In some circumstances, conventional breeding methods and the genetic material already in place could be used to create crops with higher allelopathic potential. Marker-assisted selection could expedite the process significantly since the allelopathic potential is likely inherited as a complicated trait. Researchers may think about genetic engineering if it is determined that the current genetic material is insufficient for successful allelopathy (Paltridge, 2000). In the first scenario, spontaneous genetic variability is used to create many genotypic variants with minimal impact on phenotypic features, whereas genetic transformation produces variants with significant effects. Diversity within and between species serves as a genetic pool from which crops with greater allelopathic potential can be selected. The quantitative trait loci (QTL) analysis allows the location and effect calculation of the genetic components governing any characteristic through the joint investigation of marker genotype segregation in individuals or lines (Ain et al., 2023).

Allelopathy is a condition in which a plant emits chemicals (allelochemicals) that impact other nearby plants' ability to grow and survive. Allelochemicals, specifically those related to interactions between plants and microorganisms, may prove to be a vital source of various agrochemicals for resolving numerous problems brought on by subpar conventional methods and improper handling of synthetic herbicides (López-Ráez et al., 2012). However, several traditional cover crops with allelopathic activity are crucial for controlling weeds and, consequently, for replacing synthetic pesticides that are not environmentally friendly with herbicides (Büchi et al., 2019). Furthermore, these plants can create green or organic herbicides, eliminate long-chain hazardous effects in agroecosystems, and stop catastrophic phenomena like bioaccumulation and biomagnification (Liebman et al., 2018). With the use of QTL mapping, which connected the traits in sorghum (Shehzad & Okuno, 2020) and rice (Chung et al., 2020) with several chromosomes and suggested the role of multiple allelochemicals, the genetic mechanism of allelopathy is explained molecularly. Functional genomics is used to classify genetic material related to the manufacture of certain recognized allelochemicals, which presents the possibility of enhancing allelopathy by molecular breeding (Aci et al., 2022).

By concurrently examining marker genotype segregation in individuals or lines, QTL analysis enables the position and effect calculation of the genetic components driving any attribute (Asins et al., 2009). Zhang et al. (2005) identified 13 QTLs that accounted for 5–10% of the variation in the early vigor constituents in a population obtained from a hybrid rice line derived from an indica cultivar and a non-weed-suppressing japonica cultivar. Using high-throughput SNP genotyping, Chung et al., (2020) identified QTLs associated with allelopathic characteristics of rice in 98 F8 RILs (recombinant inbred lines) produced by crossing a nonallelopathic species (Nong-an) with a species with a high allelopathic propensity (Sathi). Potential genes for allelopathic wheat breeding were found on chromosomes 1A, 2B, and 5D in wheat by QTL analysis (Zuo et al., 2012). Marker-assisted selection can be used to increase allelopathic potential (Bertholdsson & Tuvesson, 2005). Enhancing allelopathy in conjunction with plant breeding may lead to cultivars of crops that have a higher potential for suppressing weeds. Field trials can be conducted to test allelopathic activity in the field, which can be further combined with techniques like CRISPR/Cas9 and QTL mapping to modify alleles of interest for greater weed suppression.

3.2.3 Interactions with Microorganisms

One possible source of new pesticides is naturally occurring microorganisms that reduce weeds. In agricultural and natural settings, a variety of microorganisms have been developed as bioherbicides to manage invasive and weedy plants (Aneja et al., 2017). Applying the substances that these organisms produce—also known as "natural products"—is an alternative to employing live microbes to manage weeds. Weed-suppressive and allelopathic chemicals can be extracted as natural products originating from microbes and plants (Katz & Baltz, 2016). Outside of agriculture, natural products derived from microbes have significant uses. For instance, microorganisms found in soil and water were the source of more than 60% of the FDA-approved anti-infective and anti-tumorigenic drugs currently on the market (Charlop-Powers et al., 2015; Pham et al., 2019). Metagenomic techniques for the separation of microbial antibiotics and enzymes have been made possible by advances in sequencing and molecular biology. This could result in the identification of novel weed-controlling mechanisms (Kao-Kniffin et al., 2013). By encouraging weed seed degradation, preventing germination, or stopping germination, microbial bioherbicide strains can lower the amount of weed seed banks. In reduced-till or no-till cropping systems, these bioherbicides, which target the early phases of weed development, have a lot of potential (Cordeau et al., 2016). For example, effective preemergent bioherbicides have been produced from *Pseudomonas* strains that were isolated from weed rhizospheres. Green foxtail (*Setaria viridis*), an annual weedy grass, was 90% less likely to sprout when a strain of Pseudomonas fluorescens was employed as a bioherbicide (Daigle et al., 2002). The ability of several *Fusarium* strains with the potential to degrade seeds has been assessed (Müller-Stöver et al., 2016). Post-emergent bioherbicides, in contrast to these pre-emergent ones, work by lowering the number of seeds produced, which controls the weed seed bank. For instance, according to Baudoin et al. (1993), *Puccinia carduorum* reduced the yield of musk thistle (*Carduus nutans*) seeds by as much as 57%. However, the biocontrol agent *Pseudomonas* spp. decreased downy brome (*Bromus tectorum*) seed production by 64% (Kennedy et al., 1991).

Numerous commercial biocontrol products have been created using soil-isolated weed-suppressive microorganisms (Trognitz et al., 2016). There are currently just a few registered bioherbicides (made from living or dead microorganisms) in the United States (Cheng et al., 2022). Molecular biology has been used in several instances to transfer herbicide-resistant genes to target plants through microbes. However, as it is still challenging to create and regulate the spread of microorganisms in the intricate ecology of soil, the release of genetically engineered microbes into the environment would undoubtedly be controversial. A more practical strategy would be to find bacteria that are already beneficial and then control how they interact with the plant through the plant itself; that is, increase the plant's capacity to draw in and maintain a relationship with beneficial microorganisms.

3.2.4 Reducing "Fitness" in Weed Populations

The notion of introducing a gene that could increase weeds' susceptibility to herbicides into their germplasm. Adding another "unfitness factor" to weed populations would be a logical development of this strategy. Pollen from unsuitable weeds sown in the population specifically for this purpose may propagate this. Using terminator technology to disperse pollen containing the terminator gene throughout the weed population and destroy any seeds resulting from fertilization by said pollen would be a logical conclusion to the idea (Paltridge, 2000). A small adjustment to the plan might cause "termination" to begin a while after the first release of pollen containing the terminator gene. The idea of "unfitness" spreading throughout a weed population is seriously inadequate because it is improbable that any one plant in the population will acquire the unfitness element. Affected plants would eventually become less fit than their relatives of the wild type, which would cause them to be outcompeted and disappear from the population. As a result, the factor would have to be consistently delivered to the population, and it is difficult to imagine a situation in which this could be done on a commercial basis.

3.3 Biological Control

A most important role for molecular biology is in the taxonomy of weeds and their biocontrol agents, as this information helps match control agents to different weed biotypes in genetically diverse populations. Information on the genetic diversity present in weed populations is also crucial to predicting population responses to biocontrol strategies.

Like many other domains, biological control of weeds (BCW) frequently faces a taxonomic impediment caused by a shortage of taxonomic expertise on the specific species involved or by imperfect taxonomic knowledge (i.e., weed plants and their possible biocontrol agents). Any biological control project must have taxonomic and phylogenetic data for both the primary pre- and post-release phases (Briese, 2005). These data are crucial for determining the origins and identities of target weeds and control agents, identifying problems like hybrids and cryptic species, and creating host test lists. Accurate bioecological information (e.g., weed ecology, natural enemies) is essential for identifying the target weed in BCW to search for the desired target species. An example of the fundamental requirement for accurate taxonomic identification in BCW is provided by the case of weevils of the genus *Cyrtobagous Hustache* attacking the floating fern *Salvinia molesta* D.S. Mitchell (Salviniales, Salviniaceae) (Zimmerman, 1993). In addition to conventional morphological identification techniques, molecular techniques are becoming more and more significant in identifying (and/or delimiting) weed species or genotypes to enable surveys of natural enemies (Gaskin et al., 2011).

When native and invading lineages coexist with instances of hybridization, it can be crucial to identify distinct subspecies, hybrids, or haplotypes using genetic techniques in BCW. Reid et al., (2023) used molecular markers and morphological data to identify hybrids associated with the invasive weed in South Africa and determine their likely parentage. Studies to identify potential biocontrol candidates often result in inventories of natural enemies, from which potentially effective biocontrol agents are selected after the target weed has been identified (Briese, 2000). At this point, it is critical to identify the possible biocontrol agent or agents with accuracy (Andersen & Wagner, 2016). These days, molecular information is combined with data from ecological, morphological, and other sources to identify and characterize species; this process is called integrative taxonomy. For instance, Toševski et al., (2011) used an integrative taxonomic approach and a variety of evidence sources to investigate weevils in the genus *Mecinus Germar* (Coleoptera: Curculionidae), which are thought to be a good fit for biological control of two invasive toadflax species from Europe that have spread to North America. Based on morphological traits, nuclear and mitochondrial DNA markers, and bioecological data, the comparative and phylogenetic analysis results indicated that the weevils were a complex of multiple cryptic species with different host-plant preferences rather than a single species. Although useful and increasingly widespread, molecular methods for verifying species identification can only supplement the taxonomic knowledge necessary based on morphological data. Insufficient integration of molecular and ecological information in the field-tested biocontrol agents limits the relevance of laboratory results to field environments. More reliable protocols are required for species verification.

3.4 DNA Chip Technology

The identification and cloning of genes crucial to weed management utilizing the burgeoning DNA chip technology would be a remarkable contribution to the field of molecular biology (Paltridge, 2000). The genes encoding traits of plant vigor, perennial habit, and allelochemical synthesis stand out in this review as being particularly receptive to the method. Libraries of DNA representing the pertinent crop plants, grouped in a microarray, would be the instruments needed for the work. After that, these would be filtered using expressed sequences from tissues or plants that exhibited the intended phenotype, allowing hybridization patterns to be compared to those acquired from tissues or plants that did not express the desired phenotype. The timing of these screenings will rely on the commercial availability of reagents, chips, detection equipment, and software. Although currently unaffordable, optimism in the private sector indicates that systems that are more reasonably priced may become accessible (Lemieux et al., 1998).

3.5 Intellectual Property Issues

Numerous methods created in the field of molecular biology are exclusive creations. Researchers in the agriculture industry frequently overlook this, and in most situations, study findings have little practical significance in terms of commerce. Occasionally, though, a study finding is discovered that has significant business ramifications. At this point, significant issues may occur, such as patent owners requesting payments for using their technology, a cut of any earnings, or outright banning its commercialization. Under these circumstances, researchers lament not delving more into intellectual property matters or adopting a copyrighted method when a nonpatented one was accessible. Because of this, scientists are advised to think about intellectual property concerns whenever they plan to use patented technologies, including instruments used in molecular biology.

4 Potential Gene Targets for Reducing Weed Fitness

Research on weed biology and physiology aims to reduce weed fitness and colonization in farmlands by comprehending weeds' habitat, life cycle, propagation, and proliferation patterns in relation to contemporary agriculture. Herbicide use and other management-based strategies can reduce the fitness of weed populations in agricultural areas. Lowering global weed seed banks would be the long-term objective of weed management; this would effectively limit weed populations rather than address the symptoms of managing weeds as they emerge. Depending on the desired outcome, several methods that combine genomic and biotechnological technology can be used with either crop or weed species.

Different genomic tools: With an increasing amount of omics data on weed species available, researchers can gain additional insight into weedy features like dormancy, invasiveness, and herbicide tolerance/susceptibility processes. Targeted exome resequencing was employed to address glyphosate resistance in *Ipomoea purpurea*. Rather than finding mutations in the predicted glyphosate target protein of EPSPS, Van Etten et al., (2020) found selective sweeps in several genes involved in herbicide detoxification that varied by population. Thus, although the physiological basis for glyphosate resistance in this weed is identical, there are multiple populations with distinct ways of herbicide detoxification.

Understanding how herbicide-resistant weeds have developed could be useful in identifying genes and gene networks that could be changed in different crops to produce distinct herbicide tolerance mechanisms. Furthermore, it can help to clarify complex features like abiotic stressors and the various nontarget site resistance mechanisms employed by weeds (Ravet et al., 2018). Nonetheless, the creation of genomic tools and an omics database for weeds faces a variety of challenges. These include the difficulties in figuring out the biology underlying the great diversity of weed species and the variety of systems in which weeds live (Martin et al., 2019). Furthermore, the maintenance of significant genetic variability in weeds poses a challenge to obtaining the appropriate genotypes (e.g., very homogeneous lines) for the construction of genomic reference materials, in contrast to reference model plant species and other extensively researched crop species (Vigueira et al., 2013).

Marker-assisted crop breeding offers promise not only for yield enhancement and disease resistance but also for weed control strategies. It also produces notable genetic advantages for several attributes. Scientists can connect genetic element modifications to critical features for improved weed control, such as herbicide resistance or features that allow crops to outcompete weeds, by using genome-wide association studies (GWAS). This technique has been successful when used on crop species with variable levels of natural herbicide resistance, such as wheat (Shi et al., 2020), cotton (Thyssen et al., 2014, 2018), sorghum (Adhikari et al., 2020), and faba bean (Abou-Khater et al., 2022). Like this, using chemical mutagenesis to create distinct genetic diversity in mutant populations can lead to the discovery of novel allelic variations and/or novel mechanisms of action for weed management. The development of imidazolidinone tolerance in chickpeas serves as evidence of the application of this technique (Croser et al., 2021) and wheat (Newhouse et al., 1992) via means of ethyl methane sulfonate (EMS) mutant populations.

4.1 CRISPR/Cas9

The CRISPR/Cas9 approach is one of the most widely used technologies for genetically editing crops and other creatures. This versatile technique for highly targeted genome editing is mostly employed to disrupt gene function (Zhang et al., 2018).

The application of CRISPR/Cas9 in crop development has been the subject of numerous reviews (Massel et al., 2021). Modified Cas9 proteins have also enabled other techniques for precise genome editing conceivable, such as base editing and prime editing (Molla et al., 2021). There are two possible uses for the CRISPR/Cas9 system in weed management, making it an undeniably promising tool. The first would be genetically engineering the crop to outcompete weeds and increase yield. Increasing the crop's resistance to herbicides, enhancing early vigor, or reducing weed establishment using allelopathic techniques are some ways to do this. A different approach would be to alter the weed's growth or herbicide resistance by using gene editing techniques on the plant itself. To comprehensively evaluate the ecological implications of CRISPR/Cas9-based weed management strategies, it is essential to conduct carefully designed field trials, which would focus on assessing the long-term effects on biodiversity, gene flow, and ecosystem stability.

4.2 Transient Technology

Without permanently altering the genomic DNA, transient technology permits the user to influence the plant's gene expression in short bursts. This implies that results do not have to follow stringent GM laws, require a lot less time to develop, and do not depend on tissue culture methods to alter the genome. Furthermore, instead of using a method of gene drive replication of the desired modifications across the population, these technologies may frequently be used to modify features on the spot in weeds or crops that are already growing. It has been demonstrated that using double-stranded RNA (dsRNA) spray packing enhances several disease and pest control choices for a variety of plant species (Mitter et al., 2017). Using this method, a dsRNA that targets and downregulates the expression of important genes implicated in the development of disease or pests in the host plant is packaged. There is a lot of promise for using a similar strategy to reduce weeds.

RNA interference (RNAi) can be used to target herbicide-resistant weeds, preventing the expression of their tolerance mechanism and maintaining the efficacy of previously developed herbicides. Alternatively, this system could be modified to target important genes that are unique to weed species without affecting how those genes are expressed in crops; in this way, lowering gene expression could potentially lessen weed competitiveness (e.g., reduced seed set, loss of flowering, and development issues). Rapid advancements have been made in spray-on technologies, where systems for the temporary expression of genes (or RNAi machinery) through packaging into viral vectors have been devised (Torti et al., 2021). To conclude, RNAi and transient gene expression are promising biotechnological approaches to address herbicide-resistant weeds. However, the long-term efficacy and potential risks associated with these methods are areas that require thorough investigation.

4.3 Target Genes for Herbicide and the Cytochrome P450 Family

Despite the challenges in putting molecular approaches to weed management into practice—either through genome editing techniques or the dsRNA spray for transient editing—several potential genes could be targeted for knockout and knockdown in weed species to reduce fitness, or conversely, genes targeted in the crop could enhance fitness. The molecular techniques may target any of many possible gene targets. These strategies frequently aim to target genes that, when disturbed, cause a variety of downstream phenotypic changes that lower the plant's fitness and survival. For instance, modifying genes belonging to the cytochrome P450 family may enhance herbicide efficacy. The administration of herbicides has been demonstrated to cause an upregulation of these proteins (Lu et al., 2015). This behavior mitigates the effects of ACCases and acetolactate synthase (ALS) inhibitors, among other herbicides (Peterson et al., 2016). However, because these gene families are often large and diverse, it is nearly impossible to target these genes specifically. The cytochrome P450 gene family and other gene families involved in nontarget site resistance, such as glutathione-s-transferases and glycosyltransferases, are obvious candidates for genetic manipulation. (Martin et al., 2019). Consequently, features and genetic elements that might be a part of the regulatory network of these superfamilies that confer herbicide resistance may be found using genetic resources from omics research.

4.4 Genes That Regulate Plant Growth

There are two strategies to enhance weed management: increasing the crop plant's early vigor or reducing the weed species' competitiveness. Therefore, genes associated with the basic processes of growth and development may be targeted for either trait. Weeds' fitness and competitiveness would be decreased if these genes were knocked down, as this would make them more susceptible to biotic and abiotic stimuli. These genes are implicated in the first stage of the phenylpropanoid biosynthesis pathway, which results in the synthesis of precursors of various primary and secondary metabolites important for growth and responses to a range of biotic and abiotic stressors, including pathogen infection, temperature changes, and UV radiation (Feng et al., 2022).

4.5 Genes That Determine Sex and Flowering Time

Plants have a variety of sexual systems, such as apomixis, outcrossing, and obligatory selfing. Sexual systems, including hermaphroditism, dioecism, monoecism, and others, are a result of the genetics underlying the temporal and spatial development of reproductive systems (Bawa & Beach, 1981; Charlesworth, 2002). Gene targets for gene drive development and spray-on transient technology have been identified from studies of the genetic basis of sex-determining genes influencing floral and reproductive organ development in the genus *Silene* (Charlesworth, 2013) and the weed species *Amaranthus tuberculatus* and *Amaranthus palmeri* (Montgomery et al., 2019, 2021).

4.6 Dormancy of Seeds

Targeting genes linked to dormancy is another way to inhibit seed fitness. Plant control and longevity are regulated by distinct combinations of seed dormancy genes (Pipatpongpinyo et al., 2020). Frequently, these genes contain transcription factors that initiate the production of flavonoids, gibberellic acid, and ABA (Debeaujon et al., 2007). The capacity of seeds to germinate in their native environment may be restricted by using CRISPR/Cas9 to target dormant genes in seeds and incorporate them into gene drives. This would work similarly to "terminator technology" or genetic use restriction technology (GURT) to safeguard the intellectual property of genetically modified materials (Lombardo, 2014).

4.7 Shattering of Seeds

One important characteristic of weeds that separates domesticated from wild plants is seed cracking (Dong & Wang, 2015). Instead of the ripe grains in crops naturally falling off, one of the mainstays of domestication is the retention of inflorescences or pods, which enables farmers to harvest the crops. It would be crucial to eliminate the seed-shattering characteristic to stop weeds from spreading throughout agricultural areas. Both monocots and dicots use the abscission layer, which forms in the inflorescence or pods because of lignification and thickening of the cell walls, as a shared mechanism for seed breakage (Dong & Wang, 2015).

4.8 Modification of the Root Exudate Profile

In the field of weed management, research on the molecular manipulation of crop root exudate patterns to enhance crop competitiveness and decrease weed fitness is relatively new. This is already the case in the struggle between the parasitic plant *Striga hermonthica* and sorghum (Bellis et al., 2020). This is a major concern while growing this staple food in Africa (Ejeta & Gressel, 2007). Bellis et al., (2020) used CRISPR/Cas9 editing to construct a loss-of-function mutation in the low germination stimulant 1 (LGS1) gene. The fitness of the parasitic weed is believed to be impacted by this mutation, which alters the stereochemistry of strigolactones in the root exudates.

5 Conclusions

Molecular biology plays a crucial role in quantifying genetic diversity in weed populations and the many biocontrol chemicals that target them, as well as in the development of new herbicides and herbicide-resistant crops. The scientific community already acknowledges these responsibilities, and they will remain significant heading forward. Additionally, there are significant responsibilities in the breeding of traits that enhance crop competitiveness and in the creation of safe methods for herbicide-resistant farming. Even though these regions are currently being investigated, there is a lot of room to grow the endeavor. It appears unlikely that traits for enhanced crop plant vigor and/or allelopathic potential will be selected for using markers. One significant element found in this book chapter that may have broad effects on weed management is the impending revolution in gene discovery. Together with improved methods for plant transformation, the isolation of different genes influencing crop competitiveness will likely lead to the production of crops that are more weed-resistant. These crops will have a significant effect on weed control in the future, assuming that they can be cultivated commercially and are welcomed by the community.

References

Abou-Khater, L., Maalouf, F., Jighly, A., Alsamman, A. M., Rubiales, D., Rispail, N., Hu, L., Ma, Y., Balech, R., Hamwieh, A., Baum, M., & Kumar, S. (2022). Genomic regions associated with herbicide tolerance in a worldwide faba bean (*Vicia faba* L.) collection. *Scientific Reports, 12*, 158. https://doi.org/10.1038/s41598-021-03861-0

Aci, M. M., Sidari, R., Araniti, F., Lupini, A. (2022). Emerging trends in allelopathy: A genetic perspective for sustainable agriculture. *Agronomy 12(9)*, 2043. https://doi.org/10.3390/agronomy12092043.

Adhikari, P., Goodrich, E., Fernandes, S. B., Lipka, A. E., Tranel, P., & Brown, P. (2020). Genetic variation associated with PPO-inhibiting herbicide tolerance in sorghum. *PLoS One, 15*, e0233254. https://doi.org/10.1371/journal.pone.0233254

Ain, Q., Mushtaq, W., Shadab, M., et al. (2023). Allelopathy: An alternative tool for sustainable agriculture. *Physiology and Molecular Biology of Plants, 29*, 495–511. https://doi.org/10.1007/s12298-023-01305-9

Alagbo, O., Akinyemiju, O., & Chauhan, B. (2022). Weed management in rainfed upland rice fields under varied agro ecologies in Nigeria. *Rice Science, 29*, 2.

Andersen, J. C., & Wagner, D. L. (2016). Systematics and biological control. In R. G. van Driesche, D. Simberloff, B. Blossey, C. Causton, M. S. Hoddle, D. L. Wagner, C. O. Marks, K. M. Heinz, & K. D. Warner (Eds.), *Integrating biological control into conservation practice* (pp. 105–129). Wiley.

Aneja, K. R., Khan, S. A., & Aneja, A. (2017). Bioherbicides: Strategies, challenges and prospects. In T. Satyanarayana, S. Deshmukh, & B. Johri (Eds.), *Developments in fungal biology and applied mycology* (pp. 449–470). Springer. https://doi.org/10.1007/978-981-10-4768-8_23

Asins, M. J., Bernet, G. P., Villalta, I., Carbonell, E. A. (2009). QTL analysis in plant breeding. *In: Molecular techniques in crop improvement,* (pp. 3–21).

Barrett, L. G., Legros, M., Kumaran, N., Glassop, D., Raghu, S., & Gardiner, D. M. (2019). Gene drives in plants: Opportunities and challenges for weed control and engineered resilience. *Proceedings of the Royal Society B: Biological Sciences, 286*, 20191515.

Baudoin, A. B. A. M., Abad, R. G., Kok, L. T., & Bruckart, W. L. (1993). Field evaluation of *Puccinia carduorum* for biological control of musk thistle. *Biological Control, 3*, 136–152. https://doi.org/10.1006/bcon.1993.1009

Bawa, K. S., & Beach, J. H. (1981). Evolution of sexual systems in flowering plants. *Annals of the Missouri Botanical Garden, 68*, 254–274. https://doi.org/10.2307/2398798

Bellis, E. S., Kelly, E. A., Lorts, C. M., Gao, H., Deleo, V. L., & Rouhan, G. (2020). Genomics of sorghum local adaptation to a parasitic plant. *Proceedings of the National Academy of Sciences of the United States of America, 117*, 4243–4251. https://doi.org/10.1073/pnas.1908707117

Bertholdsson, N. O., & Tuvesson, S. (2005). Possibilities to use marker assisted selection to improve allelopathic activity in cereals. *In: COST SUSVAR/ECO-PB workshop on organic plant breeding strategies and the use of molecular markers, 17*, 67.

Blackshaw, R. E., Anderson, R. L., & Lemerle, D. (2007). Cultural weed management. In M. K. Upadhyaya & R. E. Blackshaw (Eds.), *Non-chemical weed management. Principles, concepts and technology* (pp. 35–47). CABI.

Briese, D. T. (2000). Classical biological control. In B. M. Sindel (Ed.), *Australian weed management systems* (pp. 161–192). R.G. and F.J. Richardson.

Briese, D. T. (2005). Translating host-specificity test results into the real world: The need to harmonize the yin and yang of current testing procedures. *Biological Control, 35*(3), 208–214. https://doi.org/10.1016/j.biocontrol.2005.02.001

Büchi, L., Wendling, M., Amossé, C., Jeangros, B., Charles, R. (2019). Cover crops to secure weed control strategies in a maize crop with reduced tillage. *Field Crops Research, 247*, 107583. https://doi.org/10.1016/j.fcr.2019.107583.

Charlesworth, D. (2002). Plant sex determination and sex chromosomes. *Heredity, 88*, 94–101. https://doi.org/10.1038/sj.hdy.6800016

Charlesworth, D. (2013). Plant sex chromosome evolution. *Journal of Experimental Botany, 64*, 405–420. https://doi.org/10.1093/jxb/ers322

Charlop-Powers, Z., Owen, J. G., Reddy, B. V. B., Ternei, M., Guimaraes, D. O., & de Frias, U. A. (2015). Global biogeographic sampling of bacterial secondary metabolism. *eLife, e05048*. https://doi.org/10.7554/eLife.05048

Chauhan, B. S. (2020). Grand challenges in weed management. *Frontiers in Agronomy, 1*, 3. https://doi.org/10.3389/fagro.2019.00003

Chauhan, B. S., & Manalil, S. (2022). Seedbank persistence of four summer grass weed species in the northeast cropping region of Australia. *PLoS One, 17*, e0262288.

Chen, K., Peng, Y., Zhang, L., Wang, L., Mao, D., & Zhao, Z. (2021). Whole transcriptome analysis resulted in the identification of Chinese sprangletop (*Leptochloa chinensis*) genes involved in cyhalofop-butyl tolerance. *BMC Genomics, 22*, 521. https://doi.org/10.1186/s12864-021-07856-z

Chen, K., Yang, H., Wu, D., Peng, Y., Lian, L., Bai, L., & Wang, L. (2024). Weed biology and management in the multi-omics era: Progress and perspectives. *Plant Communications, 5*(4), 100816.

Cheng, L., DiTommaso, A., & Kao-Kniffin, J. (2022). Opportunities for microbiome suppression of weeds using regenerative agricultural technologies. *Frontiers in Soil Science, 2*, 838595. https://doi.org/10.3389/fsoil.2022.838595

Chung, I. M., Ham, T. H., Cho, G. W., Kwon, S. W., Lee, Y., Seo, J., An, Y. J., Kim, S. Y., Kim, S. H., & Lee, J. (2020). Study of quantitative trait loci (QTLs) associated with allelopathic trait in rice. *Genes, 11*(5), 470.

Cordeau, S., Triolet, M., Wayman, S., Steinberg, C., & Guillemin, J. P. (2016). Bioherbicides: Dead in the water? A review of the existing products for integrated weed management. *Crop Protection, 87*, 44–49. https://doi.org/10.1016/j.cropro.2016.04.016

Croser, J., Mao, D., Dron, N., Michelmore, S., Mcmurray, L., & Preston, C. (2021). Evidence for the application of emerging technologies to accelerate crop improvement—A collaborative pipeline to introgress herbicide tolerance into chickpea. *Frontiers in Plant Science, 12*, 779122.

Daigle, D. J., Connick, W. J., Boyetchko, S. M. (2002). Formulating a weed–suppressive bacterium in ''*Pesta*''. *Weed Technology, 16*, 407–413.

Debeaujon, I., Lepiniec, L., Pourcel, L., & Routaboul, J. M. (2007). In K. Bradford & H. Nonogaki (Eds.), *Seedcoat development and dormancy in annual plant reviews, seed development, dormancy and germination*. Blackwell Publishing Ltd.

Dong, H., Huang, Y., & Wang, K. (2021). The development of herbicide resistance crop plants using CRISPR/Cas 9-mediated gene editing. *Genes, 12*(6), 912.

Dong, Y., & Wang, Y. Z. (2015). Seed shattering: From models to crops. *Frontiers in Plant Science, 6*, 476. https://doi.org/10.3389/fpls.2015.00476

Duke, S. O. (2012). Why have no new herbicide modes of action appeared in recent years? *Pest Management Science, 68*, 505–512. https://doi.org/10.1002/ps.2333

Ejeta, G., & Gressel, J. (2007). *Integrating new technologies for striga control: Towards ending the witch-hunt*. World Scientific Publishing Co. Pvt. Ltd.

Endo, M., & Toki, S. (2013). Creation of herbicide-tolerant crops by gene targeting. *Journal of Pesticide Science, 38*, 49–59.

Feng, Y., Huang, Q., Zhang, R., Li, J., Luo, K., & Chen, Y. (2022). Molecular characterisation of PAL gene family reveals their role in abiotic stress response in lucerne (*Medicago sativa*). *Crop and Pasture Science, 73*, 1558. https://doi.org/10.1071/CP21558

Gaines, T. A., Duke, S. O., Morran, S., Rigon, C. A. G., Tranel, P. J., Küpper, A., Dayan, F. E. (2020). Mechanisms of evolved herbicide resistance. *Journal of Biological Chemistry, 295(30)*, 10307–10330. https://doi.org/10.1074/jbc.REV120.013572.

Gaskin, J. F., Bon, M.-C., Cock, M. J., Cristofaro, M., De Biase, A., De Clerck-Floate, R., Ellison, C. A., Hinz, H. L., Hufbauer, R. A., Julien, M. H., & Sforza, R. (2011). Applying molecular-based approaches to classical biological control of weeds. *Biological Control, 58*, 1–2.

Goggin, D. E., Cawthray, G. R., & Powles, S. B. (2016). 2,4-D resistance in wild radish: Reduced herbicide translocation via inhibition of cellular transport. *Journal of Experimental Botany, 67*, 3223–3235.

Heap, I. (2022). The international herbicide-resistant weed database. Available at: https://www.weedscience.org/Pages/ChronologicalIncrease.aspx.

Holt, J.S. (1995). Plant responses to light: A potential tool for weed management. *Weed Science, 43*, 474–482.

Jabran, K., & Chauhan, B. S. (2018). Overview and significance of non-chemical weed control. In *Non-chemical weed control*. Academic Press.

Kao-Kniffin, J., Carver, S. M., & DiTommaso, A. (2013). Advancing weed management strategies using metagenomic techniques. *Weed Science, 61*, 171–184. https://doi.org/10.1614/WS-D-12-00114.1

Katz, L., & Baltz, R. H. (2016). Natural product discovery: Past, present, and future. *Journal of Industrial Microbiology & Biotechnology, 43*, 5. https://doi.org/10.1007/s10295-015-1723-5

Kennedy, A. C., Young, F. L., Elliott, L. F., & Douglas, C. L. (1991). Rhizobacteria suppressive to the weed downy brome. *Soil Science Society of America Journal, 55*, 722–727. https://doi.org/10.2136/sssaj1991.03615995005500030014x

Kumar, P., Singh, Y., & Singh, U. P. (2013). Evaluation of cultivars and herbicides for control of barnyard grass and nutsedge in boro rice Indian. *Indian Journal of Weed Science, 45*, 76–79.

Laforest, M., Soufiane, B., Simard, M. J., Obeid, K., Page, E., & Nurse, R. E. (2017). Acetyl-CoA carboxylase overexpression in herbicide-resistant large crabgrass (*Digitaria sanguinalis*). *Pest Management Science, 73*, 2227–2235.

Lan, Y., Sun, Y., Liu, Z., Wei, S., Huang, H., & Cao, Y. (2022). Mechanism of resistance to pyroxsulam in multiple-resistant *Alopecurus myosuroides* from China. *Plants, 11*, 1645.

Lemieux, B., Aharoni, A., & Schena, M. (1998). Overview of DNA chip technology. *Molecular Breeding, 4*, 277–289.

Liang, Y. L., & Richards, R. A. (1994). Coleoptile tiller development is associated with fast early vigour in wheat. *Euphytica, 80*, 119–124.

Liebman, A. M., Grossman, J., Brown, M., Wells, M. S., Reberg-Horton, S. C., Shi, W. (2018). Legume cover crops and tillage impact nitrogen dynamics in organic corn production. *Agronomy Journal, 110(3)*, 1046–1057. https://doi.org/10.2134/agronj2017.08.0474.

Lombardo, L. (2014). Genetic use restriction technologies: A review. *Plant Biotechnology Journal, 12*, 995–1005. https://doi.org/10.1111/pbi.12242

López-Ráez, J. A., Bouwmeester, H., Pozo, M. J. (2012). Communication in the rhizosphere, a target for pest management. *In: Agroecology and strategies for climate change* (pp. 109–133). Springer.

Lu, H. P., Edwards, M., Wang, Q. Z., Zhao, H. J., Fu, H. W., & Huang, J. Z. (2015). Expression of cytochrome P450 CYP81A6 in rice: Tissue specificity, protein subcellular localization, and response to herbicide application. *Journal of Zhejiang University. Science. B, 16*, 113–122. https://doi.org/10.1631/jzus.B1400168

Mahajan, G., Ramesha, M. S., & Chauhan, B. S. (2014). Response of rice genotypes to weed competition in dry direct-seeded rice in India. *The Scientific World Journal, 2014*. https://doi.org/10.1155/2014/641589

Mahajan, G., Ramesha, M. S., & Chauhan, B. S. (2015). Genotypic differences for water-use efficiency and weed competitiveness in dry direct-seeded rice (*Oryza sativa* L.). *Agronomy Journal, 107*, 1573–1583.

Manicardi, A., Milani, A., Scarabel, L., Mora, G., Recasens, J., & Llenes, J. M. (2023). First report of glyphosate resistance in an *Amaranthus palmeri* population from Europe. *Weed Research, 65*. https://doi.org/10.1111/wre.12579

Manning, A. (2021). International weed genomics consortium-collaborative effort to combat crop-threatening weeds headed by CSU Scientists. Available at: https://agsci.source.colostate.edu/international-effort-to-combat-crop-threatening-weeds-headed-by-csu-scientists/

Martin, S. L., Parent, J. S., Laforest, M., Page, E., Kreiner, J. M., & James, T. (2019). Population genomic approaches for weed science. *Plants, 8*, 354. https://doi.org/10.3390/plants8090354

Massel, K., Lam, Y., Wong, A. C. S., Hickey, L. T., Borrell, A. K., & Godwin, I. D. (2021). Hotter, drier, CRISPR: The latest edit on climate change. *Theoretical and Applied Genetics, 134*, 1691–1709. https://doi.org/10.1007/s00122-020-03764-0

Mitter, N., Worrall, E. A., Robinson, K. E., Li, P., Jain, R. G., & Taochy, C. (2017). Clay nanosheets for topical delivery of RNAi for sustained protection against plant viruses. *Nature Plants, 3*, 16207. https://doi.org/10.1038/nplants.2016.207

Molla, K. A., Sretenovic, S., Bansal, K. C., & Qi, Y. (2021). Precise plantgenome editing using base editors and prime editors. *Nature Plants, 7*, 1166–1187. https://doi.org/10.1038/s41477-021-00991-1

Montgomery, J. S., Sadeque, A., Giacomini, D. A., Brown, P. J. & Tranel, P. J. (2019). Sex-specific markers for waterhemp (*Amaranthus tuberculatus*) and palmer amaranth (*Amaranthus palmeri*). *Weed Science, 67*, 412–418.

Montgomery, J. S., Giacomini, D. A., Weigel, D., & Tranel, P. J. (2021). Male-specific Y-chromosomal regions in waterhemp (*Amaranthus tuberculatus*) and Palmer amaranth (*Amaranthus palmeri*). *The New Phytologist, 229*, 3522–3533. https://doi.org/10.1111/nph.17108

Moretti, M. L., & Hanson, B. D. (2017). Reduced translocation is involved in resistance to glyphosate and paraquat in *Conyza bonariensis* and *Conyza canadensis* from California. *Weed Research, 57*, 25–34.

Müller-Stöver, D., Nybroe, O., Baraibar, B., Loddo, D., Eizenberg, H., & French, K. (2016). Contribution of the seed microbiome to weed management. *Weed Research, 56*, 335–339. https://doi.org/10.1111/wre.12218

Newhouse, K. E., Smith, W. A., Starrett, M. A., Schaefer, T. J., & Singh, B. K. (1992). Tolerance to imidazolinone herbicides in wheat. *Plant Physiology, 100*, 882–886. https://doi.org/10.1104/pp.100.2.882

Ogawa, T., Kawahigashi, H., Toki, S., & Handa, H. (2008). Efficient transformation of wheat by using a mutated rice acetolactate synthase gene as a selectable marker. *Plant Cell Reports, 27*, 1325–1331.

Paltridge, N. G. (2000). Applications for molecular biology in weed management. *Plant Protection Quarterly, 15*(2).

Papapanagiotou, A. P., Damalas, C. A., Bosmali, I., Madesis, P., Menexes, G., & Eleftherohorinos, I. (2022). Multiple resistance of silky windgrass to acetolactate synthase- and acetyl-CoA synthase–inhibiting herbicides. *Weed Technology, 36*, 334–343. https://doi.org/10.1017/wet.2022.24

Perotti, V. E., Larran, A. S., Palmieri, V. E., Martinatto, A. K., & Permingeat, H. R. (2020). Herbicide resistant weeds: A call to integrate conventional agricultural practices, molecular biology knowledge and new technologies. *Plant Science, 290*, 110255.

Peterson, M. A., Mcmaster, S. A., Riechers, D. E., Skelton, J., & Stahlman, P. W. (2016). 2,4-D past, present, and future: A review. *Weed Technology, 30*, 303–345. https://doi.org/10.1614/WT-D-15-00131.1

Pham, J., Yilma, M. A., Feliz, A., Majid, M. T., & Maffetone, N. (2019). A review of the microbial production of bioactive natural products and biologics. *Frontiers in Microbiology, 10*, 1404. https://doi.org/10.3389/fmicb.2019.01404

Pipatpongpinyo, W., Korkmaz, U., Wu, H., Kena, A., Ye, H., Feng, J., Gu, X. Y. (2020). Assembling seed dormancy genes into a system identified their effects on seedbank longevity in weedy rice. *Heredity 124*, 135–145. https://doi.org/10.1038/s41437-019-0253-8.

Ramesh, K., Matloob, A., Aslam, F., Florentine, S. K., & Chauhan, B. S. (2017). Weeds in a changing climate: Vulnerabilities, consequences, and implications for future weed management. *Frontiers in Plant Science, 8*, 95.

Ravet, K., Patterson, E. L., Krähmer, H., Hamouzová, K., Fan, L., & Jasieniuk, M. (2018). The power and potential of genomics in weed biology and management. *Pest Management Science, 74*, 2216–2225. https://doi.org/10.1002/ps.5048

Reid, M. K., Paterson, I. D., Coetzee, J. A., Gettys, L. A., & Hill, M. P. (2023). Know the enemy: Investigating genetic contributions from putative parents of invasive *Nymphaea mexicana* hybrids in South Africa as part of efforts to develop biological control. *Biological Control, 84*, 105291. https://doi.org/10.1016/j.biocontrol.2023.105291

Saito, K., Azoma, K., & Rodenburg, J. (2010). Rodenburg plant characteristics associated with weed competitiveness of rice under upland and lowland conditions in West Africa. *Field Crops Research, 116*, 308–317.

Sammons, R. D., & Gaines, T. A. (2014). Glyphosate resistance: State of knowledge. *Pest Management Science, 70*, 1367–1377.

Sardana, V. (2016). Role of competition in managing weeds: An introduction to the special issue. *Crop Protection, 95*, 1.

Schwartz-Lazaro, L. M., & Copes, J. T. (2019). A review of the soil seedbank from a weed scientists' perspective. *Agronomy, 9*, 369.

Sen, M. K., Hamouzová, K., Mikulka, J., Bharati, R., Košnarová, P., & Hamouz, P. (2021). Enhanced metabolism and target gene overexpression confer resistance against acetolactate synthase-inhibiting herbicides in *Bromus sterilis*. *Pest Management Science, 77*, 2122–2128. https://doi.org/10.1002/ps.6241

Sen, M. K., Bhattacharya, S., Bharati, R., Hamouzová, K., & Soukup, J. (2023). Comprehensive insights into herbicide resistance mechanisms in weeds: A synergistic integration of transcriptomic and metabolomic analyses. *Frontiers in Plant Science, 14*, 1280118.

Shaner, D. L. (2000). The impact of glyphosate-tolerant crops on the use of other herbicides and on resistance management. *Pest Management Science, 6*, 320–326.

Shehzad, T., Okuno, K. (2020). Genetic analysis of QTLs controlling allelopathic characteristics in sorghum. *PLoS ONE, 15(7)*, e0235896. https://doi.org/10.1371/journal.pone.0235896.

Shi, C., Zheng, Y., Geng, J., Liu, C., Pei, H., & Ren, Y. (2020). Identification of herbicide resistance loci using a genome-wide association study and linkage mapping in Chinese common wheat. *Crop Journal, 8*, 666–675. https://doi.org/10.1016/j.cj.2020.02.004

Smith, A. E., & Secoy, D. M. (1975). Forerunners of pesticides in classical Greece and Rome. *Journal of Agricultural and Food Chemistry, 23*, 1050–1055.

Sudianto, E., Beng-Kah, S., Ting-Xiang, N., Saldain, N. E., Scott, R. C., & Burgos, N. R. (2013). Clearfield rice: Its development, success, and key challenges on a global perspective. *Crop Protection, 49*, 40–51.

Thyssen, G., Mccarty, J. C., Li, P., Jenkins, J. N., & Fang, D. D. (2014). Genetic mapping of non-target-site resistance to a sulfonylurea herbicide (Envoke®) in upland cotton (*Gossypium hirsutum* L.). *Molecular Breeding, 33*, 341–348. https://doi.org/10.1007/s11032-013-9953-6

Thyssen, G. N., Naoumkina, M., Mccarty, J. C., Jenkins, J. N., Florane, C., & Li, P. (2018). The P450 gene CYP749A16 is required for toleranceto the sulfonylurea herbicide trifloxysulfuron sodium in cotton (*Gossypium hirsutum* L.). *BMC Plant Biology, 18*, 186. https://doi.org/10.1186/s12870-018-1414-2

Torti, S., Schlesier, R., Thümmler, A., Bartels, D., Römer, P., & Koch, B. (2021). Transient reprogramming of crop plants for agronomic performance. *Nature Plants, 7*, 159–171. https://doi.org/10.1038/s41477-021-00851-y

Toševski, I., Caldara, R., Jovic, J., Hernández-Vera, G., Baviera, C., Gassmann, Emerson, B. C. (2011). Morphological, molecular and biological evidence reveal two cryptic species in Mecinus janthinus Germar (Coleoptera, Curculionidae), a successful biological control agent of Dalmatian toadflax, Linaria dalmatica (Lamiales, Plantaginaceae). *Systematic Entomology, 36(4)*, 741–753. https://doi.org/10.1111/j.1365-3113.2011.00593.x.

Trognitz, F., Hackl, E., Widhalm, S., & Sessitsch, A. (2016). The role of plant-microbiome interactions in weed establishment and control. *FEMS Microbiology Ecology, 92*, 138. https://doi.org/10.1093/femsec/fiw138

Van Etten, M., Lee, K. M., Chang, S. M., & Baucom, R. S. (2020). Parallel and nonparallel genomic responses contribute to herbicide resistance in Ipomoea purpurea, a common agricultural weed. *PLoS Genetics, 16*, e1008593. https://doi.org/10.1371/journal.pgen.1008593

Varanasi, A., Vara Prasad, P. V., & Jugulam, M. (2016). Impact of climate change factors on weeds and herbicide efficacy. *Advances in Agronomy, 135*, 107–146. https://doi.org/10.1016/bs.agron.2015.09.002

Vigueira, C. C., Olsen, K. M., & Caicedo, A. L. (2013). The red queen in the corn: Agricultural weeds as models of rapid adaptive evolution. *Heredity, 110*, 303–311. https://doi.org/10.1038/hdy.2012.104

Wong, A. C. S., Massel, K., Lam, Y., Hintzsche, J., & Chauhan, B. S. (2022). Biotechnological road map for innovative weed management. *Frontiers in Plant Science, 13*, 887723. https://doi.org/10.3389/fpls.2022.887723

Wu, H., Pratley, J., Lemerle, D., & Haig., T. (1999). Crop cultivars with allelopathic capability. *Weed Research, 39*, 171–180.

Zabala-Pardo, D., Gaines, T., Lamego, F. P., & Avila, L. A. (2022). RNAi as a tool for weed management: Challenges and opportunities. *Advances in Weed Science, 40*, e020220096.

Zhang, Z. H., Yu, S. B., Yu, T., Huang, Z., & Zhu, Y. G. (2005). Mapping quantitative trait loci (QTLs) for seedling-vigor using recombinant inbred lines of rice (*Oryza sativa* L.). *Field Crops Research, 91*(2–3), 161–170.

Zhang, Y., Massel, K., Godwin, I. D., & Gao, C. (2018). Applications and potential of genome editing in crop improvement. *Genome Biology, 19*, 210. https://doi.org/10.1186/s13059-018-1586-y

Zhang, W., Miao, Z., Li, N., He, C., & Sun, T. (2022). Review of current robotic approaches for precision weed management. *Current Robotics Reports, 3*, 139–151.

Zhou, H., Liu, B., Weeks, D. P., Spalding, M. H., & Yang, B. (2014). Large chromosomal deletions and heritable small genetic changes induced by CRISPR/Cas9 in rice. *Nucleic Acids Research, 42*, 10903–10914.

Zhu, G., Wang, H., Gao, H., Liu, Y., Li, J., & Feng, Z. (2023). Multiple resistance to three modes of action of herbicides in a single italian ryegrass (*Lolium multiflorum* L.) population in China. *Agronomy, 13*, 216. https://doi.org/10.3390/agronomy13010216

Zimmerman, E. C. (1993). *Australian weevils (Coleoptera: Curculionoidea)* (Volume III. Nanophyidae, Rhynchophoridae, Erirhinidae, Curculionidae: Amycterinae, literature consulted, 854 pp). CSIRO.

Zuo, S., Liu, G., & Li, M. (2012). Genetic basis of allelopathic potential of winter wheat based on the perspective of quantitative trait locus. *Field Crops Research, 135*, 67–73.

Chapter 10
Advances in Omics Approaches for Weed Management

Romesh Salgotra, Prerna Johar, Monika Sood, and Bhagirath Singh Chauhan

Abstract In the past decade, the field of weed science has seen significant advancements through the application of omics techniques, leading to a deeper understanding of traits associated with weeds, including their invasiveness, hybridization, and resistance. The rapid progress in multi-omics research has been largely driven by advancements in sequencing technologies and analytical methodologies. Omics approaches offer extensive opportunities to uncover the molecular and physiological processes that contribute to the emergence of resilient weed species. The adoption of new "omics" technologies, such as genomics, proteomics, metabolomics, and other modern high-throughput molecular methods, in their application in weed biology research has notably increased. In weed science, these advanced omics platforms are increasingly being used to investigate the mechanisms underlying herbicide resistance, the evolution of weed resistance, and crop–weed interactions. The growing use of these sophisticated analytical tools not only deepens our understanding of weed biology but also aids in developing more effective management strategies. As research continues to evolve, the potential for these technologies to revolutionize weed science is substantial, paving the way for innovative solutions to ongoing agricultural challenges.

Keywords High-throughput analysis · Omics techniques · Weed biology · Weed management

R. Salgotra (✉) · P. Johar
Institute of Biotechnology, Sher-e-Kashmir University of Agricultural Sciences & Technology of Jammu, Jammu, J & K, India

M. Sood
Division of PHM, Sher-e-Kashmir University of Agricultural Sciences & Technology of Jammu, Jammu, J & K, India

B. S. Chauhan
Queensland Alliance for Agriculture and Food Innovation (QAAFI), The University of Queensland, Gatton, QLD, Australia

B. S. Chauhan, G. Mahajan (eds.), *Recent Advances in Weed Science*,
https://doi.org/10.1007/978-3-031-91014-2_10

1 Introduction

Weeds represent a fascinating and unique category of plants that significantly impact agricultural practices and ecological systems. However, many aspects of their biological, ecological, and genetic characteristics remain underexplored (Chen et al., 2024). These plants occupy a distinctive niche within plant biology, recognized for their exceptional ability to rapidly adapt to abrupt and severe changes in selective pressures imposed by human agricultural activities (Ravet et al., 2018). The creation of a public genomics database, which includes reference genomes for over 50 weed species, masks a pivotal advancement in research aimed at deepening our understanding of weed biology and evolutionary processes (Sheth & Thaker, 2014). Nevertheless, further efforts are necessary to enhance the accuracy of annotations, particularly in relation to complex gene families associated with herbicide resistance, structural variations, and mobile genetic elements. Current findings indicate that adaptive genetic diversity in weeds frequently arises from structural modifications (Patterson et al., 2019). The availability of reference genome assemblies opens avenues for developing standardized and cost-efficient methods for genotypic data collection, aiding in the identification of genetic variants related to key traits. Furthermore, there is an urgent need to enhance techniques focused on functional gene validation and hypothesis testing in weed species to substantiate the effects of genetic variants identified in these studies (Werck-Reichhartet et al., 2000). This includes advancing systems for transformation, gene editing, and transient gene silencing and expression (Patterson et al., 2019).

Globally, agricultural producers and landscape managers invest substantial financial resources and labor to combat weedy plants and mitigate their adverse effects (Oerke, 2006). Weeds threaten native flora by outcompeting them, thereby reducing biodiversity and disrupting ecosystem balance (Royal Botanic Gardens Kew, 2017). Additionally, they can negatively impact soil health and water availability. The presence of weeds can lead to significant reductions in crop yields, escalating production costs and lowering the quality of agricultural products (Oerke, 2006). Furthermore, weeds can develop resistance to herbicides through genetic mutations and the pressures of natural selection. They may adopt new resistance strategies, such as modifying their metabolic processes or altering their growth characteristics, which can undermine the effectiveness of chemical control methods.

The strategies employed to manage these challenges range from traditional, low-tech methods like manual weed removal to advanced, high-tech solutions, such as utilizing computer vision for targeted herbicide applications (Salgotra & Chauhan, 2021; Chaudhary et al., 2023). Regardless of the level of technological advancement, these control strategies impose significant selection pressures on weed populations, often leading to the rapid evolution of resistance (Maroli et al., 2016). This dynamic creates an ongoing arms race between humans and weeds, with humans continuously innovating and refining control methods while weeds adapt and evolve to counteract these efforts. Genomic strategies are emerging as a pioneering approach in weed management, leveraging advances in genetic research to address the complexities of controlling weed populations (Ravet et al., 2018). The advent of sequencing technologies and analytical methodologies has greatly advanced the field of multi-omics research, highlighting the importance of "omics" in various biological disciplines. The application of omics methodologies to examine different biological aspects provides a broad array of opportunities for elucidating the molecular and physiological processes that underlie the development of resilient phenotypes (Chen et al., 2024).

Over the past two decades, the term "omics" has been integral to numerous biological disciplines, reflecting a significant evolution in the field (Brunetti et al., 2018). Genomics primarily investigates the functional aspects of the genome, while transcriptomics and proteomics focus on the qualitative and quantitative characteristics of the transcriptome and proteome, respectively, within specific biological contexts. Complementary omics approaches, such as metabolomics, phenomics, and lipidomics, enhance traditional methodologies by offering a holistic perspective on cellular mechanisms (Patterson et al., 2019). The emergence of high-throughput functional omics technologies has significantly advanced our comprehension of the complex molecular and environmental interactions that shape biological systems (Kitano, 2002). The utilization of omics methodologies in exploring various biological dimensions presents remarkable potential for uncovering the molecular and physiological foundations that contribute to the development of resilient phenotypes. Importantly, omics technologies have achieved significant advancements, leading to the sequencing of an increasing number of plant genomes, which has, in turn, supported the development of plants with targeted agronomic traits and biological functions (Kantar et al., 2017; Nelson et al., 2018). The foundational components of omics platforms encompass genomics, transcriptomics, and proteomics (Fig. 10.1), which have been pivotal in enhancing our understanding of biological systems and remain essential in investigating diverse biological processes (Palsson, 2002; Rochfort, 2005).

In the past decade, the application of omics methodologies in weed science has increased significantly to focus on the biology and physiology of weeds. Molecular analyses utilizing DNA markers, such as restriction fragment length polymorphisms (RFLPs), random amplified polymorphic DNAs (RAPDs), simple sequence repeats (SSRs), amplified fragment length polymorphisms (AFLPs), cleaved amplified polymorphic sequences (CAPS), and inter-simple sequence repeats (ISSRs), have developed our understanding of various traits in weeds, including their resilience, dormancy, invasiveness, genetic diversity, and hybridization among species (Delye, 2013; Keith et al., 2017; Salgotra & Stewart, 2022). Beyond genomics, a range of other omics methodologies has been applied to critical areas within weed science, encompassing stress responses, invasiveness, herbicide resistance, and genetic diversity. However, the complex interplay between molecular and environmental

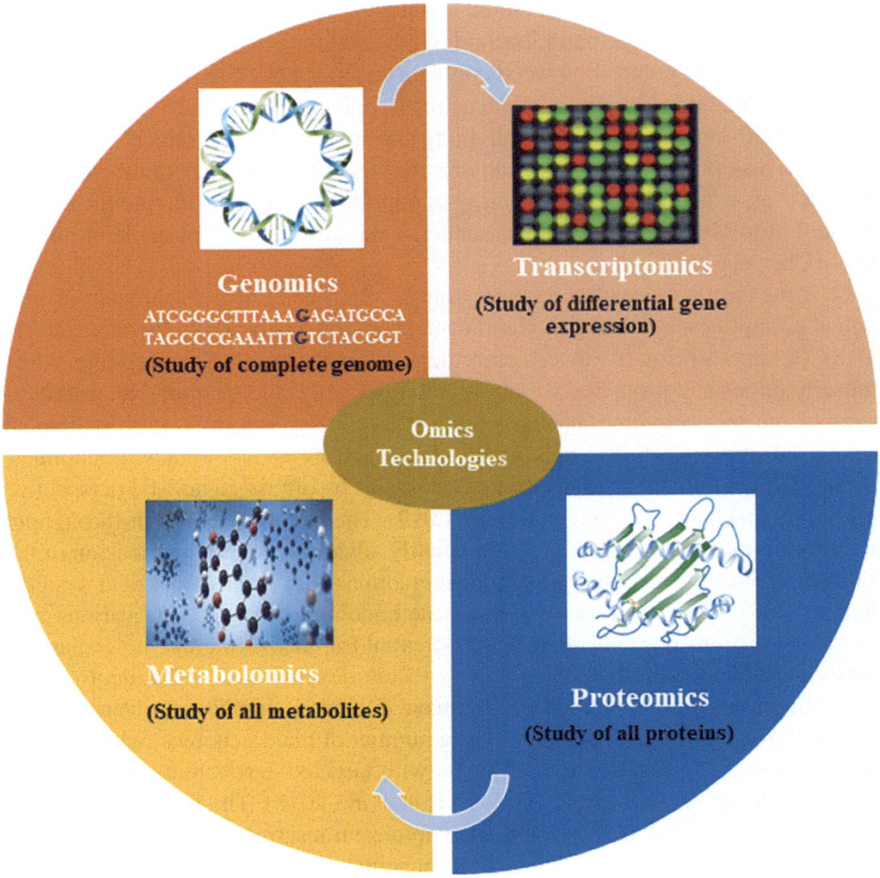

Fig. 10.1 Omics techniques used in weed management

factors means that no single omics approach can fully capture the intricacies of fundamental physiological processes (Liberman et al., 2012). Therefore, a comprehensive systems biology framework is essential for accurately characterizing the molecular, biochemical, and physiological conditions of target organisms.

The integration of omics approaches not only aids in annotating unknown genes but also in identifying their regulatory networks and the metabolic pathways they influence. This, in turn, enhances our understanding of the genotype–phenotype relationship and improves current weed management practices in agriculture (Perez-Alonso et al., 2018). While numerous omics platforms exist, this chapter will focus on the omics strategies employed to explore various aspects of agriculturally significant weeds that have yet to be thoroughly investigated, particularly the mechanisms underlying herbicide resistance and the identification of potential herbicidal phytochemicals through omics methodologies.

2 Genomics

Weeds pose considerable economic and environmental challenges on a global scale. Over the past 10 years, there has been a marked increase in the number of weed species whose genomes have been sequenced (Fig. 10.2) and assembled de novo, with over 26 species now documented (Huang et al., 2023). The genomic sizes of these species vary significantly, ranging from 270 Mb for *Barbarea vulgaris* to nearly 4.4 Gb for *Aegilops tauschii*. Notably, polyploid species often exhibit enhanced adaptive advantages compared to diploid species (te Beest et al., 2012). With the advancement of sequencing technologies, the genomes of polyploid weed species have been successfully sequenced, such as *Leptochloa chinensis, Echinochloa oryzicola, Capsella bursa-pastoris,* and *Echinochloa crus-galli.* Interestingly, some weed species are closely related to crop species, representing various subspecies or accessions of the same species, which allows the crop genome to serve as a useful reference for the weeds. For example, the genome of *Echinochloa colona* var. *frumentacea* has been referenced for *Echinochloa colona* var. *colona,* while other crops such as sorghum, pearl millet, rye, sugar beet, and rice have also been utilized as reference genomes for their weedy relatives (Wu et al., 2022). The draft genomes of *Echinochloa crus-galli* species, *Echinochloa crus-galli,* and

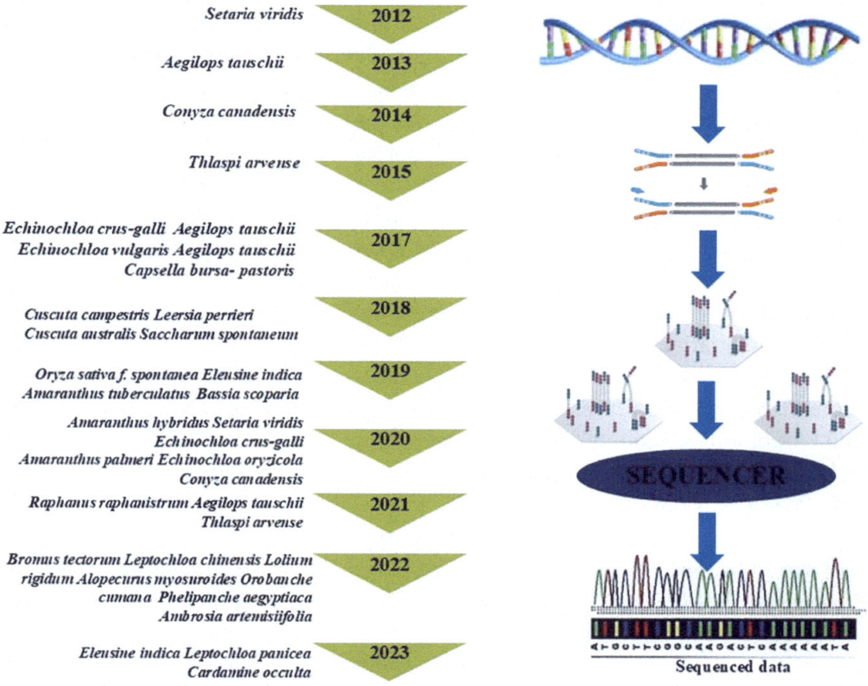

Fig. 10.2 Timeline of weed genome sequencing/population genomics studies

Echinochloa oryzicola, which thrive in paddy fields and compete with rice crops, were assembled and anchored to chromosomes using data from chromosome conformation capture techniques (Wu et al. 2022). The genome of *Oryza sativa f. spontanea* was sequenced and assembled at the chromosome level (Sun et al., 2019). Additionally, the genomes of tetraploid *Leptochloa chinensis* have been successfully assembled (Wang et al., 2022). The genome of *Thlaspi arvense*, an invasive species prevalent in wheat fields, was first assembled in 2015 and later anchored to chromosomes in subsequent studies (Dorn et al., 2015; Geng et al., 2021; Nunn et al., 2022).

2.1 Weed Genome Sequencing

The extent of genome sequence data and genetic resources available for model plants, including *Arabidopsis thaliana, Medicago truncatula*, and *Brachypodium distachyon*, as well as various dicot and monocot crops, is considerably greater than that for agronomic weed species. Nevertheless, the genomes of several agriculturally important weeds, particularly the highly heterozygous *Amaranthus* species, such as *Amaranthus tuberculatus, Amaranthus hybridus,* and *Amaranthus palmeri*, have been sequenced and organized into chromosome-level genomes (Montgomery et al., 2020). The substantial genomic information generated from these investigations has greatly propelled research in weed management and biology, especially regarding the understanding of their origins and evolutionary processes (Huang et al., 2023). The genetic and genomic resources obtained from these weed genomes have become crucial assets for advancing crop improvement initiatives.

Genomics is an exceptionally powerful tool for understanding the intricate structures of plant genomes. This discipline plays a crucial role in investigating the origins and evolutionary paths of weeds, as well as in examining their genomic frameworks and identifying the genetic markers that enhance their remarkable adaptability to various environmental conditions (Maroli et al., 2016). Recently, the weed science community formed the International Weed Genomics Consortium to foster the integration of genomics into weed science research.

The sequencing of the *Arabidopsis thaliana* genome in 2000 was a landmark accomplishment, as it represented the first complete genome sequence of a higher plant (Arabidopsis Genome Initiative, 2000). In contrast, research on weed genomes has historically lagged; for example, the first weed genome *Setaria viridis* (Brutnell et al., 2012), was only published in 2012, followed by *Aegilops tauschii* (Jia et al., 2013). The release of the third version of the *Aegilops tauschii* genome in 2017 was a notable milestone, as it provided the first chromosome-level genome for a weed (Zhao et al., 2017). While crop genomics has advanced rapidly, only three weed genomes had been assembled before 2017. However, subsequent advancements in sequencing technologies have enabled the publication of 32 weed genomes (47 versions), with 31 of these at the chromosome level (Table 10.1). Among these, the

Table 10.1 Important list of genome-sequenced weeds

Weed species	Genome sequencing completed	Associated crops	Year of released	Focus of genomic studies	Key findings/applications	References
Setaria viridis	Yes	Various cereal crops (e.g., maize, millet)	2012	Comparative genomics, stress responses	Model for grass genomics; insights into weed biology and evolution	Brutnell et al. (2012)
Aegilops tauschii	Yes	Wheat	2013	Whole genome sequence	Identification of wheat genes associated with disease resistance	Jia et al. (2013)
Amaranthus palmeri	Yes	Soybean, cotton, maize, sorghum	2014	Herbicide resistance (glyphosate)	Identified resistance genes and mechanisms; aids in managing herbicide-resistant populations	Gaines et al. (2014)
Conyza canadensis	Yes	Cotton, maize, soybean	2014	Genomic patterns were detected that were associated with glyphosate-resistant or glyphosate-susceptible biotypes	To understand weediness and the evolution of herbicide resistance and to devise new management strategies	Peng et al. (2014)
Thlaspi arvense	Yes	Wheat	2015	Comparative genomic analyses	Identifying gene targets whose manipulation can be predicted to result in improvements for pennycress weed	Dorn et al. (2015)
Echinochloa crusgalli	Yes	Rice, maize, soybean	2017	Herbicide resistance (ALS inhibitors)	Provided insights into resistance mechanisms and control strategies.	Guo et al. (2017)
Capsella bursapastoris	Yes	Various crops (e.g., lettuce, radish)	2015	Comparative genomics, adaptation	Evolutionary insights and adaptation to herbicide treatments	Kasianov et al. (2017)
Aegilops tauschii	Yes	Wheat	2017	Advanced sequencing and mapping techniques used to produce a reference genome of *Aegilops tauschii*	Providing a valuable resource for comparative genetic studies	Luo et al. (2017); Zhao et al. (2017); Kasianov et al. (2017)
Barbarea vulgaris	Yes	Lawn	2017	Comparative genomic analysis determines genetic differences between them and using genetic analysis	Identification of genes to detect the evolutionary and functional changes and their defense metabolites	Byrne et al. (2017)

(continued)

Table 10.1 (continued)

Weed species	Genome sequencing completed	Associated crops	Year of released	Focus of genomic studies	Key findings/applications	References
Cucumis myriocarpus	Yes	Cucurbits (e.g., melon, cucumber)	2021	Invasiveness, stress adaptation	Studied for adaptation mechanisms in tropical environments	Ling et al. (2021)
Cuscuta australis	Yes	Fabaceae	2018	High-quality reference genome of *Cuscuta australis*	Genes that are normally involved in root development of *Cuscuta australis*	Sun et al. (2018)
Saccharum spontaneous	Yes	Poaceae	2018	Whole genome sequence	New knowledge and resources to accelerate sugarcane improvement	Zhang et al. (2018)
Eleusine indica	Yes	Fabaceae	2019	Whole genome sequence	Identify all major herbicide-resistance-related genes	Zhang et al. (2019)
Oryza sativa f. spontanea	Yes	Rice	2019	de novo assembly of a high-quality weedy rice genome	Identified novel quantitative trait loci underlying weedy-specific traits and the genomic modifications underlying weedy rice evolution	Sun et al. (2019)
Amaranthus tuberculatus	Yes	Cotton, maize, soybean	2019	Glyphosate resistance	To understand the evolutionary mechanisms driving the spread of resistance	Kreiner et al. (2019)
Sorghum halepense	Yes	Sorghum, maize, cotton	2020	Herbicide resistance (glyphosate)	Insights into glyphosate resistance and control methods	Paterson et al. (2020)
Plantago lanceolata	Yes	Various crops (e.g., grasslands, pastures)	2023	Stress responses, herbicide adaptation	Genetic adaptations to environmental stress and herbicides	Lyu et al. (2023)
Abutilon theophrasti	Yes	Soybean, maize, cotton	2021	Herbicide resistance (glyphosate and other herbicides)	Provides insights into resistance mechanisms and adaptation	Lv et al. (2021)

Poaceae family of weeds has been the most extensively sequenced, likely due to their significant impact as pests on Poaceae crops such as rice, wheat, and maize, which are vital food sources for a large segment of the global population.

3 Transcriptome

RNA sequencing (RNA-Seq) is a powerful methodology that quantifies the abundance of transcripts from a genome at a specific time point, yielding data applicable to various analyses, such as identifying differentially expressed transcripts across varying conditions, examining genetic variations, and characterizing alternative splicing events (Chen et al., 2017). Unlike microarray-based transcriptomics, which offers relative quantification, Next Generation Sequencing (NGS) techniques provide absolute quantification of transcript expression along with the complete sequence of all transcripts present in a sample. This approach allows for exploring gene expression changes, including upregulation or downregulation that may enhance herbicide metabolism, and facilitates the investigation of potential nonsynonymous mutations in genes associated with resistance (Patterson et al., 2019). The identification of transcripts with unique regulatory patterns or mutations serves as a foundation for hypotheses that can be rigorously tested and validated. Transcriptomic analysis serves several important functions in biological research:

(i) It allows for comparing gene expression levels across various conditions, treatments, or genotypes, such as contrasting herbicide-resistant and sensitive accessions. By examining the complete transcriptomes of these differing accessions, researchers can pinpoint genes and pathways that are differentially expressed, thereby enhancing the understanding of the adaptive and evolutionary mechanisms of weeds in response to herbicide application. Additionally, transcriptomics facilitates the investigation of gene expression patterns, uncovering specific profiles that vary across different conditions or temporal stages.

(ii) It also supports functional enrichment analysis, which identifies biological processes, pathways, or molecular functions that are disproportionately represented among genes of interest. Furthermore, co-expression network analysis aids in the discovery of modules or clusters of co-regulated genes that may share functional relationships.

(iii) The advent of Next Generation Sequencing (NGS) technologies has significantly advanced transcriptome sequencing, with over 50 studies utilizing these methods to elucidate the mechanisms of herbicide resistance in weeds. Techniques such as RNA-Seq have provided robust and precise approaches for transcriptome analysis in nonmodel species, including weeds, without the necessity of a fully assembled genome.

The digital nature of RNA-Seq permits the detection of transcripts, enabling the quantification of transcripts even at low expression levels (Chen et al., 2017). To substantiate these hypotheses, it is essential to conduct subsequent validation experiments utilizing forward genetics, which play a pivotal role in confirming gene functionality (Neeraj et al., 2023). This validation may encompass several strategies, such as investigating the cosegregation of a molecular marker with resistance traits, assessing the presence of this marker across unrelated populations within the same species, and ideally performing expression or knockout analyses in model organisms like *Arabidopsis* or yeast. Once validation is achieved, differential expression can be evaluated in the validation samples through qRT-PCR analysis (Kohlhase et al., 2019). This thorough methodology ensures that the insights gained from RNA-Seq studies are meticulously tested and confirmed, thereby enhancing our comprehension of gene function and its significance in resistance mechanisms. Subsequent analyses of transcriptome data identify enriched pathways that contain differentially expressed transcripts (Lee & Tranel, 2008; Horvath, 2010). RNA-Seq is especially advantageous for studying complex gene families associated with metabolic resistance, such as cytochrome P450s, glutathione-S-transferases, glucosyl transferases, and ABC transporters. Nonetheless, it is crucial to recognize that the results obtained from RNA-Seq experiments should be cautiously interpreted to establish causal relationships for potential resistance mechanisms.

The utilization of RNA-seq technology within the realm of weed science has become increasingly prominent, primarily due to its ability to measure gene expression levels and detect sequence variations across various samples (Zargar et al., 2022). However, the absence of reference genome sequences for many weed species necessitates a continued reliance on de novo transcriptome assembly techniques (Majeed et al., 2022). Additionally, the significant heterozygosity and frequent polyploidy observed in many weed species pose challenges that can negatively impact the accuracy and quality of these de novo assemblies (Liang et al., 2021). Although RNA-seq is widely used to explore herbicide resistance, the presence of genetic variations between resistant and susceptible biotypes may compromise its reliability, increasing the risk of misidentifying resistance genes. Therefore, it is crucial to consider and address genetic background differences to improve the precision and reliability of RNA-seq analyses aimed at identifying genuine herbicide-resistant genes. Furthermore, conventional bulk analysis methods often overlook rare cell types and fail to capture the heterogeneity within cell populations adequately (Majeed et al., 2022).

Transcriptomics has emerged as a prominent multi-omics strategy in weed science, significantly contributing to the understanding of herbicide resistance mechanisms and responses to herbicides (Salas-Perez et al., 2018). The majority of research in weed transcriptomics has relied on de novo assembly due to the lack of available reference genomes, which introduced potential inaccuracies and constraints (Chen et al., 2024). Nevertheless, recent advancements in weed genomics have led to a rise in studies utilizing reference genomes. Although many weed species still do not have well-defined reference genomes, necessitating the continued use of de novo transcriptome analysis methods. These methods complicate the accurate interpretation and annotation of transcriptomic data (Ravet et al., 2018). Nevertheless, ongoing efforts in weed genomics are continuously expanding the availability of reference genomes, which will facilitate more comprehensive transcriptomics studies in the future.

The foundational transcriptomic data for the weed species *Conyza canadensis* was reported by Yang et al. (2021). Since then, researchers have sequenced and compiled at least 22 additional weed transcriptomes, including species prevalent in agricultural settings, turfgrass, and those classified as invasive (Yang et al., 2021). Notably, the *Echinochloa genus* has been extensively studied through RNA-Seq (Table 10.2), with over ten investigations focusing on herbicide resistance in species such as *Echinochloa crus-galli* and *E. colona*. These studies have rigorously examined various herbicides, particularly glyphosate, which is the most widely applied herbicide globally, recognized for its broad-spectrum efficacy against both annual and perennial plants.

Transcriptomic analyses have successfully identified numerous genes associated with herbicide resistance (Chen et al., 2024). By employing RNA-Seq, researchers have compared the transcriptomic profiles of glyphosate-resistant and sensitive *Echinochloa crus-galli,* leading to the discovery of key genes, including aldo-keto reductase, ABC transporter, and CYP450 genes, implicated in glyphosate resistance mechanisms. Subsequent experiments involving the overexpression of candidate genes such as *EcABCC8, EcCYP81A68*, and *EcAKR4-1* in transgenic plants have

Table 10.2 Transcriptomic studies of different weed populations

Weedy species	Associated crops	Year of released	Region	Reference
Oryza sativa f. *spontanea*	Rice	2017	China, USA, Japan, Korea	Babineau et al. (2017); Qiu et al. (2017); Han et al. (2022)
Echinochloa crus-galli	Rice	2019	China	Ye et al. (2019); Wu et al. (2022)
Amaranthus tuberculates	Maize, soyabean	2019	USA	(Kreiner et al. (2019)
Digitaria longiflora	–	2020	Africa	Abrouk et al. (2020)
Digitaria exilis	–	2020	Africa	Abrouk et al. (2020)
Thlaspi arvense	–	2021	China	Geng et al. (2021)
Secale cereale subsp. *segetale*	Rice	2021	All over the world	Sun et al. (2021)
Echinochloa walter	Rice	2022	USA	Wu et al. (2022)
Echinochloa colona var. *colona*	Rice	2022	All over the world	Wu et al. (2022)
Echinochloa oryzicola	Rice	2022	All over the world	Wu et al. (2022)
Leptochloa chinensis	Rice	2022	All over the world	Wu et al. (2022)
Ambrosia artemisiifolia	Tomato	2022	USA	Bieker et al. (2022)
Cardamine occulta	Rapeseed, mustard	2023	China	Li et al. (2023)

validated their contributions to glyphosate resistance. Specifically, *EcCYP81A68* and *EcAKR4-1* have been shown to metabolize herbicides within the plant, while *EcABCC8* was found to facilitate the transport of glyphosate into vesicles, thereby protecting the plant from herbicide toxicity (Pan et al., 2022). Furthermore, *EcABCC8* exhibits significant homology with *ABCC8* genes in rice, soybean, and maize, and the overexpression of these related genes has also been linked to glyphosate resistance (Pan et al., 2021).

Numerous investigations within the domain of weed science have been undertaken to explore the transcriptomic characteristics associated with herbicide resistance traits. These investigations have concentrated on both target-site resistance mechanisms and non-target site resistance (NTSR) mechanisms (Iwakami et al., 2014). The thorough examination of the complete transcriptome has been particularly instrumental in elucidating NTSR mechanisms, which often involve a complex interplay of multiple genes and gene families. For example, RNA-Seq methodologies have been employed to study NTSR in various grass species, including resistance to acetyl-CoA carboxylase (ACCase) inhibitors in *Lolium rigidum*, and acetolactate synthase (ALS) inhibitors in *Alopecurus myosuroides* (Cai et al., 2023). Additionally, research has assessed both NTSR and target-site mechanisms in *Echinochloa crus-galli*, leveraging next-generation sequencing technologies to pinpoint candidate genes linked to ALS inhibitor and synthetic auxin resistance. The dynamics of glyphosate and glyphosate resistance have also been investigated through RNA-Seq in tall morning glory (*Ipomoea purpurea* (L.) and kochia (*Bassia scoparia* (L.). Differential expression analysis was carried out to identify a range of candidate genes involved in metabolism, signaling, and defense, which may help clarify the variations in glyphosate response across different populations (Leslie & Baucom, 2014). Similarly, Wiersma et al. (2015) employed RNA-Seq to validate the overexpression of 5-enolpyruvylshikimate-3-phosphate synthase (EPSPS) in glyphosate-resistant *Bassia scoparia*, attributing this to gene amplification. These findings also indicated that no other genes within the shikimate pathway, aside from EPSPS, displayed differential expression between resistant and susceptible populations, and it was confirmed that the EPSPS sequence from the resistant population did not harbor any mutations linked to resistance (Wiersma et al., 2015). Deep sequencing was employed to identify mutations that might be expressed at low levels and go undetected by conventional sequencing methods. Recently, a mutation linked to dicamba resistance in *B. scoparia* was discovered through transcriptomics, followed by functional validation using forward genetics and expression in heterologous systems (LeClere et al., 2018).

4 Metabolomics

The realization that genes, transcripts, and proteins alone do not adequately explain the complex physiological and biochemical responses that link gene expression to resultant phenotypic traits. In this context, metabolomics serves as an essential approach, facilitating the thorough identification and quantification of low-molecular-weight metabolites, collectively referred to as the metabolome, thus offering a powerful strategy for molecular phenotyping (Fiehn, 2002; Chen et al., 2024). These metabolites are typically found within biological fluids, cells, or tissues and play a crucial role in various metabolic processes. The field of metabolomics focuses on the identification and quantification of these low-molecular-weight metabolites, providing insights into the biochemical state of an organism. By examining the metabolomic profile, researchers can gain a deeper understanding of metabolic pathways and their implications for health and disease. A standard workflow in metabolomics typically involves several stages, including metabolite extraction, chromatographic separation, detection, data analysis, metabolite characterization, and statistical validation. Generally, plant metabolites are separated frequently using either liquid chromatography or gas chromatography in conjunction with mass spectrometry for detection purposes (De Vos et al., 2007; Haggarty & Burgess, 2017; Maroli et al., 2017).

In the last 10 years, metabolomics has emerged as a valuable tool for exploring the mechanisms through which both synthetic and natural herbicides affect various model plant species, such as maize (Aranîbar et al., 2001), sterile oat (Aliferis & Chrysayi-Tokousbalides, 2006), and *Arabidopsis* (Jaini et al., 2017). Nevertheless, there remains a notable lack of research utilizing metabolomics to analyze weed physiology in relation to herbicide applications, and the mechanisms underlying herbicide resistance, including NTSR mechanisms like detoxification and metabolic processes. Earlier studies have demonstrated that genomics and transcriptomics play a crucial role in uncovering herbicide resistance mechanisms in specific weed species. Recently, metabolomics has been applied to assess the changes occurring in the synthesis of different metabolites under stresses such as: (i) effects of chemical stress on weeds, (ii) identifying additional glyphosate resistance mechanisms, (iii) evaluating glyphosate-induced physiological changes in both glyphosate-resistant and glyphosate-tolerant weeds, and (iv) investigating herbicide metabolism in resistant weed populations (Fernández-Escalada et al. 2016). Studies by Fernández-Escalada et al. (2016) and Maroli et al. (2017) have focused on the metabolic alterations in weeds following exposure to sublethal glyphosate doses. These studies underscore the significance of metabolomics in clarifying the intricate relationships between herbicides and weed physiology, thereby illuminating potential resistance mechanisms and the metabolic pathways associated with herbicide tolerance.

Metabolomics serves as a robust analytical approach that enables the instant assessment of metabolite concentrations through the direct measurement of static metabolite pools. Nevertheless, the intricate and dynamic characteristics of metabolic processes, which exhibit rapid turnover rates, necessitate a comprehensive quantitative analysis of intracellular fluxes to fully comprehend metabolic networks (Fernie & Morgan, 2013). The quantification of metabolic fluxes across individual reactions within these networks can only be achieved through indirect estimations utilizing isotopically labeled metabolic tracers (Gaudin et al., 2014). Recent advancements in fluxomics, particularly stable isotope–resolved metabolomics

(SIRM), have emerged as effective methodologies for assessing fluxes within complex metabolic networks (Maroli et al., 2016; Srivastava et al., 2016).

In weed science, the application of flux-based omics techniques to explore competitive physiology remains relatively limited (Maroli et al., 2016). Implementing SIRM experiments with stable isotope–labeled metabolic precursors, or tracers, is an optimal strategy for investigating metabolic fluxes in weeds, as it involves enriching growth media with labeled nutrients that can be monitored throughout the metabolic network (Gaudin et al., 2014; Maroli et al., 2016). By observing the movement of these labeled nutrients within the metabolic network as part of endogenous metabolic processes, flux rates can be indirectly inferred from variations in metabolite levels and the distribution of isotopes across the network. For example, the increase in amino acid levels in glyphosate-susceptible plants after glyphosate treatment is a frequently noted phenomenon (Fernández-Escalada et al., 2017). Fernández-Escalada et al. (2017) demonstrated that biotypes of *Amaranthus palmeri* susceptible to glyphosate accumulate amino acids at significantly higher levels than their resistant counterparts. This phenomenon of increased amino acid accumulation in susceptible biotypes following glyphosate application is often linked to the process of proteolysis, as noted in earlier research. Nevertheless, Maroli et al. (2016) employed SIRM to offer a more nuanced understanding of the mechanisms behind the heightened amino acid levels in susceptible *Amaranthus palmeri* biotypes. The research revealed that the accumulation of amino acids in these susceptible biotypes is attributable to a combination of proteolysis (catabolism) and the de novo synthesis of certain amino acids. In contrast, the glyphosate-resistant biotype predominantly displayed amino acid concentrations resulting from de novo synthesis (anabolism). This investigation highlights the importance of contemporary omics technologies in clarifying the interplay between the metabolome and metabolic pool dynamics, thereby shedding light on the relationship between glyphosate's action and the de novo synthesis of amino acids.

In another study, Fernández-Escalada et al. (2017) employed genetic, biochemical, and metabolic analyses to explore the physiological differences between glyphosate-resistant and glyphosate-susceptible populations of *Amaranthus palmeri*, providing new insights into the physiological responses associated with evolved glyphosate resistance. They observed that aromatic amino acids do not play a significant role in regulating the EPSPS protein and suggested that maintaining a consistent pool of free amino acids, including aromatic amino acids, is crucial for enhancing glyphosate resistance via EPSPS gene amplification. Similarly, Maroli et al. (2015) found that glyphosate resistance in *Amaranthus palmeri* may be supported by an increased antioxidant capacity, as several metabolites with known antioxidant properties were found to be elevated in the resistant biotype compared to its susceptible counterpart. Furthermore, Serra et al. (2015) subjected *Lolium perenne* to various chemical stressors, including glyphosate and its degradation product AMPA, at subtoxic concentrations, revealing that these stressors triggered distinct physiological disruptions and intricate metabolic changes through multiple modes of action. Collectively, these studies underscore the necessity of monitoring metabolic disturbances following herbicide application to better understand the sequence of cellular phytotoxic events.

5 Proteomics

Plant proteomics is a specialized field of research focused on the comprehensive analysis of proteins within plants, encompassing aspects such as protein content, expression levels, structural characteristics, functional roles, interaction networks, and post-translational modifications. In the context of weed science, the application of proteomics has a longer history than that of genomics and transcriptomics (Yang et al. 2017a, b). Yajima et al. (2004) investigated the proteomes of both herbicide-susceptible and herbicide-resistant *Sinapis arvensis* to identify specific proteins that may contribute to resistance against the herbicide dicamba. Initially, the focus of proteomics in this field was primarily on understanding the responses of weeds to herbicidal treatments. However, advancements in proteomic technologies have facilitated a shift toward using these methods to enhance our understanding of weed classification and adaptability to various ecological environments. This evolution led to a growing number of studies aimed at deciphering how weeds adjust to different environmental conditions through proteomic analysis. As a result, the development of proteomics within weed science has expanded research horizons, fostering a deeper understanding of the mechanisms underpinning weed resistance and adaptability. By investigating the proteomic profiles of diverse weed species, scientists can clarify the molecular pathways associated with herbicide resistance and identify potential targets for effective weed management strategies (Chen et al., 2024). Ultimately, the incorporation of proteomics into weed science has paved the way for novel explorations of the intricate relationships between weeds and their surrounding ecosystems.

6 Role of Omics Techniques in Weed Management

6.1 Genetic Diversity to Study the Spread of Weeds

The exploration of genetic diversity among and within ecotypes has significantly advanced through the utilization of genomic tools, which provide a comprehensive framework for genotype characterization (Salgotra & Zargar, 2020). The effectiveness of molecular markers in distinguishing closely related genotypes and revealing the genetic diversity and structural characteristics inherent in various plant species is well documented (Salgotra & Stewart, 2020). This knowledge is instrumental in understanding the underlying mechanisms that contribute to the resilience and adaptability of these plants across different environments. By analyzing population structures, researchers can uncover genetic variation patterns that may shed light on the ecological interactions and survival strategies employed by these weeds. Ultimately, such insights are essential for formulating effective weed management strategies (Chauhan et al., 2022). Studies on genetic diversity can develop more focused and sustainable methods for controlling weed populations, thereby improving agricultural productivity.

The emergence of herbicide resistance in agricultural weeds results directly from significant selection pressure (Baucom, 2019). As resistance alleles spread within these populations, they generate a diverse range of effects, impacting both individual organisms and entire populations. The transmission of herbicide-resistance genes via pollen and seeds is intricately linked to the degree of connectivity among different populations (Hawkins et al., 2019). The likelihood of new herbicide resistance traits arising from pre-existing genetic variation depends on the presence of such variation within the population (Dlugosch & Parker, 2008). Additionally, the spread of herbicide resistance genes through reproductive means is shaped by the degree of interconnectedness among populations. The potential for new resistance traits to emerge from existing genetic diversity within a population is reliant on the availability of that diversity. Moreover, the genetic variation present in a novel environment may be restricted, as it is anticipated that only a fraction of the total genetic diversity of the species will be introduced to that area (Warwick, 1990).

6.2 Improving Herbicide Resistance

Improvements in the diagnosis of weed resistance, along with historical field data, will aid in developing of precision weed control strategies. Moreover, enhanced diagnostics and precise mapping may either bolster or complicate the enforcement of zero-tolerance policies regarding newly discovered herbicide resistance mutations that present significant challenges to agricultural practices. Currently, the International Survey of Herbicide Resistant Weeds (ISHRW) acts as the main repository for data on herbicide-resistant weeds, providing a thorough database encompassing recent resistance findings and genomic DNA sequences pertinent to herbicide targets across various weed species (Heap 2019). In the field of antibiotic resistance, the Resistance Map offers an interactive resource that outlines global trends in antibiotic usage and resistance, while the Comprehensive Antibiotic Resistance Database (CARD) further compiles data on antibiotic resistance genes, their related proteins, and mechanisms (CDDEP 2019). This database also forecasts potential antibiotic resistance genes from unannotated contigs and evaluates their occurrence in sequenced genomes. Such comprehensive data resources are crucial for weed scientists, allowing them to explore essential inquiries regarding the emergence and spread of resistance. Additionally, there is increasing acknowledgment of the need for platforms like weedscience.org to expand their scope, integrating more in-depth reporting on resistance mechanisms, including target-site resistance and NTSR (Iwakami et al., 2014). This expansion could thereby enhance the overall understanding of resistance dynamics in weed populations (Jia et al., 2017).

6.3 Understanding Herbicide Resistance Mechanisms

The increase in sequencing projects focused on uncovering resistance mechanisms has led to the identification of potential driver genes linked to resistant phenotypes. Nevertheless, many of these studies do not follow up with functional validation of the candidate genes identified, leaving the fundamental reasons for resistance inadequately addressed. In cases where validation has been successfully conducted, researchers have utilized *Agrobacterium tumefaciens*-mediated transformation methods across various model plant systems, such as *Arabidopsis thaliana* (Saika et al., 2014), tobacco (*Nicotiana benthamiana*) (Li et al., 2019), rice *calli* (*Oryza sativa*) (Iwakami et al., 2014; Chu et al., 2018), transgenic rice (Chen et al., 2019; Pan et al., 2019), and budding yeast (*Saccharomyces cerevisiae*) (Iwakami et al., 2019).

Despite the progress made in sequencing and gene identification, the discipline of weed science frequently overlooks the functional validation of these genes. This oversight can be linked to insufficient investment in the development of stable plant transformation techniques specifically designed for weed species (Patterson et al., 2019). Consequently, valuable insights that could arise from a deeper understanding of the genetic foundations of resistance in weeds remain largely unexamined. Developing methodologies for plant transformation is crucial, especially in the context of weeds, where conventional methods may not be readily applicable. Bridging this gap could lead to a more thorough comprehension of resistance mechanisms and allow researchers to validate candidate genes more effectively. Ultimately, improving the ability to achieve stable transformation in weed species will be vital for advancing the field and tackling the challenges presented by resistant phenotypes.

6.4 Functional Genomics in Gene Prediction

The identification of gene functions plays a crucial role in developing effective weed management strategies (Patterson et al., 2019). Researchers can explore plant gene functions through transient expression systems, which allow for the targeted knockout or overexpression of specific gene variants. In the context of non-model species, various methodologies, including RNA interference (RNAi) and virus-induced gene silencing (VIGS), are utilized to investigate gene functions (Becker and Lange 2010; Lange et al., 2013). For example, VIGS has been effectively applied to silence particular genes, such as *CYP749A16* in cotton varieties tolerant to trifloxysulfuron, as well as a cluster of GST genes in cotton resistant to Verticillium wilt (Thyssen et al., 2018).

6.4.1 RNAi

Beyond VIGS, an alternative strategy involves inoculating plants with engineered viruses that promote the production of anti-sense RNA, which subsequently leads to the degradation of specific mRNA targets (Patterson et al., 2019). This technique, notably illustrated by the barley stripe mosaic virus system in cereal crops, serves as a viable method for mRNA suppression (Bennypaul et al., 2012). Additional methodologies include using anti-sense silencing oligonucleotides, such as small interfering RNAs (siRNAs) (Numata et al., 2014), which can be delivered via a protein carrier, as well as the high-pressure applications of double-stranded RNA (dsRNA) (Dalakouras et al., 2016) or the direct introduction of long dsRNA (Dubrovina et al., 2019).

In contrast to reverse genetics approaches that primarily aim to silence gene expression, the assessment of gain-of-function resulting from specific gene variants can be conducted through transient expression in plant systems. This can be accomplished by employing promoter-targeted RNA-directed DNA methylation, particularly in instances where DNA methylation plays a critical role in regulating gene transcription (Shibuya et al., 2009). Additionally, transient infection with *Agrobacterium* can facilitate the expression of a candidate gene, offering valuable insights into the functional implications of the gene variant being studied (Yoo et al., 2007).

6.4.2 Gene Editing

The exploration of gene functions in model organisms has predominantly utilized alternative transfer of DNA (t-DNA) or transposon insertional mutagenesis to create mutant plants, commonly referred to as gene knock-outs or phenotype screening. Following this initial phase, plants that display noteworthy phenotypes are subjected to further investigation to pinpoint the affected gene(s). These approaches require establishing and maintaining a considerable amount of germplasm, alongside a significant allocation of resources. Such endeavors are generally feasible only when a robust scientific community is engaged in the study of a specific species, exemplified by *Arabidopsis*. In contrast, researchers focused on weed science may find a more efficient strategy in employing gene editing technologies, such as zinc finger nucleases (ZFNs) or transcription activator-like effector nucleases (TALEN) (Li et al., 2012). Furthermore, the application of gene editing methods like clustered regularly interspaced short palindromic repeats/CRISPR-associated protein 9 (CRISPR/Cas9), which utilizes small RNA for sequence-specific DNA cleavage, offers a swift and precise means of gene modification. The CRISPR system has proven to be both effective and specific, operating with high efficiency in either transient or stable contexts. Within the CRISPR/Cas9 framework, the guide RNA (gRNA) sequence is crucial for the successful knock in and knockout of targeted genes (Feng et al., 2013). The weed science community stands to benefit greatly from the implementation of these innovative techniques to confirm the functionality of candidate genes. However, as with other methodologies for gene function analysis, it is essential to invest in plant transformation techniques tailored to weed species to enable thorough gene editing (Mushtaq et al. 2018). By adopting these state-of-the-art technologies, researchers in weed science can deepen their understanding of gene functions and expedite progress within the discipline.

6.5 Allelopathic Effects of Weeds

Allelochemicals, which are secondary metabolites synthesized by plants, play a crucial role in modulating growth and developmental processes. These bioactive substances have attracted significant interest as sustainable alternatives for weed management in agricultural systems due to their environmentally friendly characteristics (Khanh et al., 2007; Jabran et al., 2015). The sequencing and assembly of the genome of *E. crusgalli* have provided important insights into its genetic structure, particularly highlighting the existence of multiple copies of a biosynthetic gene cluster responsible for producing 2,4-dihydroxy-7-methoxy-1,4-benzoxazin-3-one. This compound has been shown to inhibit the growth of adjacent rice plants, thereby enhancing our understanding of the allelopathic relationships between *E. crusgalli* and rice (Guo et al., 2017). As a significant member of the plant benzoxazinoids, 2,4-dihydroxy-7-methoxy-1,4-benzoxazin-3-one represents a category of protective and allelopathic secondary metabolites governed by the *Bx* biosynthetic gene cluster. A comparative genomic study conducted by Wu et al. (2022) involving 41 grass species suggested that the *Bx* cluster likely emerged from gene duplication or chromosomal translocation of *Bx* genes. Additionally, their research indicated that the ancestral *Bx* cluster was acquired by the Triticeae lineage through horizontal gene transfer from the Panicoideae lineage (Wu et al., 2022).

To explore allelopathic interactions between crops and weeds, researchers frequently utilize transcriptomics and metabolomics techniques. Zhang et al. (2023) revealed reciprocal inhibition during the co-cultivation of rice and weeds. Through comprehensive transcriptomic and metabolomic analyses, the study identified several genes and metabolites that exhibited differential expression and abundance. Importantly, these genes and metabolites were significantly enriched within the phenylpropanoid and flavonoid biosynthesis pathways. The application of transcriptomics and metabolomics methodologies in investigating allelopathy offers significant insights into the molecular dynamics that govern plant interactions. The study elucidates the complex interplay of genes and metabolites involved in the allelopathic interactions between agricultural crops and competing weeds. By focusing on the biosynthetic pathways of phenylpropanoids and flavonoids, this research enhances our understanding of the biochemical mechanisms that facilitate allelopathic responses within plant ecosystems (Zhang et al., 2023).

6.6 Management of Herbicide Resistance

The emergence of herbicide resistance in weed populations represents a critical and urgent challenge in weed management. To date, 251 weed species have evolved resistance to 162 distinct herbicides, with resistance mechanisms identified in 23 of the 26 molecular sites of action associated with herbicides (Heap 2019). This phenomenon of herbicide resistance has highlighted a pervasive and intensifying problem that undermines the efficacy of existing weed control strategies. Given this scenario, it is essential to thoroughly investigate the fundamental factors that contribute to the rise of herbicide resistance across various weed species (Neve et al., 2014).

The mechanisms facilitating herbicide resistance can be classified into two primary categories: target-site resistance and NTSR (Powles & Yu, 2010). Target-site resistance is characterized by mutations in specific genes that alter the binding sites or structures of target proteins, thereby diminishing the effectiveness of herbicides. This distinction illustrates the intricate nature of herbicide resistance and the necessity for a detailed understanding of the genetic and biochemical mechanisms involved (Chen et al., 2024). As resistance spreads across a wide array of species and herbicide action modes, a comprehensive understanding of these mechanisms is crucial for formulating effective strategies to combat herbicide resistance in weed populations.

Herbicide resistance in weeds extends beyond mere target site modifications, encompassing a variety of mechanisms that enable these plants to survive herbicide exposure. These mechanisms include changes in gene expression, gene duplications, and chromosomal rearrangements that affect critical processes such as herbicide absorption, distribution, and metabolic breakdown (Gaines et al., 2020). A deeper comprehension of these NTSR mechanisms is essential for devising more effective management strategies for herbicide-resistant weed populations (Chauhan et al., 2022). The emergence of multi-omics methodologies has greatly facilitated the exploration of herbicide resistance in weeds, enabling the analysis of diverse omics data, including genomics, transcriptomics, proteomics, and metabolomics. This integrative approach allows for a thorough investigation of the molecular underpinnings of herbicide resistance, leading to the identification of pivotal genes, proteins, and metabolic pathways associated with NTSR, ultimately fostering the development of novel strategies to address the challenge of herbicide-resistant weeds (Gaines et al., 2010; Patterson et al., 2019).

Enzyme families such as cytochrome P450 (CYP450) and glutathione S-transferase (GST) play crucial roles in the metabolism and detoxification of herbicides. Recent research has indicated notable expansions of CYP450 and GST gene families within the genomes of prominent rice weed species, including *E. crus-galli* (Guo et al., 2017) and *L. chinensis* (Wang et al., 2022). Comparative genomic studies contrasting herbicide-resistant and sensitive populations of *Amaranthus tuberculatus* have illuminated the evolutionary trajectories linked to herbicide resistance, providing valuable insights into the underlying mechanisms of this adaptive phenomenon (Kreiner et al., 2019). Additionally, genomic analyses of *Amaranthus palmeri* have uncovered that herbicide resistance may be facilitated by extrachromosomal circular DNA harboring essential enzymes that are the targets of herbicides, such as glyphosate. By decoding the complete genome sequences of weeds, researchers can better pinpoint the specific genes responsible for herbicide resistance (Koo et al., 2018; Gaines et al., 2014). This approach allows for the identification and confirmation of mutations in target genes or differentially expressed genes that regulate herbicide resistance, providing valuable information for developing strategies to combat herbicide-resistant weeds effectively.

6.7 Use of Nanoherbicides in Weed Management

The evolution of nanotechnology has significantly broadened the scope of applying nanomaterials in agriculture. Among these innovations, nanoherbicides have surfaced as viable substitutes for conventional chemical herbicides, demonstrating comparable or superior efficacy while requiring reduced application quantities (Zargar et al., 2023). Despite their potential, the effects of these nanoherbicides on agricultural crops remain inadequately explored, necessitating further research. To bridge this knowledge gap, a series of investigations employing metabolomics and transcriptomics methodologies have been conducted to evaluate the metabolic responses of various crops, including cucumber, maize, and spinach, to $Cu(OH)_2$ nanopesticide exposure (Chaud et al., 2021). The findings revealed notable alterations in the metabolite profiles of the examined plants. Specifically, spinach exhibited diminished levels of metabolites related to antioxidants and defense mechanisms; maize showed a decrease in chlorophyll content and biomass alongside elevated concentrations of potassium, phosphorus, and precursors of phenolic acids; while cucumber revealed increased mRNA expression of genes linked to antioxidant activity and detoxification in response to the nanopesticide. These investigations underscore the utility of omics technologies in enhancing our understanding of the intricate interactions between crops and their environmental contexts, thereby shedding light on the implications of nanopesticide application in agricultural practices (Shangguan et al., 2024). Nonetheless, comprehensive studies are essential to fully elucidate the ramifications of nanopesticides on crop species and their surrounding ecosystems.

7 Conclusion

Over the past decade, significant progress has been made in weed genomics; however, it is clear that this area still lags behind the more mature domain of crop genomics. The utilization of genomics, transcriptomics, and various high-throughput screening methods is crucial for advancing our understanding of essential weed biology. These innovative approaches yield critical information regarding the responses of weeds to biotic and abiotic stressors, as well as their interactions with cultivated crops. By integrating these omics technologies, researchers can more swiftly gather insights into the genetics, biology, and physiology of weeds. The knowledge acquired through these sophisticated platforms plays a vital role in shaping sustainable weed management strategies. A deeper understanding of weed biology and adaptive strategies enables scientists to develop more efficient and ecologically sound methods for controlling weed populations in the long run. This advancement is essential for improving agricultural productivity while reducing the environmental footprint of weed management practices.

References

Abrouk, M., Ahmed, H. I., Cubry, P., et al. (2020). Fonio millet genome unlocks African orphan crop diversity for agriculture in a changing climate. *Nature Communications, 11*, 4488.

Aliferis, K. A., & Chrysayi-Tokousbalides, M. (2006). Metabonomic strategy for the investigation of the mode of action of the phytotoxin (5*S*, 8*R*, 13*S*, 16*R*) - (-) - pyrenophorol using ^{1}H nuclear magnetic resonance fingerprinting. *Journal of Agricultural and Food Chemistry, 54*(5), 1687–1692.

Araníbar, N., Singh, B. K., Stockton, G. W., & Ott, K. H. (2001). Automated mode-of-action detection by metabolic profiling. *Biochemical and Biophysical Research Communications, 286*(1), 150–155.

Babineau, M., Mahmood, K., Mathiassen, S. K., Kudsk, P., & Kristensen, M. (2017). *De novo* transcriptome assembly analysis of weed *Apera spica-venti* from seven tissues and growth stages. *BMC Genomics, 18*, 128.

Baucom, R. S. (2019). Evolutionary and ecological insights from herbicide-resistant weeds: what have we learned about plant adaptation and what is left to uncover? *The New Phytologist, 223*, 68–82.

Becker, A., & Lange, M. (2010). VIGS–genomics goes functional. *Trends in Plant Science, 15*, 1–4.

Bennypaul, H. S., Mutti, J. S., Rustgi, S., Kumar, N., Okubara, P. A., & Gill, K. S. (2012). Virus-induced gene silencing (VIGS) of genes expressed in root, leaf, and meiotic tissues of wheat. *Functional & Integrative Genomics, 12*(1), 143–156.

Bieker, V. C., Battlay, P., Petersen, B., Sun, X., Wilson, J., Brealey, J. C., Bretagnolle, F., Nurkowski, K., Lee, C., Barreiro, F. S., et al. (2022). Uncovering the genomic basis of an extraordinary plant invasion. *Science Advances, 8*, eabo5115.

Brunetti, L., Gundry, M. C., Sorcini, D., Guzman, A. G., Huang, Y. H., Ramabadran, R., Gionfriddo, I., et al. (2018). Mutant NPM1 Maintains the Leukemic State through HOX Expression. *Cancer Cell, 34*(3), 499–512.

Brutnell, T. P., Wang, L., Swartwood, K., Goldschmidt, A., Jackson, D., Zhu, X. G., Kellogg, E., & Van Eck, J. (2012). *Setaria viridis*: A model for C_4 photosynthesis. *Plant Cell, 22*(8), 2537–2544.

Byrne, S. L., Erthmann, P. O., Agerbirk, N., Bak, S., Hauser, T. P., Nagy, I., Paina, C., & Asp, T. (2017). The genome sequence of *Barbarea vulgaris* facilitates the study of ecological biochemistry. *Scientific Reports, 7*, 40728.

Cai, L., Comont, D., MacGregor, D., Lowe, C., Beffa, R., Neve, P., & Saski, C. (2023). The black-grass genome reveals patterns of non-parallel evolution of polygenic herbicide resistance. *The New Phytologist, 237*(5), 1891–1907.

Center for Disease Dynamics, Economics and Policy, Resistance Map (CDDEP) (2019). Available online: https://resistancemap.cddep.org/AntibioticResistance.php/. Accessed 16 July 2024.

Chaud, M., Souto, E. B., Zielinska, A., Severino, P., Batain, F., Oliveira-Junior, J., & Alves, T. (2021). Nanopesticides in Agriculture: Benefits and challenge in agricultural productivity, toxicological risks to human health and environment. *Toxics, 9*(6), 131.

Chaudhary, N., Salgotra, R. K., & Chauhan, B. S. (2023). Genetic enhancement of cereals using genomic resources for nutritional food security. *Genes (Basel), 14*(9), 1770.

Chauhan, B. S., Kaur, V., & Salgotra, R. K. (2022). Genetic diversity and population structure analysis to study the evolution of herbicide resistance in *Echinochloa colona* ecotypes in Australia. *Acta Physiologiae Plantarum, 44*(3), 32.

Chen, J., Chu, Z., Han, H., Goggin, D. E., Yu, Q., Sayer, C., & Powles, S. B. (2019). A Val-202-Phe α-tubulin mutation and enhanced metabolism confer dinitroaniline resistance in a single *Lolium rigidum* population. *Pest Management Science, 76*(2), 645–652.

Chen, J., Huang, H., Wei, S., Huang, Z., Wang, X., & Zhang, C. (2017). Investigating the mechanisms of glyphosate resistance in goosegrass (*Eleusine indica* (L.) Gaertn.) by RNA sequencing technology. *The Plant Journal, 89*(2), 407–415.

Chen, K., Yang, H., Wu, D., Peng, Y., Lian, L., Bai, L., & Wang, L. (2024). Weed biology and management in the multi-omics era: Progress and perspectives. *Plant Communications, 5*(4), 100816.

Chu, Z., Chen, J., Nyporko, A., Han, H., Yu, Q., & Powles, S. (2018). Novel α-tubulin mutations conferring resistance to dinitroaniline herbicides in *Lolium rigidum*. *Frontiers in Plant Science, 9*, 97.

Dalakouras, A., Wassenegger, M., McMillan, J. N., Cardoza, V., Maegele, I., Dadami, E., Runne, M., Krczal, G., & Wassenegger, M. (2016). Induction of silencing in plants by high-pressure spraying of in vitro-synthesized small RNAs. *Frontiers in Plant Science, 7*, 1327.

De Vos, R. C., Moco, S., Lommen, A., Keurentjes, J. J., Bino, R. J., & Hall, R. D. (2007). Untargeted large-scale plant metabolomics using liquid chromatography coupled to mass spectrometry. *Nature Protocols, 2*(4), 778.

Delye, C. (2013). Unravelling the genetic bases of non-target-site-based resistance (NTSR) to herbicides: a major challenge for weed science in the forthcoming decade. *Pest Management Science, 69*(2), 176–187.

Dlugosch, K. M., & Parker, I. M. (2008). Founding events in species invasions: Genetic variation, adaptive evolution, and the role of multiple introductions. *Molecular Ecology, 17*(1), 431–449.

Dorn, K. M., Fankhauser, J. D., Wyse, D. L., & Marks, M. D. (2015). A draft genome of field pennycress (*Thlaspi arvense*) provides tools for the domestication of a new winter biofuel crop. *DNA Research, 22*(2), 121–131.

Dubrovina, A. S., Aleynova, O. A., Kalachev, A. V., Suprun, A. R., Ogneva, Z. V., & Kiselev, K. V. (2019). Induction of transgene suppression in plants via external application of synthetic dsRNA. *International Journal of Molecular Sciences, 20*(7), 1585.

Feng, Z., Zhang, B., Ding, W., Liu, X., Yang, D. L., Wei, P., Cao, F., Zhu, S., Zhang, F., Mao, Y., & Zhu, J. K. (2013). Efficient genome editing in plants using a CRISPR/Cas system. *Cell Research, 23*(10), 1229–1232.

Fernández-Escalada, M., Zulet-Gonzalez, A., Gil-Monreal, M., Zabalza, A., Ravet, K., Gaines, T., & Royuela, M. (2017). Effects of EPSPS copy number variation (CNV) and glyphosate application on the aromatic and branched chain amino acid synthesis pathways in *Amaranthus palmeri*. *Frontiers in Plant Science, 8*, 1970.

Fernie, A. R., & Morgan, J. A. (2013). Analysis of metabolic flux using dynamic labelling and metabolic modelling. *Plant, Cell & Environment, 36*(9), 1738–1750.

Fiehn, O. (2002). Metabolomics—The link between genotypes and phenotypes. *Plant Molecular Biology, 48*, 155–171.

Gaines, T. A., Zhang, W., Wang, D., Bukun, B., Chisholm, S. T., Shaner, D. L., Nissen, S. J., Patzoldt, W. L., Tranel, P. J., Culpepper, A. S., Grey, T. L., et al. (2010). Gene amplification confers glyphosate resistance in *Amaranthus palmeri*. *Proceedings of the National Academy of Sciences of the USA, 107*(3), 1029–1034.

Gaines, T. A., Shaner, D. L., Ward, S. M., Leach, J. E., Preston, C., & Westra, P. (2014). Mechanism of resistance of evolved glyphosate-resistant Palmer amaranth (*Amaranthus palmeri*). *Journal of Agricultural and Food Chemistry, 59*(11), 5886–5889.

Gaines, T. A., Duke, S. O., Morran, S., Rigon, C. A. G., Tranel, P. J., Kupper, A., & Dayan, F. E. (2020). Mechanisms of evolved herbicide resistance. *The Journal of Biological Chemistry, 295*(30), 10307–10330.
Gaudin, Z., Cerveau, D., Marnet, N., Bouchereau, A., Delavault, P., Simier, P., & Pouvreau, J. B. (2014). Robust method for investigating nitrogen metabolism of 15N labeled amino acids using AccQ•Tag ultra performance liquid chromatography-photodiode array-electrospray ionization-mass spectrometry: application to a parasitic plant-plant interaction. *Analytical Chemistry, 86*(2), 1138–1145.
Geng, Y., Guan, Y., Qiong, L., Lu, S., An, M., Crabbe, M. J. C., Qi, J., Zhao, F., Qiao, Q., & Zhang, T. (2021). Genomic analysis of field pennycress (*Thlaspi arvense*) provides insights into mechanisms of adaptation to high elevation. *BMC Biology, 19*(1), 143.
Guo, L., Qiu, J., Ye, C., et al. (2017). *Echinochloa crus-galli* genome analysis provides insight into its adaptation and invasiveness as a weed. *Nature Communications, 8*, 1031.
Haggarty, J., & Burgess, K. E. (2017). Recent advances in liquid and gas chromatography methodology for extending coverage of the metabolome. *Current Opinion in Biotechnology, 43*, 77–85.
Han, B., Cui, D., Ma, X., Cao, G., Zhang, H., Koh, H. J., & Han, L. (2022). Evidence for evolution and selection of drought-resistant genes based on high-throughput resequencing in weedy rice. *Journal of Experimental Botany, 73*(7), 1949–1962.
Hawkins, N. J., Bass, C., Dixon, A., & Neve, P. (2019). The evolutionary origins of pesticide resistance. *Biological Reviews, 94*, 135–155.
Heap, I. (2019). *The International Survey of Herbicide Resistant Weeds*. Available online: http://www.weedscience.org. Accessed on 2 April 2019.
Horvath, D. (2010). Genomics for weed science. *Current Genomics, 11*(1), 47–51.
Huang, Y., Wu, D., Huang, Z., Li, X., Merotto, A., Jr., Bai, L., & Fan, L. (2023). Weed genomics: yielding insights into the genetics of weedy traits for crop improvement. *aBIOTECH, 4*(1), 20–30.
Iwakami, S., Endo, M., Saika, H., Okuno, J., Nakamura, N., Yokoyama, M., Watanabe, H., Toki, S., Uchino, A., & Inamura, T. (2014). Cytochrome P450 CYP81A12 and CYP81A21 are associated with resistance to two acetolactate synthase inhibitors in *Echinochloa phyllopogon*. *Plant Physiology, 165*(2), 618–629.
Iwakami, S., Kamidate, Y., Yamaguchi, T., Ishizaka, M., Endo, M., Suda, H., Nagai, K., Sunohara, Y., Toki, S., Uchino, A., Tominaga, T., & Matsumoto, H. (2019). CYP81A P450s are involved in concomitant cross-resistance to acetolactate synthase and acetyl-CoA carboxylase herbicides in *Echinochloa phyllopogon*. *The New Phytologist, 221*(4), 2112–2122.
Jabran, K., Mahajan, G., Sardana, V., & Chauhan, B. S. (2015). Allelopathy for weed control in agricultural systems. *Crop Protection, 72*, 57–65.
Jaini, R., Wang, P., Dudareva, N., Chapple, C., & Morgan, J. A. (2017). Targeted metabolomics of the phenylpropanoid pathway in *Arabidopsis thaliana* using reversed phase liquid chromatography coupled with Tandem Mass Spectrometry. *Phytochemical Analysis, 28*(4), 267–276.
Jia, J., Zhao, S., Kong, X., Li, Y., Zhao, G., He, W., Appels, R., Pfeifer, M., Tao, Y., et al. (2013). Aegilops tauschii draft genome sequence reveals a gene repertoire for wheat adaptation. *Nature, 496*(7443), 91–95.
Jia, B., Raphenya, A. R., Alcock, B., Waglechner, N., Guo, P., Tsang, K. K., Lago, B. A., Dave, B. M., Pereira, S., et al. (2017). CARD 2017: expansion and model-centric curation of the comprehensive antibiotic resistance database. *Nucleic Acids Research, 45*(D1), D566–D573.
Kantar, M. B., Nashoba, A. R., Anderson, J. E., Blackman, B. K., & Rieseberg, L. H. (2017). The genetics and genomics of plant domestication. *Bioscience, 67*, 971–982.
Kasianov, A. S., Klepikova, A. V., Kulakovskiy, I. V., Gerasimov, E. S., Fedotova, A. V., Besedina, E. G., Kondrashov, A. S., Logacheva, M. D., & Penin, A. A. (2017). High-quality genome assembly of *Capsella bursa-pastoris* reveals asymmetry of regulatory elements at early stages of polyploid genome evolution. *The Plant Journal, 91*(2), 278–291.
Keith, B. K., Burns, E. E., Bothner, B., Carey, C. C., Mazurie, A. J., Hilmer, J. K., Biyiklioglu, S., Budak, H., & Dyer, W. E. (2017). Intensive herbicide use has selected for constitutively

elevated levels of stress-responsive mRNAs and proteins in multiple herbicide-resistant *Avena fatua* L. *Pest Management Science, 73*, 2267–2281.

Khanh, T. D., Xuan, T. D., & Chung, I. (2007). Rice allelopathy and the possibility for weed management. *The Annals of Applied Biology, 151*, 325–339.

Kitano, H. (2002). Systems biology: a brief overview. *Science (NY), 295*(5560), 1662–1664.

Kohlhase, D. R., O'Rourke, J. A., MDK, O., & Graham, M. A. (2019). Using RNA-seq to characterize responses to 4-hydroxyphenylpyruvate dioxygenase (HPPD) inhibitor herbicide resistance in waterhemp (*Amaranthus tuberculatus*). *BMC Plant Biology, 19*, 182.

Koo, D.-H., Molin, W. T., Saski, C. A., Jiang, J., Putta, K., Jugulam, M., Friebe, B., & Gill, B. S. (2018). Extrachromosomal circular DNA based amplification and transmission of herbicide resistance in crop weed *Amaranthus palmeri*. *Proceedings of the National Academy of Sciences of the USA, 115*(13), 3332–3337.

Kreiner, J. M., Giacomini, D. A., Bemm, F., Waithaka, B., Regalado, J., Lanz, C., Hildebrandt, J., Sikkema, P. H., Tranel, P. J., Weigel, D., Stinchcombe, J. R., & Wright, S. I. (2019). Multiple modes of convergent adaptation in the spread of glyphosate-resistant *Amaranthus tuberculatus*. *Proceedings of the National Academy of Sciences, 116*(42), 21076–21084.

Lange, M., Yellina, A. L., Orashakova, S., & Becker, A. (2013). Virus-induced gene silencing (VIGS) in plants: an overview of target species and the virus-derived vector systems. *Methods in Molecular Biology (Clifton, NJ), 975*, 1–14.

LeClere, S., Wu, C., Westra, P., & Sammons, R. D. (2018). Cross-resistance to dicamba, 2, 4-D, and fluroxypyr in *Kochia scoparia* is endowed by a mutation in an AUX/IAA gene. *Proceedings of the National Academy of Sciences of the USA, 115*(13), E2911–E2920.

Lee, R. M., & Tranel, P. J. (2008). Utilization of DNA microarrays in weed science research. *Weed Science, 56*(2), 283–289.

Leslie, T., & Baucom, R. S. (2014). De novo assembly and annotation of the transcriptome of the agricultural weed Ipomoea purpurea uncovers gene expression changes associated with herbicide resistance. *G3 (Bethesda), 4*, 2035–2047.

Li, T., Liu, B., Spalding, M. H., Weeks, D. P., & Yang, B. (2012). High-efficiency TALEN-based gene editing produces disease-resistant rice. *Nature Biotechnology, 30*(5), 390–392.

Li, Z. K., Chen, B., Li, X. X., Wang, J. P., Zhang, Y., Wang, X. F., Yan, Y. Y., et al. (2019). A newly identified cluster of glutathione S-transferase genes provides Verticillium wilt resistance in cotton. *The Plant Journal, 98*(2), 213–227.

Li, L. Z., Xu, Z. G., Chang, T. G., Wang, L., Kang, H., Zhai, D., Zhang, L. Y., Zhang, P., Liu, H., Zhu, X. G., & Wang, J. W. (2023). Common evolutionary trajectory of short life-cycle in Brassicaceae ruderal weeds. *Nature Communications, 14*, 290.

Liang, Y., Tabien, R. E., Tarpley, L., Mohammed, A. R., & Septiningsih, E. M. (2021). Transcriptome profiling of two rice genotypes under mild field drought stress during grain-filling stage. *AoB Plants, 13*(4), plab043.

Liberman, L. M., Sozzani, R., & Benfey, P. N. (2012). Integrative systems biology: an attempt to describe a simple weed. *Current Opinion in Plant Biology, 15*(2), 162–167.

Ling, J., Xie, X., Gu, X., Zhao, J., Ping, X., Li, Y., et al. (2021). High-quality chromosome-level genomes of *Cucumis metuliferus* and *Cucumis melo* provide insight into *Cucumis* genome evolution. *The Plant Journal, 107*(1), 136–148.

Luo, M. C., Gu, Y. Q., Puiu, D., Wang, H., Twardziok, S. O., Deal, K. R., Huo, N., et al. (2017). Genome sequence of the progenitor of the wheat D genome *Aegilops tauschii*. *Nature, 551*(7681), 498–502.

Lv, Y., Yi, Y., Han, R., Tong, X., & Takahashi, H. (2021). The complete chloroplast genome of *Abutilon theophrasti* medic (Malvaceae). *Mitochondrial DNA Part B: Resources, 6*(3), 912–913.

Lyu, S., Mei, Q., Liu, H., Wang, B., Wang, J., Lambers, H., Wang, Z., Dong, B., Liu, Z., & Deng, S. (2023). Genome assembly of the pioneer species Plantago major L. (Plantaginaceae) provides insight into its global distribution and adaptation to metal-contaminated soil. *DNA Research, 30*(4), dsad013.

Majeed, A., Johar, P., Raina, A., Salgotra, R. K., Feng, X., & Bhat, J. A. (2022). Harnessing the potential of bulk segregant analysis sequencing and its related approaches in crop breeding. *Frontiers in Genetics, 13*, 944501.

Maroli, A. S., Nandula, V. K., Dayan, F. E., Duke, S. O., Gerard, P., & Tharayil, N. (2015). Metabolic profiling and enzyme analyses indicate a potential role of antioxidant systems in complementing glyphosate resistance in an *Amaranthus palmeri* biotype. *Journal of Agricultural and Food Chemistry, 63*(41), 9199–9209.

Maroli, A., Nandula, V., Duke, S., & Tharayil, N. (2016). Stable isotope resolved metabolomics reveals the role of anabolic and catabolic processes in glyphosate-induced amino acid accumulation in *Amaranthus palmeri* biotypes. *Journal of Agricultural and Food Chemistry, 64*(37), 7040–7048.

Maroli, A. S., Nandula, V. K., Duke, S. O., Gerard, P., & Tharayil, N. (2017). Comparative metabolomic analyses of Ipomoea lacunosa biotypes with contrasting glyphosate tolerance captures herbicide-induced differential perturbations in cellular physiology. *Journal of Agricultural and Food Chemistry, 66*(8), 2027–2039.

Montgomery, J. S., Giacomini, D., Waithaka, B., Lanz, C., Murphy, B. P., Campe, R., et al. (2020). Draft Genomes of *Amaranthus tuberculatus*, *Amaranthus hybridus*, and *Amaranthus palmeri*. *Genome Biology and Evolution, 12*(11), 1988–1993.

Mushtaq, M., Bhat, A.K., Zahoor, M.A., Sakina, A., Ali, S., Singh, A.K., Tyagi, A., Salgotra, R.K., Dar, A.A. & Bhat, R. (2018). CRISPR/Cas approach: A new way of looking at plant-abiotic interactions. *Journal of Plant Physiology, 224–225*, 156–162, https://doi.org/10.1016/j.jplph.2018.04.001

Nelson, R., Wiesner-Hanks, T., Wisser, R., & Balint-Kurti, P. (2018). Navigating complexity to breed disease-resistant crops. *Nature Reviews. Genetics, 19*(1), 21–33.

Neeraj, C., Kumar, S. R. & Singh, C.B. (2023). Genetic enhancement of cereals using genomic resources for nutritional food security. *Genes, 14*, 1770. https://doi.org/10.3390/genes14091770

Neve, P., Busi, R., Renton, M., & Vila-Aiub, M. M. (2014). Expanding the eco-evolutionary context of herbicide resistance research. *Pest Management Science, 70*(9), 1385–1393.

Numata, K., Ohtani, M., Yoshizumi, T., Demura, T., & Kodama, Y. (2014). Local gene silencing in plants via synthetic dsRNA and carrier peptide. *Plant Biotechnology Journal, 12*(8), 1027–1034.

Nunn, A., Rodriguez-Arevalo, I., Tandukar, Z., Frels, K., Contreras-Garrido, A., Carbonell-Bejerano, P., Zhang, P., Ramos Cruz, D., Jandrasits, K., Lanz, C., et al. (2022). Chromosome-level *Thlaspi arvense* genome provides new tools for translational research and for a newly domesticated cash cover crop of the cooler climates. *Plant Biotechnology Journal, 20*(5), 944–963.

Oerke, E. C. (2006). Crop losses to pests. *The Journal of Agricultural Science, 144*, 31–43.

Palsson, B. (2002). In silico biology through "omics". *Nature Biotechnology, 20*(7), 649–650.

Pan, L., Yu, Q., Han, H., Mao, L., Nyporko, A., Fan, L., Bai, L., & Powles, S. (2019). Aldo-keto reductase metabolizes glyphosate and confers glyphosate resistance in *Echinochloa colona*. *Plant Physiology, 181*(4), 1519–1534.

Pan, L., Yu, Q., Wang, J., Han, H., Mao, L., Nyporko, A., Maguza, A., Fan, L., Bai, L., & Powles, S. (2021). An ABCC-type transporter endowing glyphosate resistance in plants. *Proceedings of the National Academy of Sciences of the USA, 118*(16), e2100136118.

Pan, L., Guo, Q., Wang, J., Shi, L., Yang, X., Zhou, Y., Yu, Q., & Bai, L. (2022). CYP81A68 confers metabolic resistance to ALS and ACCase-inhibiting herbicides and its epigenetic regulation in *Echinochloa crus-galli*. *Journal of Hazardous Materials, 428*, 128225.

Paterson, A. H., Kong, W., Johnston, R. M., Nabukalu, P., Wu, G., Poehlman, W. L., Goff, V. H., Isaacs, K., Lee, T. H., Guo, H., Zhang, D., et al. (2020). The evolution of an invasive plant, *Sorghum halepense* L. (*'Johnsongrass') Frontiers in Genetics, 11*, 317.

Patterson, E. L., Saski, C., Kupper, A., Beffa, R., & Gaines, T. A. (2019). Omics potential in herbicide-resistant weed management. *Plants (Basel), 8*(12), 607.

Peng, Y., Lai, Z., Lane, T., Nageswara-Rao, M., Okada, M., Jasieniuk, M., et al. (2014). De novo genome assembly of the economically important weed horseweed using integrated data from multiple sequencing platforms. *Plant Physiology, 166*(3), 1241–1254.

Perez-Alonso, M. M., Carrasco-Loba, V., Medina, J., Vicente-Carbajosa, J., & Pollmann, S. (2018). When transcriptomics and metabolomics work hand in hand: A case study characterizing plant CDF transcription factors. *High Throughput, 7*(1), 7.

Powles, S. B., & Yu, Q. (2010). Evolution in action: plants resistant to herbicides. *Annual Review of Plant Biology, 61*, 317–347.

Qiu, J., Zhou, Y., Mao, L., et al. (2017). Genomic variation associated with local adaptation of weedy rice during de-domestication. *Nature Communications, 8*, 15323.

Ravet, K., Patterson, E., Krahmer, H., Hamouzova, K., Fan, L., et al. (2018). The power and potential of genomics in weed biology and management. *Pest Management Science, 74*(10), 2216–2225.

Rochfort, S. (2005). Metabolomics reviewed: a new "omics" platform technology for systems biology and implications for natural products research. *Journal of Natural Products, 68*(12), 1813–1820.

Royal Botanic Gardens Kew, State of the World's Plants. (2017). Available online: https://stateoftheworldsplants.com. Accessed 10 August 2024.

Saika, H., Horita, J., Taguchi-Shiobara, F., et al. (2014). A novel rice cytochrome P450 gene, CYP72A31, confers tolerance to acetolactate synthase-inhibiting herbicides in Rice and Arabidopsis. *Plant Physiology, 166*(3), 1232–1240.

Salas-Perez, R. A., Saski, C. A., Noorai, R. E., Srivastava, S. K., Lawton-Rauh, A. L., Nichols, R. L., & Roma-Burgos, N. (2018). RNA-Seq transcriptome analysis of *Amaranthus palmeri* with differential tolerance to glufosinate herbicide. *PLoS One, 13*, e0195488.

Salgotra, R. K., & Stewart, C. N., Jr. (2020). Functional markers for precision plant breeding. *International Journal of Molecular Sciences, 21*(13), 4792.

Salgotra, R. K. & Zargar, S. M. (2020). Rediscovery of genetic and genomic resources for future food security. ISBN 978-981-15-0155-5, ISBN 978-981-15-0156-2 (eBook). https://doi.org/10.1007/978-981-15-0156-2

Salgotra, R., & Chauhan, B. S. (2021). The first report of target-site resistance to glyphosate in sweet summer grass (*Moorochloa eruciformis*). *Plants, 10*(9), 1885. https://doi.org/10.3390/plants10091885.

Salgotra, R. K., & Stewart, C. N., Jr. (2022). Genetic Augmentation of Legume Crops Using Genomic Resources and Genotyping Platforms for Nutritional Food Security. *Plants, 11*(14), 1866. https://doi.org/10.3390/plants11141866

Serra, A. A., Couee, I., Renault, D., Gouesbet, G., & Sulmon, C. (2015). Metabolic profiling of *Lolium perenne* shows functional integration of metabolic responses to diverse subtoxic conditions of chemical stress. *Journal of Experimental Botany, 66*(7), 1801–1816.

Shangguan, W., Chen, H., Zhao, P., et al. (2024). Scenario-oriented nanopesticides: Shaping nanopesticides for future agriculture. *Advanced Agrochem, 3*, 265. https://doi.org/10.1016/j.aac.2024.07.002

Sheth, B. P., & Thaker, V. S. (2014). Plant systems biology: Insights, advances and challenges. *Planta, 240*, 33–54.

Shibuya, K., Fukushima, S., & Takatsuji, H. (2009). RNA-directed DNA methylation induces transcriptional activation in plants. *Proceedings of the National Academy of Sciences of the USA, 106*(5), 1660–1665.

Srivastava, A., Kowalski, G. M., Callahan, D. L., Meikle, P. J., & Creek, D. J. (2016). Strategies for extending metabolomics studies with stable isotope labelling and fluxomics. *Metabolites, 6*(4), 32.

Sun, G., Xu, Y., Liu, H., et al. (2018). Large-scale gene losses underlie the genome evolution of parasitic plant *Cuscuta australis*. *Nature Communications, 9*(1), 2683.

Sun, J., Ma, D., Tang, L., et al. (2019). Population genomic analysis and *De Novo* assembly reveal the origin of weedy rice as an evolutionary game. *Molecular Plant, 12*(5), 632–647.

Sun, Y., Shang, L., Zhu, Q.-H., Fan, L. & Guo, L. (2021). Twenty years of plant genome sequencing: achievements and challenges. *Trends in Plant Science, 27*, 391–401.

te Beest, M., Le Roux, J. J., Richardson, D. M., et al. (2012). The more the better? The role of polyploidy in facilitating plant invasions. *Annals of Botany, 109*(1), 19–45.

The Arabidopsis Genome Initiative. (2000). Analysis of the genome sequence of the flowering plant *Arabidopsis thaliana*. *Nature, 408*, 796–815. https://doi.org/10.1038/35048692

Thyssen, G. N., Naoumkina, M., McCarty, J. C., Jenkins, J. N., Florane, C., Li, P., & Fang, D. D. (2018). The P450 gene CYP749A16 is required for tolerance to the sulfonylurea herbicide trifloxysulfuron sodium in cotton (*Gossypium hirsutum* L.). *BMC Plant Biology, 18*(1), 186.

Wang, L., Sun, X., Peng, Y., Chen, K., Wu, S., Guo, Y., et al. (2022). Genomic insights into the origin, adaptive evolution, and herbicide resistance of *Leptochloa chinensis*, a devastating tetraploid weedy grass in rice fields. *Molecular Plant, 15*(6), 1045–1058.

Warwick, S. I. (1990). Allozyme and life history variation in five northwardly colonizing North American weed species. *Plant Systematics and Evolution, 169*, 41–54.

Werck-Reichhart, D., Hehn, A., & Didierjean, L. (2000). Cytochromes P450 for engineering herbicide tolerance. *Trends in Plant Science, 5*(3), 116–123.

Wiersma, A. T., Gaines, T. A., Preston, C., Hamilton, J. P., Giacomini, D., Robin Buell, C., Leach, J. E., & Westra, P. (2015). Gene amplification of 5-enol-pyruvylshikimate-3-phosphate synthase in glyphosate-resistant *Kochia scoparia*. *Planta, 241*(2), 463–474.

Wu, D., Shen, E., Jiang, B., Feng, Y., Tang, W., Lao, S., et al. (2022). Genomic insights into the evolution of Echinochloa species as weed and orphan crop. *Nature Communications, 13*(1), 689.

Yajima, W., Hall, J. C., & Kav, N. N. (2004). Proteome-level differences between auxinic-herbicide susceptible and -resistant wild mustard (*Sinapis arvensis* L.). *Journal of Agricultural and Food Chemistry, 52*(16), 5063–5070.

Yang, X., Zhang, Z., Gu, T., Dong, M., Peng, Q., Bai, L., & Li, Y. (2017a). Quantitative proteomics reveals ecological fitness cost of multi-herbicide resistant barnyardgrass (*Echinochloa crus-galli* L.). *Journal of Proteomics, 150*, 160–169.

Yang, C., Xu, L., Zhang, N., Islam, F., Song, W., Hu, L., Liu, D., Xie, X., & Zhou, W. (2017b). iTRAQ-based proteomics of sunflower cultivars differing in resistance to parasitic weed *Orobanche cumana*. *Proteomics, 17*, 13–14.

Yang, Y., Gardner, C., Gupta, P., Peng, Y., Piasecki, C., Millwood, R. J., Ahn, T. H., & Stewart, C. N., Jr. (2021). Novel Candidate Genes Differentially Expressed in glyphosate-treated horseweed (*Conyza canadensis*). *Genes (Basel), 12*(10), 1616.

Ye, C. Y., Tang, W., Wu, D., Jia, L., Qiu, J., Chen, M., Mao, L., Lin, F., Xu, H., Yu, X., Lu, Y., Wang, Y., Olsen, K. M., Timko, M. P., & Fan, L. (2019). Genomic evidence of human selection on Vavilovian mimicry. *Nature Ecology & Evolution, 3*(10), 1474–1482.

Yoo, S. D., Cho, Y. H., & Sheen, J. (2007). Arabidopsis mesophyll protoplasts: a versatile cell system for transient gene expression analysis. *Nature Protocols, 2*(7), 1565–1572.

Zargar, S. M., Mir, R. A., Ebinezer, L. B., Masi, A., Hami, A., Manzoor, M., Salgotra, R. K., Sofi, N. R., Mushtaq, R., Rohila, J. S., & Rakwal, R. (2022). Physiological and Multi-omics approaches for explaining drought stress tolerance and supporting sustainable production of rice. *Frontiers in Plant Science, 12*, 803603.

Zargar, M., Bayat, M., Saquee, F. S., Diakite, S., Ramzanovich, N. M., & Akhmadovich, K. A. S. (2023). New advances in nano-enabled weed management using Poly(Epsilon-Caprolactone)-based nanoherbicides: A review. *Agriculture, 13*(10), 2031.

Zhang, J., Zhang, X., Tang, H., Zhang, Q., Hua, X., Ma, X., et al. (2018). Allele-defined genome of the autopolyploid sugarcane *Saccharum spontaneum* L. *Nature Genetics, 50*(12), 1565–1573.

Zhang, H., Hall, N., Goertzen, L. R., Bi, B., Chen, C. Y., Peatman, E., Lowe, E. K., Patel, J., & McElroy, J. S. (2019). Development of a goosegrass (*Eleusine indica*) draft genome and application to weed science research. *Pest Management Science, 75*(10), 2776–2784.

Zhang, L., Chen, K., Li, T., Yuan, S., Li, C., Bai, L. & Wang, L. (2023). Metabolomic and transcriptomic analyses of rice plant interaction with invasive weed Leptochloa chinensis. *Frontiers of Plant Science, 25*(14),1271303. https://doi.org/10.3389/fpls.2023.1271303

Zhao, G., Zou, C., Li, K., et al. (2017). The *Aegilops tauschii* genome reveals multiple impacts of transposons. *Nature Plants, 3*, 946–955.

Index

B. S. Chauhan, G. Mahajan (eds.), *Recent Advances in Weed Science*,
https://doi.org/10.1007/978-3-031-91014-2

The manufacturer's authorised representative in the EU is Springer Nature Customer Service Centre GmbH, Europaplatz 3, 69115 Heidelberg, Germany. If you have any concerns regarding our products, please contact ProductSafety@springernature.com

Printed and bound by CPI Group (UK) Ltd, Croydon, CR0 4YY
30/09/2026
02225590-0009